RESTRUCTURING ASIAN ECONOMICS FOR THE NEW MILLENNIUM

RESEARCH IN ASIAN ECONOMIC STUDIES

Series Editor: M. Dutta

RESEARCH IN ASIAN ECONOMIC STUDIES VOLUME 9-A

RESTRUCTURING ASIAN ECONOMICS FOR THE NEW MILLENNIUM

EDITED BY

J. R. BEHRMAN

University of Pennsylvania, Philadelphia, PA, USA

M. DUTTA

Rutgers University, New Brunswick, NJ, USA

S. L. HUSTED

University of Pittsburgh, Pittsburgh, PA, USA

P. SUMALEE

Chulalongkorn University, Bangkok, Thailand

C. SUTHIPHAND

Chulalongkorn University, Bangkok, Thailand

P. WIBOONCHUTIKULA

Chulalongkorn University, Bangkok, Thailand

Emerald

United Kingdom – North America – Japan
India – Malaysia – China

Emerald Group Publishing Limited
Howard House, Wagon Lane, Bingley BD16 1WA, UK

Reprints and permission service
Contact: booksandseries@emeraldinsight.com

British Library Cataloguing in Publication Data
A catalogue record for this book is available from the British Library

ISBN: 978-0-76-230709-8
ISSN: 1047-126X (Series)

Printed and bound by CPI Group (UK) Ltd, Croydon, CR0 4YY

CONTENTS

LIST OF CONTRIBUTORS *ix*

FOREWORD
Thienchay Kiranandana *xi*

PREFACE
M. Dutta *xiii*

SECTION ONE

1. OPTIMAL MONETARY AND EXCHANGE POLICY UNDER GLOBAL FINANCIAL CRISIS
M. R. Chatu Mongol Sonakul *3*

2. A MACROECONOMIC CORE OF OPEN ECONOMIES IN ASIA FOR PROGRESSIVE INDUSTRIALIZATION IN THE NEW MILLENNIUM
M. Dutta *9*

3. THE EURO, DOLLAR, AND INTERNATIONAL MONETARY SYSTEM
Jürgen Schröder *25*

4. THE ARCHITECTURE OF THE MULTILATERAL ORGANIZATIONS
Peter J. Lloyd *37*

5. CONTAGION IN THE EAST ASIAN CURRENCY CRISIS
Mardi Dungey and Vance L. Martin *69*

6. LONG RUN EQUILIBRIUM EXCHANGE RATE BEHAVIOUR: ASIAN/YEN EXCHANGE RATES BEFORE THE CRASH
Steven Husted *85*

SECTION TWO

7. SCHOOLING IN ASIA: SELECTED MICRO EVIDENCE ON DETERMINANTS AND EFFECTS AND POLICY IMPLICATIONS
Jere R. Behrman 105

8. GROWTH, CRISIS & POVERTY INCIDENCE IN SOUTHEAST ASIA
Peter G. Warr 165

9. MACROECONOMIC FUNDAMENTALS OF POSTWAR ECONOMIC GROWTH IN JAPAN – A GREAT SUCCESS AND RECENT FLUCTUATION LESSONS IN ASIAN ECONOMIES
Hiromitsu Ishi 177

10. KOREAN FINANCIAL CRISIS DURING 1997–1998: CAUSES AND CHALLENGES
Jang-Hee Yoo and Chul Woo Moon 197

11. THE CURRENT ECONOMIC SITUATION AND TRENDS IN CHINA WITH A DISCUSSION OF EMPLOYMENT-RELATED ISSUES IN THE PROCESS OF INDUSTRIALIZATION
Fan-Zhang Huang 215

12. THE EVOLUTION OF COMMERCIAL BANKING AND FINANCIAL MARKETS IN TAIWAN
Tzong-Shian Yu 229

13. A LONG AND BUMPY ROAD TO ECONOMIC RECOVERY: THE CASE OF INDONESIA
Anwar Nasution 249

14. ECONOMICS FOR THE MILLENNIUM
Pasuk Phongpaichit 257

15. CHALLENGES TO HUMAN RESOURCE DEVELOPMENT IN ASIA
M. G. Quibria 261

SECTION THREE

16. EQUILIBRIUM REAL EXCHANGE RATE, REAL EXCHANGE RATE MISALIGNMENT, AND THE CURRENT ACCOUNT: AN ECONOMETRIC ANALYSIS OF THAILAND'S EXPERIENCES
Shuji Kasajima and Sukanda Lewis 281

17. APPROPRIATENESS OF FLEXIBLE EXCHANGE RATE RÉGIME IN ECONOMIC ADJUSTMENTS: AN ASIAN EXPERIENCE
Chartchai Parasuk and Wimonrat Sukcharoen 323

18. MONETARY CONDITIONS & MONETARY POLICY IN SMALL OPEN ECONOMIES: EMPIRICAL RESULTS OF THAILAND UNDER FLOATING RATE RÉGIME
Rungsun Hataiseree 341

19. THE OPTIMAL-CURRENCY-AREA INDEX: SELECTED ASIAN COUNTRIES
Chayodom Sabhasri and Gene Janevathanavitya 363

20. THE ASIAN FINANCIAL CRISIS AND THE BALANCE SHEETS OF BANKING AND CORPORATE SECTORS: AN EMPIRICAL EVALUATION
Sung Yeung Kwack 375

21. THE THAI FINANCIAL CRISIS AND THE ROLE OF PUBLIC FINANCIAL INSTITUTIONS
Masumi Kishi 395

LIST OF CONTRIBUTORS

Jere R. Behrman	University of Pennsylvania, Philadelphia, PA, USA
Mardi Dungey	LaTrobe University, Victoria, Australia
M. Dutta	Rutgers University, New Brunswick, NJ, USA
Rungsun Hataiseree	Bank of Thailand, Bangkok, Thailand
Fan-zhang Huang	formerly of State Planning Commission of PRC, Beijing, China
Hiromitsu Ishi	Hitotsubashi University, Kunitachi, Tokyo, Japan
Gene Janevathanavitya	Chulalongkorn University, Bangkok, Thailand
Shuji Kasajima	Thammasat University, Bangkok, Thailand
Masumi Kishi	Chuo University, Tokyo, Japan
Sung Yeung Kwack	Howard University, Washington, DC, USA
Sukanda Lewis	Thammasat University, Bangkok, Thailand
Peter J. Lloyd	The University of Melbourne, Victoria, Australia
Vance L. Martin	Melbourne University, Melbourne, Australia
Chul Woo Moon	EWHA University, Seoul, Korea
Anwar Nasution	University of Indonesia, Jakarta, Indonesia

Chartchai Parasuk	National Finance Public Company Limited, Bangkok, Thailand
Pasuk Phongpaichit	Chulalongkorn University, Bangkok, Thailand
M. G. Quibria	The Economic Analysis and Research Division of the Asian Development Bank
Chayodom Sabhasri	Chulalongkorn University, Bangkok, Thailand
Jürgen Schröder	Universitat Mannheim, Mannheim, Germany
M. R. Chatu Mongol Sonakul	Governor, Bank of Thailand, Bangkok
Wimonrat Sukcharoen	National Finance Public Company Limited, Bangkok, Thailand
Peter G. Warr	The Australian National University, Canberra, Australia
Jang-Hee Yoo	EWHA University, Seoul, Korea
Tzong-Shian Yu	Chinese Institute of Economics and Business, Taipei, Taiwan

FOREWORD

The forty chapters in six sections in RESTRUCTURING ASIAN ECONOMIES FOR THE NEW MILLENNIUM present the revised versions of presentations made at the international conference on "*A Macroeconomic Core of An Open Economy for Progressive Industrialization and Development of Asian Economies in the New Millennium*", jointly organized by the Faculty of Economics, Chulalongkorn University and the American Committee on *Asian Economic Studies (ACAES)*, on December16–18, 1998 in Bangkok. Conferees included a large number well-known economists, experts in economic policy-making and researchers in allied fields, who presented papers, served on panel discussions and participated in discussions and deliberations regarding the Asian financial crisis and strategies for industrialization and development in the future. Invited speakers, panelists, and discussants came from the United States, Germany, Australia, Japan, Korea, People's Republic of China, Hong Kong, Taiwan, Indonesia, the Philippines, Singapore, Vietnam, India, Bangladesh joined their peers from Thailand at this three-day event. This book is the final output of the scholarly exercises at the conference which was a superb success.

The Asian economic crisis and future development is a hot subject among academics and policy-makers worldwide. Several countries in Asia came to face severe economic recession in the wake of their financial and foreign exchange crisis which began in 1997. The recession was deep and the challenges for managers of respective economies for an optimum specification of monetary and fiscal policy guideliens became all too great. Many came to wonder whether the so-called "*Asian Miracle*" all of a sudden became a part of history. If so, how will Asia recover from the crisis and experience high growth and continuing development as in the past quarter of the century? Can such a crisis be forewarned? Which development paths should Asia seek so that the crisis will not be repeated in the future? In other words, the challenge is to formulate rational economic policies for Asia'a newly industrialized/industrializing economies so that their industrialization and development can sustainably progress as Asia enters the Twenty-First Century.

The book addresses the above issues by covering many topics including financial meltdown and industrial development, money and financial markets, exchange rates, capital flows, economic reforms, human capital development, impact of the financial crisis on growth and poverty incidence, as well as other development problems in Asia-Pacific countries. This conference featured a keynote speech by M. R. Chatu Momgol SONAKUL, Governor of the Bank of Thailand on "Optimal monetary and exchange policy under global financial crisis" (see chapter one section). A roundtable discussion on "*Asian Economies in the New Millennium*", chaired by KIM Duk-choong of Seoul (now Minister for Education, Government of Korea) was joined by Juergen SCHROEDER of University of Mannheim, Mannheim, Hiromitsu ISHI of Hitotsubashi University, Tokyo, HUANG Fan-zhang of State Planning Commission of PRC, Beijing, Peter J. LLOYD of University of Melbourne, Victoria,

M. G. QUIBRIA of Asian Development Bank, Manila, Jang-Hee YOO and Chul –Woo MOON of EWHA University, Seoul, Tzong-shiana YU of Chinese Institute of Economics and Business, Taipei, Anwar Nasution of University of Indonesia, Jakarta (now Deputy Governor of Bank Indonesia), and Pasuk PHONGPAICHIT ofChulalongkorn University, Bangkok plus a dinner lecture for Thailand's Ministry of Finance on policies and measures to cope with the Thai financial crisis were special features of the conference program. The subject of this book is definitely very timely and should be of great interest to scholars, policy-makers, and the general public who follow the economic development in Asia. RESTRUCTURING ASIAN ECONOMIES FOR THE NEW MILLENNIUM should contribute to a broader perspective and at the same time, a deeper understanding of the subject.

Thienchay Kiranandana, Ph.D.
Professor of Economics, and
President of Chulalongkorn University
Bangkok, Thailand

PREFACE

RESTRUCTURING ASIAN ECONOMIES FOR THE NEW MILLENNIUM

Since the 1970s researchers in economics have been engaged in studying Asia's newly industrialized/industrializing economies and literature in economics is very rich reporting their findings. As Asian economies in the Northeast and Southeast Asia recorded spectacularly high rates of growth of their respective gross domestic products with high rates of savings and investments as percentages of their respective gross domestic products with reasonably moderate rates of inflation, many hailed it as "The Asian Miracle". In the late 1990s, came the now-famous Asian financial crisis and challenged several of Asia's success stories, and many scholars came to be overwhelmed. Few wrote a treatise forecasting the crisis. There were hardly any warning from national and/or international think tanks. Again a miracle? Happily, financial crisis in most affected Asian economies has been contained.

There is a need for sustained, in-depth research of economic happenings in Asian economies and that is a challenge to all students of economics, be they in Asia or in any other continent. Let there be no search for what one may call the final truth. Let us search for the truth as it affects the economies of Asia we study. The traditional agriculture-dominated Asian economies have accepted the challenge of industrialization, and that too at an accelerated rate. Their economic agenda is to be an able and competitive actor in the world market as they adopt an economic program of industrialization and internationalization. They want to import physical capital form the relatively capital-abundant mature industrialized economies and pay for them by way of exporting a share of their newly manufactured industrial products to the world market, and thus earning export-revenue in convertible currencies. This is an import-export led growth model, not an export-led one, as some have called it. The economic gains are shared by partners at both ends, indeed a win-win situation. In the new millennium Asian economies, indeed economies of all continents, will face newer challenges and we must pursue our studies to find our best answers to them.

In two parts, A and B, RESTRUCTURING ASIAN ECONOMIES FOR THE NEW MILLENNIUM presents 40 research papers in six sections with a foreword by Thienchay Kiranandana, President of Chulalongkorn University, Bangkok, Thailand, himself a distinguished economist.

Section one includes six chapters. Chapter 1 presents a learned discourse by Governor M. R. Chatu Mongol Sonakul of Bank of Thailand on optimal monetary and exchange policy under global financial crisis. M. Dutta follows in chapter 2 with a discussion on a macroeconomic core of open economies in Asia for progressive industrialization in the new millennium. In chapter 3, Jürgen Schröder makes a presentation on The Euro, Dollar and International Monetary System inviting his readers to examine the eventful economic impacts of the emerging new global money market and to be ready for the impacts that there will be. Peter J. Lloyd in chapter 4 presents a forceful discussion of the architecture of the multilateral organizations, the IMF, OECD, G-7, World Bank, BIS, WTO and poses the question if there are too many of them. In chapter 5, Mardi Dungey and Vance L. Martin specify a model and present their statistical findings regarding the contagion in the East Asian Currency Crisis, and in chapter 6, Steven Husted writes on the long-run equilibrium exchange rate behavior with focus on Asian/Yen exchange rates before the crash.

Section two consists of nine chapters. In chapter 7, Jere R. Benhrman presents a discussion of "Schooling in Asia" analyzing micro evidence on determinants and effects and policy implications, relative to selected countries in South and Southeast Asia. Chapter 8 adds a discourse by Peter G. Warr on growth, crisis and poverty in Southeast Asia. In chapter 9, Hiromitsu Ishi writes on macroeconomic fundamentals of postwar economic growth of Japan with an analysis of the story of a great success and recent economic frustrations in Asian economies. In chapter 10, Jang-Hee Yoo and Chul Woo Moon discuss the causes and challenges of the Korean financial crisis during 1997–1998. Fan-zhang Huang in chapter 11 dwells on the current economic situation and trends in China with an analysis of employment related issues in the process of industrialization. Tzhong-shian Yu follows in chapter 12 writing on the evolution of commercial banking and financial markets in Taiwan and pointing to the challenges of the financial crisis for all Asian economies. In chapter 13, Anwar Nasution writes on Indonesia and argues that a long and bumpy road to Indonesia's economic recovery lies ahead. Chapter 14 is a thematic presentation on the challenges for the Asian economies in the new millennium by Pasuk Phongpaichit. M. G. Quibria in Chapter 15 writes on challenges to human resource development in Asia.

Section three presents six studies. Chapter 16 offers an econometric study of Thailand's experiences relative to equilibrium real exchange rate, real exchange rate misalignment and the current account by Shuji Kasajima and Sukanda Lewis. In chapter 17, Chartchai Parasuk and Wimonrat Sukcharoen analyze appropriateness of flexible exchange rate regime in economic adjustments for Asian economies. Rungsun Hataiseree follows in chapter 18 examining monetary conditions and monetary policy in small open economies with empirical results of Thailand under floating rate regime. Chayodom Sabhasri and Gene Janevathanavitya in chapter 19 report a study on the optimum currency index in the context of selected Asian countries. Sung Yeung Kwack presents in chapter 20 an empirical evaluation of the Asian financial crisis relative to the balance sheets of banking and corporate sectors. In chapter 21, Masumi Kishi offers a study of the Thai financial crisis and the role of public financial institutions.

Section four includes 8 chapters. Chapter 22 presents a study on microeconomic foundations of macroeconomic crisis with risk analysis of Asian non-Financial firms with comparisons to Latin America by Stephen Reynolds, Somchai Ratanakomut and James P. Gander. In chapter 23, Sunti Tirapat and Nittayagasetwat Aekkachai predict a firm's financial distress using micro and macro variables for Thailand. Teerana Bhongmakapat, in chapter 24, offers an analysis of the management of the Thai crisis with focus on asset bubbles, globalization and macroeconomic policy. Chapter 25 presents a study of China's macroeconomic stabilization policies in the process of high growth, system transition and opening to the outside world by Changsheng Xu. Do Loc Diep follows in chapter 26 with a discourse on foreign exchange policy in the context of monetary-financial crisis in Asia and economic renewal of Vietnam. Chapter 27 presents a study by Chyau Tuan and Linda Fund Yee Ng on the post-industrialization economic growth of Hong Kong from the perspectives of a mega-city. Jung-Chao Liu and Li-Lai Xu in chapter 28 analyze China's foreign exchange rate policy in the context of the Asian financial crisis. In chapter 29, Saleem M. Khan offers an analytic study on the concept of a new regionalism in South Asia.

Section five consists of 5 chapters on studies on foreign investment. In chapter 30, Yang-Taek Lim presents a much comprehensive analysis of the role of foreign direct investment for Korea's economic recovery and future directions in foreign investment promotion policy. Amit S. Ray offers in chapter 31 a study based on the evidence from the Indian chemical industry and has posed the question if foreign firms are technologically more active and progressive as they invest in less developed countries. Shujiro Urata and Hiroki

Kawai in chapter 32 offers an empirical study on locational determinants of Japanese foreign direct investment. In chapter 33, Momtaz Uddin Ahmed analyzes trends, patterns and policies of foreign direct investment and technology flows in Asia and the Pacific region in the 1990s. Wiboonchutikula Paitoon, Tubitimtong Bangorn, Lakshmi Raut and Chaivichayachat Bundit follow in chapter 34 with an exhaustive and comprehensive study of Thailand's capital flows.

In conclusion, section six has 6 chapters. In chapter 35, Hiro Lee, David Roland-Hoist and Dominique van der Mensbrugghe present a study on general equilibrium assessments of trade liberalization in APEC countries. Juanjai Ajanant, in chapter 36, projects an economic perspective of trade and industrialization of Thailand in 2020. In chapter 37, Romeo M. Bautista, Sherman Robinson and Moataz El-Said make use of SAM and CGE analytic frameworks and present alternative industrial development paths for Indonesia. Robert H. Aten follows in chapter 38 with an extended study of the Indonesian economy in the context of imperfect competition and progression of techonolgical advancement. In chapter 39, Wottana S. Janjaroen and Siripen Supakankunti study health services systems and the consequences from the general agreement of trade in services (GATS). In the concluding chapter 40, Piyasiri Wickramasekara presents a comprehensive study of the implications for employment, labor markets and poverty as the economic crisis confronts the economies. in the Asia-Pacific region.

M. DUTTA, EDITOR

SECTION ONE

1. OPTIMAL MONETARY AND EXCHANGE POLICY UNDER GLOBAL FINANCIAL CRISIS

M. R. Chatu Mongol Sonakul

Professor Thienchay, Professor Dutta, Dean Sumalee, Ladies and Gentlemen,

I am glad to have this chance of addressing the 22nd CAES conference on Economic Development in the New Millennium. It gives me the chance to reflect on what Thailand has done and what it still has to do. But let me keep my talk strictly confined to our financial problems. There is an abundance of other speakers who are better qualified to speak on the other areas of the economy who are discussants in the conference.

Let me break my statement into two parts.

Firstly, on what Thailand has done in monetary policy and what we still have to do.

And secondly, what is needed in the new world financial system to minimise the risk of the present world crisis from occurring again and again.

Let me begin by restating what we have done in Thailand. We floated the exchange rate to allow normal responses to the interest rate adjustment to occur. We raised the interest rate to slow down the economy, to keep the domestic funds from flowing out, and to attract some foreign funds to come in, through also the guarantee of all liabilities of the financial institutions in the country. When this worked, we lowered the interest rate and allowed the economy to begin to heal itself. Exchange rate strengthened and stabilised, interest rate fell, inflation fell because these were the right medication.

Restructuring Asian Economics for the New Millennium, Volume 9A, pages 3–8.
2001 by Elsevier Science B.V.
ISBN: 0-7623-0709-9

At the same time, we managed to stabilise the political environment. We aided the low income working group and farmers via the programme of social safety net, job creation and productivity improvement. These two critical areas seem to be well enough stabilised such that it seems possible that the boiling point will not be reached. This is about as much as anyone can hope for in these areas.

Thailand is then ready to expand and progress again. But like all developments created by years and years of mismanagement and bad decisions, there are still legacies.

The biggest legacy is probably the non performing loans. The rising NPL portfolio which is now at 39% of all loans as of September needs to be contained and reduced.

The other big legacy is probably that some financial institutions are still not stabilised. That is to say, that their future viability has still not been ensured but an umbrella scheme announced on August the 14th to inject tier 1 and tier 2 capital is available. We will have to see the details of how the Financial Restructuring Advisory Committee who were established to handle the scheme will tackle this problem. Since they were set up to advise the Governor of the Central Bank and the Minister of Finance, I have at this point restrained from making a statement of how it should be done. So that they will be giving me advise rather than another way round.

Since these major problems are only temporal and are our own and the topic of the Conference is on the next Millennium, let me not go on to explain how far advanced we are in solving these problems. Let me go on instead to give my view on what is the optimal monetary and exchange rate policy in the new Millennium which, without some changes, looks like it's going to be constantly under global financial crises.

The way to solve it probably involves extra territoriality which takes a long time and naturally is not easy.

Thailand herself has a particularly challenging decision. Despite the crisis, development-wise we are at a cross-road. The crisis will have to be overcome then we still have to answer the question of "What is the optimal monetary and exchange rate policy in normal time for a country which has just reached lower-middle income and is in the group of 21 top financial countries of the world."

Traditionally, we have used a financial regime where we try to fix the exchange rate, fix the interest rate, control inflation rate and operate monetary policy through a combination of cash deposit requirements at the central bank and little of open market operations because there were no markets to operate in. When you reach middle income, a country perhaps can go on doing that but a better way might well be to do open market operations because there are now markets to operate in.

The stock market in Thailand exists and used to be quoted internationally. It will be quoted again when things are normal because it is a big market as Thailand is a big market and Thailand is a big country.

The foreign exchange markets are now developing very fast off-shore and on-shore and will develop more. We have a master plan to develop the baht fixed income market or the debt market to be of minimal international standards beginning with the FIDF bond issued to finance losses in the financial sector.

Today we have probably finished building a yield curve of risk free baht interest rate by selling for the first time 10 year bonds in the open market. Having yield curve and benchmarks, the bond market is ready for development in term of mechanical systems like DVP, moving the repurchase market from the government run to the over-the-counter system operated by the private sector and such likes.

We are in the very advance stage of negotiation with the financial institutions in the country in order to operate a system well supported by all sectors rather than being government dictated in design.

The new governance in Thailand calls for the government to take the initiative, but not the action for doing things that are much better done by the private sector. Sorry I cannot resist that as an aside because my main interest now in good governance in all sectors and hopefully in the financial sector.

The question being asked though is what do we do with the rudimentary instruments of financial and monetary control. The passage from using government dictated instruments such as cash deposit by the financial institutions at the central bank to one of open market operations is a difficult transition.

Can we really operate without cash requirements and its multiplier effect on lending or money supply? What do we do with the capital adequacy ratio. Do we just use it to ensure the security of the financial system or do we use it also to manage liquidity?

Can we build a debt market that offers liquidity such that the government need not provide that service? Can't we build a last resort operation of the central bank such that it piggy backs on the private sector system rather that in place of the financial sector system as at present?

These questions are being looked at closely by the staff at the central bank, partly because a better financial system means a less costly financial system and less costly loans to the private sector, which will also help to alleviate our immediate problem of high NPL.

Less costly loans make projects more viable and the NPL will in itself go down somewhat while serving the long term benefit of having a better financial system for the country.

Doing all this when everything is falling down around you must be very difficult. I do not know whether other countries are doing this but it has often been noted that solving the long term problem is rather a crucial element to solving the immediate crisis that we are facing today.

Tackling it as a short term crisis probably can't solve the problem as the basic structure is at odds with the solution to the problem.

Now, in Thailand we have achieved overall stability so we can look at these problem with sufficient time to consider it rather than in haste as in the past. We have time to solve the problems one by one, but it is still like walking on thin ice. It is very comfortable when the ice doesn't break but nobody knows when the ice might break. One can only hope to walk all the way and then end up dry land soon.

Let me now talk a little bit about the world financial system. In my mind, the world financial system, the minimum that is needed is quite obvious and we have taken the matter in our own hand until that minimum is achieved.

We now claim to have sufficient information on our the financial system. We feel that we know, or could if we wish, instantly know the price, amount, and identities of counterparty if not the owner of any financial transactions in Thailand in both baht and the foreign exchange market.

With this information on a daily basis and capital account information at most on the delay-of-a-week basis, we now know what's going on in our markets. Knowing what's going on means you can do something about it, and this is one of the reason why the baht has been stable.

A stable currency normally means an appreciating currency because there is nothing like stability to make a good platform for progress. It is difficult to make progress of any kind when the ground under your feet is shaking.

But what about the off-shore markets. We don't know anything. We only know the price that the baht is being traded at the time. That's it. You really can't know how much was traded or who traded, even if we were the counter party of trade. All the off-shore market need to tell us is that somebody traded. We can't do anything about it because it is not under our jurisdiction. Nor it seems under anybody's real observation. Electronic information moves instantaneously over practically everywhere in the world including countries that don't do anything else except allowing people to use them as a base for trading for its own purpose; tax avoidance, financial control, illegal money, or whatever.

Even with countries with good financial supervision and there are quite a few of them, the attitude has been that organisations which do not affect their own interest need not be controlled or observed as they cause problems only to other countries, and the other countries should try manage their own system well enough not to be taken advantage.

The Asian, the Russian and the South American crises have proven that the whole world is inter-linked. Most countries are beginning to look alike, if not in physical characteristics, but certainly in the mainstream operations of its economic, financial and social system. There are no central planing economies worth talking about anymore. All countries now have free markets trading with the world and subject to the vagaries of the world as I have mentioned.

Of course now the advanced countries realise that their organisations of left uncontrolled are going to grow and might be a problem. Organisations freely allowed to take advantage of any system can grow big.

When things go wrong, it affects both the countries being taken advantage of and it affects the countries who allow their organisations to do that. So now I am more hopeful that major countries will realise that there is no longer the chance that just because you are rich and control your system well you will not be affected by allowing your own people to do whatever they like to less well controlled or well endowed people.

One of the problem of course is that major countries fear that doing anything may force money to move to unregulated countries. But not to do anything is inevitably to fail the world.

We have only one world to live in rather than one country. Countries no longer go to another country to colonise it and take away all its resources without proper payments. Less developed countries are now beginning to accept that countries which have more money and dominate the world look on the world as just their one world. They will invest wherever it looks like they can make money and they will stay there putting more money in if it looks like they will make even more money if they keep on staying there. They cannot just go there and carry away resources like in the past.

This goes back to my comment on extra territoriality that the world does not allow people in other countries to kill each other. It doesn't allow emission of carbon dioxide gas because the world will get too hot. It makes rules for how countries may fish or share in the international water resources.

The world cannot allow countries to operate its financial systems without due regards to the rest of the world. It must make rules.

To go back to the case of Thailand, we are actually getting by monitoring the baht in the country and at the same time restricting off-shore baht. In my mind if the world would establish a system where it monitors the financial transactions in the world and information may be called up by the participating countries or authorities, that is probably enough. We probably do not need to tax world financial transactions. We probably do not need a supervisory rule making body. Monitoring would at least allow us to evaluate much better what we really need. Who should have this responsibility? The World Bank, the IMF, a new organisation, a super power, organisations like the UN, let me not discuss this aspect as it will take too long.

With or without a new organisation, the system of monitoring world's transaction is the minimum in the new millennium. Without this then most countries will have to ban their currencies to international trading, probably making the U.S. dollar and the Euro into world currencies, just as English will become the world language in the new millennium.

If that is wanted, and it may not be such a bad idea, then Thailand is quite happy. Then we probably have started the new millennium right. We don't allow the baht to be remitted out of the country without underlying real transactions which means that the baht can be used off-shore only for trade in goods, services or capital, but not for speculation. Without a new world order in the financial system that is probably how it's going to remain. Not, as I have said, because it might even be a good idea, but because people who can do something to improve the world monetary system will not do it.

2. A MACROECONOMIC CORE OF OPEN ECONOMIES IN ASIA FOR PROGRESSIVE INDUSTRIALIZATION IN THE NEW MILLENNIUM

M. Dutta

I. ECONOMIC RATIONALE – NO MIRACLE

Before I turn to many recent economic news stories 1970–1998, first of the Asian miracle and then of the Asian debacle and/or depression, I am tempted to remind you that the term European miracle has very recently been introduced in news literature on economics.

In 1992 at an international economic conference at this campus, some of you may recall that I referred to the emerging phenomenon of one currency based European Union (EU). As we meet here in 1998, EU of 15 member economies in Western Europe has become an eventful fact of economic regionalization, and that EU's one currency – EURO – will become a reality in eighteen days, not only for the eleven EU member-economies who have signed for it, but also for EU's remaining four members and for the rest of the world.

Many of our fellow economists have found no word other than European Miracle to describe the event. Our preconceived notion of sovereign nation-state based macroeconomic systems with its respective sovereign currencies therefore warrants a careful review. The Central Banks of the eleven EU

Restructuring Asian Economics for the New Millennium, Volume 9A, pages 9–24.

ISBN: 0-7623-0709-9

member-economies have just introduced a concurrent reduction of rate of interest – the rate of interest member banks charge for overnight intra-bank loans. In 1992 you expressed your serious reservations and questioned if it were too soon for me to speak for the forthcoming EU experiment. In 1998, you cannot ignore the happenings in EU.

The concept of sovereign nation-state based macroeconomies with their respective independent sovereign currencies may be outdated (Dutta, 1995). The concept of a sovereign state, to paraphrase a noted historian, owes its origin to accidents of history. An intra-regional macroeconomic core of member-economies of the region with well-structured multilateral monetary and fiscal policy coordination may be the new order (Dutta, 1998) in the new millennium.

It is no European *miracle*. It is a rational economic decision by 15 member-nations of EU toward the objective of optimization of real economic gains for microeconomic units – households as well as business organization – in all member-economies of the Union. All available evidences point to that achievement. The region providing the base for EU is not defined by the old-fashioned imperial order, but by a map-of-the-world view of it. The member-economies' belonging to the map of the region is the core issue. In the absence of the ideal of a global UNION of all sovereign member nations of the world, a regional economic UNION is the second best, satisfying the Pareto Optimality condition (Dutta, 1998). To quote Saburo Okita, regionalism and globalism will be the core of the new framework of international economic relations.

I now return to the Asian miracle about which so much we read in the last quarter of the century. As of 1997, the story is about Asian debacle/depression/bust/crisis. In either case there has been insufficient rational economic discourse. I argue that happenings in Asia in the recent decades can and must be explained by laws of economics, no *miracle*, nor by reference to terms such as "giants", "dragons", or "tigers" of Asia.

By late 1960s and through 1970s, Asia's agriculture-based, traditional economies recognized that they could not achieve economic advancement without accelerated industrialization of their respective economies. They had relative abundance of labor (N) – not only ordinary unskilled labor (N.1) but also skilled (N.2) labor, often with comfortable endowments of natural resources. They had very limited indigenous supply of physical capital goods – equipment and technology (K). They elected to import physical capital goods (K) from relatively capital-abundant economies.

The Asian economies progressively elected to move away from their erstwhile economic policy of import substitution and opted for open economic policy regimes. An obvious experiment in the theory of comparative advantage in its many and varied forms a la Ricardo, Hecksher-Ohlin. They offered to import capital goods (K) from the advanced industrialized economies and pay for it by way of exporting some of the new manufactured products they were able to make with the imported capital goods. The straightforward Cobb-Douglas Production function was their immediate choice. Aggregate level of output (Q) is a non-linear (or log-linear) function of aggregate stock of labor (N) and aggregate stock of capital goods (K). They seemed to be not too anxious to include the role of socially owned inputs, environment and issues relative to non-homogenous labor market.

To begin with these economies in Asia were low income and had limited supply of savings (S) at home. Hence the choice was to import (K) with the inflow of foreign savings form relatively savings-surplus economies and augment the availability of indigenous and inflow of foreign savings to match the accelerated investment (I). These inflows of foreign investment were to be paid for by sharing the revenues from the export earnings which generated a pool of hard currency. The profit repatriation home by foreign investors thus became a way of operation, which in turn induced additional inflow of foreign investments. To optimize a newly industrializing Asian economy's export-earning ability, their export products needed to be competitive in the world market in terms of both cost and quality parameters. An appropriate adaptation of the imported capital goods to their respective indigenous economies rich in labor, both skilled and unskilled, and/or resources must have met this competitiveness of their export products in the world market, which we called "an adaptive innovation paradigm" (Dutta & Tantum, 1988).

Some economists in advanced industrialized countries referred to it as export-led economic development model and ignored the role of exports of capital goods from the advanced industrialized economies to these newly industrializing economies. Some economists in Asia argued against the import of capital goods from advanced economies as they argued such imports would facilitate exploitation of the under-developed economies. Many economists at both ends noted it to be a two-way process contributing to the mutual economic gain of all participating economies. It has indeed been *an import-export-led economic growth paradigm* (Klein, 1990), a rational two-way buy-and-sale market. It was by no means an economic *miracle* unless one subscribed to the preconceived notion that industrial revolution was never to be successful in Asia.

The story of newly industrialized/industrializing economies in Asia successfully progressed until late 1990s. In 1997–1998 came the financial meltdown in several Asian economies. Some economists called it the Asian bust or Asia's

Great Depression and the end of the story of industrialization of Asian economies. Again there has been a failure to appreciate the operation of the free market economy. Following several decades of economic growth, growth of gross domestic product (GDP) at an average annual rate of 9–10% in Northeast and 7–9% in Southeast Asia, a business cycle should have been anticipated. Indeed, business cycles in a cyclical economic growth path have been a part of economic experience in advanced industrialized economies.

In addition, periods of high rate of economic growth have been known to encourage "irrational expectation" for investors who engage in highly speculative financial investments, contributing to the de-linking of the basic economic relationship between the two markets, let us call them – the Wall Street and the Main Street. Short term capital movements with a goodly number of innovative products in the financial market, such as, hedge funds, junk bonds, derivatives and the like add to the game of speculation. Much of the 1997–1998 economic happenings in Korea, Thailand, and Indonesia may be explained as such. The financial meltdown in the United states in October 1987 witnessed all of this – a high rate of economic growth from December1982 through August 1987 encouraged what we now may call "irrational expectation" of the financial world in the form of junk bonds and Savings and Loan Association debacle, hostile mergers and takeovers, swallowing of poison pills. The meltdown was followed by years of low growth and then a brief but painful recession in the third quarter of 1990. Only in 1992 economic recovery became real.

Will Korea and Thailand experience recovery relatively sooner? As of December 9, The New York Times, December 10, reports that Korea is immediately paying back some U.S.$3 billion she received in emergency loans from the IMF. The IMF loan was of U.S.$58 billion of which Korea used only a part of it. Michel Camdessus, Fund's Managing Director, called the repayment an "important watershed in the process of Korea's emergence from last year's foreign exchange crisis." The situation in Indonesia merits more careful scrutiny. Malaysia has adopted her own game plan. Taiwan has been immune from the financial meltdown in Asia. Singapore and Hong Kong are very special economies as they continue to remain too open to the rest of the world. Be it further noted that People's Republic of China which adopted the open economic policy in late 1970s with an economic game plan of accelerated industrialization has been able to successfully manage its exposure to financial meltdown in her neighboring economies. Economies of Russia, now an APEC member, of India and other South Asian economies who are not members of APEC, have very limited trade exposure to the rest of the world, and their economic problems can hardly be related to the meltdown of financial markets in their neighboring Asian economies. They merit independent analyses.

The economic happenings in the process of industrialization of Asian economies are certainly *neither miracles nor debacles/depressions*. They must be understood in the broader context of experiences of advanced industrialized economies. Other economists who have been known to question the essential logic of the paradigm of open market economy – industrialization and internationalization – will, of course, elect not to join the discourse.

The meltdown of financial markets in selected Asian economies must be viewed not in the exclusive context of Asia but in its regional context. On September 5, 1983, *The New York Times* reported the passing of "a milestone" in 1982 when the United States for the first time in its history did more trade on the trans-Pacific routes than on the trans-Atlantic routes. Indeed, the crossing of the "milestone", occurred one year sooner in 1981 (Dutta, 1987, also 1985). The sustained growth in the trans-Pacific trade – a two-way economic relationship – eventually led to the institutionalization of Asia-Pacific Economic Cooperation (APEC) in 1989 with eighteen Asia-Pacific member-economies. [As of November 1997, APEC membership has been increased to twenty-one.] The United States, Canada, Japan, Australia and New Zealand, as five members of the OPTAD, led the APEC movement (Dutta, 1998), thanks to the pioneering works of Kiyoshi Kojima, and of Kojima and Drysdale. The Asia-Pacific intra-regional economic cooperation remains a fact. The success of the latest APEC Summit in Kuala Lumpur in November 1998 can be noted by the fact that an Asian Fund was constituted. An intra-regional approach to correct the meltdown of financial markets in several Asian economies is in order.

Chairman Alan Greenspan of Federal Reserve System, American Central Bank, very rightly emphasized that no single national economy can remain a "mosaic" of economic prosperity when Asian economies collapse. Many of us recall reading Joan Robinson's celebrated critique of the "beggar thy neighbor" theory.

II. THE INDUSTRIAL REVOLUTION IN ASIA BEYOND JAPAN?

Krugman (1994), reviewing the works of Lau and Kim (1994), has questioned if the industrial revolution in Asia beyond Japan is at all feasible. Did this prompt speculative investors in Asia-Pacific move their hedge funds elsewhere? One will never be able to answer the question.

Harberger (1996) has used a two-deflator approach, yielding results easier to interpret and has presented results against the Krugman hypothesis. I have argued that "estimates of residuals" by a specification of any variation of the Cobb-Douglas-family, a single equation specification of a production function, appropriately homogenous of degree, is very likely to be biased and inconsistent, especially in economies with targeted economic plans for accelerated industrialization. I have argued for a multi-variate unit cost analysis with special emphasis on skill-based disaggregation of labor (Dutta, 1996, see p. 361). Leontief's pioneering effort with an exposition of skill-based disaggregation of labor input in the Input-Output Table framework has forcefully articulated the debate. Using the U.S. data, Dale Jorgenson has done extensive research by further gender-race-education-age specific disaggregation of labor input in industry-specific production functions.

I have earlier referred to the fact of availability of skilled and professional labor in most Asian economies. I have suggested that these Asian economies successfully completed the important first step of the new industrial revolution by an interface of import-export-led growth paradigm and that Asia's new industrial revolution is expected to initiate phases of progressive "deepening" by reaching out to higher technological frontiers. [In personal communications Professors Klein and Herberger approved of this proposition.] Concurrently, the process of "deepening" will be reinforced by the "widening" of these economies to other labor/resource-abundant, yet-to-be-industrialized economies on the Asian map.

Let it be further noted that with the rate of unemployment at an all-time low level in many industrially advanced economies, especially in the United States, the skilled labor has become a scarce input in the global market. The inflow of capital goods from these economies to the skill-labor abundant Asian economies will continue to be a rational economic proposition. The paradigm of import-export-led growth that was successful in the 1970s and 1980s will come to play its role in the new millennium. The sooner an intra-regional approach will be able to fix the present meltdown of the financial markets in several Asian economies, the better will be the prospect for the shared trans-Pacific economic prosperity.

This paradigm has been responsible for accelerated industrialization and rapid rate of growth of GDP in several Asian economies. At the other end, exports of the new manufactured products from these newly industrialized/ industrializing countries to the advanced industrialized countries not only facilitated their payments for the imported capital, but also contributed to the high growth and employment, concurrently holding the rate of inflation very low in the advanced industrialized countries. The question often has been raised if the rate of unemployment in the U.S. has been as low as 4.4% in 1998, will the rate of inflation remain as low as 2%? Indeed the open economic policy of the U.S., call it globalization, has made its contributions to this rare economic experience – high growth, low unemployment rate – low rate of inflation. Contributions of the trans-Pacific trade in this regard must be emphasized and estimated.

III. THE CASE FOR AN INTRA-REGIONAL APPROACH

The meltdown of the financial markets in several Asian economies warrants immediate attention. Any newly industrialized/industrializing individual economy in Asia, given its economic magnitude, can hardly be expected to cope with the situation. Each member economy's competitive ability is limited. The two parameters we use are: (a) the share of its GDP of the world GDP (GDPi/GDPw) and (b) the share of its exports of the world export (Xi/Xw).

For the U.S., the challenge of the financial meltdown of 1987 became no less overwhelming. Given its very large economic base with its shares of world GDP roughly at 20%, and its share of world trade at about the same magnitude, the institution of a special restitution fund became a solution as it struggled to work over the years.

Let us consider a simple model: The ith country's exports (Xi) are functionally dependent on the level of word GDP (GDPw), the price level at home in the ith country (Ph), the price level in the world market (Pw) and the level of export in the world market (Xw). Similarly, the ith country's imports (Mi) will be functionally dependent on the level of GDP in the ith country (GDPi), the level of exports of the ith country (Xi), the price level at home market of the ith country (Ph) and the price level in the world market (Pw) (Dutta, 1998).

We rewrite the two equations,

$$Xi = f(GDPw, Xw, Ph, Pw) \qquad (1)$$

$$Mi = g(GDPi, Xi, Ph, Pw) \qquad (2)$$

In equilibrium, and taking Ph/Pw on the left hand side, the Ph/Pw will be functionally dependent on

$$Ph/Pw = k(GDPw, Xw, GDPi, Xi) \qquad (3)$$

It follows that if the two variables, GDPi and Xi were competitive shares of GDPw and Xw, respectively, the ith country would have a competitive advantage. The relative stability of the home and world price ratio would be a comfortable base for a stable exchange rate for its currency. Foreign investors, exporting capital goods to this economy and/or importing manufactured products from the ith economy would have comfortably less exposure to exchange rate fluctuation risk. We limit our exercise to 18 member economies of APEC.

We present (Table 1) the shares GDPapec and of Xapec of each member country, i.e. GDPi/GDPapec and Xi/Xapec. The U.S. has a dominant position on both counts. Japan is the second richest economy in the world, its shares, GDPj/GDPapec and Xj/Xapec must be taken note of. We may note that globally, the EU with its 15 members (11 of the 15 joining EURO as of January 1, 1999) will have competitive shares of GDPw and Xw. The U.S. dollar and the EURO may emerge as the two competing global currencies.

As we review the 18 APEC member-countries the picture points to the non-competitive position of most of them. Hence the choice for a regional economic game plan is pronounced. Given the fact that most of these country's have their major trade in the Asia-Pacific region, the use of GDPapec and Xapec for GDPw and Xw, respectively, may substantively relate to the competitive positions of these economies.

Table 1. Estimates of GDPi/GDPapec and Xi/Xapec – 1995.

Member-Economies	GNPi/GDPapec	Xi/Xapec
Australia	0.022	0.0235
Brunei Darussalam	0.0003	0.0011
Canada	0.037	0.0085
Chile	0.0004	0.0071
PRC	0.048	0.0660
Hong Kong	0.009	0.0770
Indonesia	0.012	0.0201
Japan	0.319	0.1962
Korea	0.028	0.0554
Malaysia	0.005	0.0328
Mexico	0.021	0.0353
New Zealand	0.0033	0.0061
Papua New Guinea	0.0003	0.0012
Philippines	0.0046	0.0078
Singapore	0.0051	0.0524
Taipei	0.017	0.0495
Thailand	0.0126	0.0250
U.S.A.	0.4563	0.2586

Source: Data on GDPi and Xi from APEC Secretariat Compilations: World Development Report 1996; Brunei Darassalam Key Indicators 1995/1996; Statistical Data Book 1996, Chinese Taipei.
Note: Hong Kong and Brunei GDP, and GNP for all others. See also Dutta, 1998.

IV. DIRECT FOREIGN INVESTMENT (DFI)

I have argued that DFI, direct foreign investment with 100% foreign ownership, "wholly foreign owned" as it is referred to in the People's Republic of China, is a preferred mode vis-à-vis the financial capital inflows (see also Fry, 1996; Lloyd, 1996; Petri, 1993). DFI has also been in operation with joint ventures with indigenous entrepreneurs in several Asian countries. The terms of agreement between the foreign and domestic investors have, of course, been varied.

My arguments in support DFI, wholly foreign owned, are:

(a) Assumption of risk – profit or loss – 100% as is the case when it is wholly foreign owned;
(b) Optimum technology import enabling the foreign investors to manufacture cost and quality competitive products for the world market, generating export earnings in hard currency – a potential base for repatriation of profit home on terms as per agreement;
(c) Optimum human capital convergence (superior skill labor, engineers, scientists, managers) enabling the foreign investor to minimize cost of importing them from the home market, and assuring, first creation and then progressive expansion of the market in the country of investment;
(d) Optimum access to the world market based on his expertise and operation in the multinational program (brand name, marketing know-how, advertisement budget on a global scale), thus ensuring exports of a share of the products from foreign investment;
(e) Optimization of the time-horizon for investment, as is natural for direct investment, and minimization of volatility attributable to easy-in and easy-out of the movement of financial capital.

In addition, three indirect points merit attention:

Firstly, DFI will introduce a framework of competition in the economy prompting the indigenous entrepreneurs, who will be prompted to be competitively cost effective by way of introduction of necessary innovative measures.

Secondly, the broadening of access to the hard currency earning world market of products manufactured by DFI in the economy will earn a good-will for the country and eventually help expand its share of the world trade.

Third, and finally, to the extent DFI will help expand the economy's export earnings, its current account balance will be balanced and the pressure on its foreign exchange rate will be eased.

China's foreign exchange reserve has risen to U.S.$ 144 billion. This fact adds to the confidence of international investors' in the Chinese economy, thus is Asian economies. With all that hefty foreign exchange reserve, China recently (December 9) offered to raise funds in the Wall Street with a bond offering which was oversubscribed, helping China to raise twice as much funds as had been expected. "When the offering was priced, China had to pay investors an interest rate only 2.8% higher than the United States Treasury pays on its 10-year Treasury securities," reports David E. Sanger, The New York Times, December 10, 1998.

V. AN OPEN ECONOMY AND FREE FLOW OF FINANCIAL CAPITAL

With all that has been said about DFI, free flow of financial capital must constitute a part, an integral part, of open economic regimes. The inflow of funds from foreign investors – individuals as well as financial institutions must be allowed. If the funds are allowed to come in, it must have the freedom to move out. The rules of free market must be allowed to work. This is where the issue of a macroeconomic core of an open economy merits careful evaluation.

Asia's newly industrialized/industrializing economies have been known to lack a well-structured macroeconomic core which must be defined by monetary and fiscal guidelines which must be *transparent* to all concerned at home and abroad. Investors, foreign as well as indigenous, must be able to comprehend and abide by it. Let us focus on the monetary guidelines. With the notable exception of Taiwan, recent meltdown in financial markets in several Asian economies very much point to the absence of an organized monetary guidelines in the ith economy.

Independence and accountability of the central bank in the ith economy is crucial. It must command optimum independence with sufficient freedom from interference by the leaders of political establishment. However, the head of the central bank individually can and must be allowed to be a member of a political party. The degree of freedom of the office can be optimized if the tenure of the office is defined by law. Accountability of the office can and must be manifest in the periodic reports on the economic status of the ith economy, prepared and presented by the Head of the central bank.

Should there be regional central banks in a given economy? It is increasingly becoming evident that such an arrangement helps the formulation of the economic agenda of the economy and may be helpful for investors, domestic and foreign.

I argued both the above points (Dutta, 1995) in the context of China's economy. The U.S. Central Banking Act of 1913 set up such a structure. In January 1998 China acted to reform her central banking system by setting up regional central banks. In January 1998 a detailed plan to that end is expected to be announced. However, China's central bank continues to lack independence and accountability in the strict term we have stated above. I venture to suggest that when the European Monetary Institute (EMI) evolves to the European Central Bank, the central banks of its 15 or 11 member-economies will effectively play the role of regional central banks. The central bank will draw upon the inputs of the central banks of each member-economy, who in turn will relate their activities to the guidelines set by the European Central Bank. Even if the ith country in Asia is relatively small in magnitude the case for regional central banks merit consideration. They relate to the various regions of the economy and collate inputs toward developing a successful national monetary policy.

An effective central bank can function effectively only when the structure of commercial banks and related financial institutions are well-structured. In the 1970s and 1980s, too much of cooperation amongst the leaders of political, industrial and money-markets have been observed by many scholars and often hailed as "a happy marriage" between private capitalism in free market and government-led socially planned economies. Such an economic game plan was certainly politically correct. Were they economically vulnerable? The question remains to be answered.

Free movement of financial capital – inflow into and outflow from – the ith economy can prove too costly an experience for it in the absence of a well-organized money market, led by a properly institutionalized central bank. The recent experience of a giant-size long-term capital management fund in the U.S. with two Nobel Laureates on its Board points to the gravity of the situation. Thanks to the effective leadership of the U.S. central bank which helped minimize the potential adverse economic effects.

Asian economies, caught in recent financial crisis, have experienced the truth with much distress. An effort to restructure their respective money markets under proper leadership of central banks must be in order. Asian economies must be institutionally ready to enable their respective economies permit free flow of financial capital.

VI. A GLOBAL INSTITUTIONAL ARRANGEMENT

Familiarity of scholars in economics with the post-World-War II global institutions – the International Monetary Fund (IMF), The World Bank, and the

World Trade Organization (WTO) must be taken for granted. Of course, these institutions have a specific role to play in a financial crisis anywhere in the world. They have been engaged in the Asian financial crisis and they have helped ease the situation to a given extent. The IMF in particular has provided rescue funds to several economies in Asia and also in the Americas. We focus our attention to Asia.

In a crisis availability of rescue funds has been known to be an important economic action. On occasions in a given economic crisis, such rescue operations have been undertaken by effective mobilization of monetary and fiscal policies from within the nation. In the U.S. such arrangements in recent times will include special loans to an automobile corporation, a hefty rescue fund for the Savings and Loan Association Debacle and so forth. Asia's need for a rescue fund is no exception. However, Asia's problem is that no rescue fund for an economic crisis from within the ith economy is absent. Hence the need for the availability of a rescue fund from outside. The IMF is an world institution to deliver the necessary rescue funds to any of its member-economy.

Arguments against this institutional arrangement warrants serious consideration.

Firstly, these large global institutions are too large and thus unable to move with promptness. A financial crisis in Thailand or Korea or Indonesia calls for a prompt action. The 182-member IMF can hardly be expected to deliver the necessary funds to any of these individual economies in crisis so promptly.

Secondly, whatever rescue funds are made available by the IMF, it comes after the fact. Indeed, in the 1970s and 1980s, the IMF and also the WB joined the chorus of Asian miracle and continued to publish eloquent reports of dynamic growth of many of these Asian economies, supported by a relevant data bank.

Third and finally, the IMF prescription to the economies in financial crisis fail to note that the real sectors of these economies merit support toward the core agenda of accelerated industrialization to which they made a commitment. To persuade these economies to undertake steps to cut down and limit their investment schedules is a prescription to reverse their growth paradigm. Thus, the IMF prescription which is anti-growth contributes to the growth of unemployment rate and sub-optimal utilization of capital goods in these economies. The parental discipline code of "TIME OUT" is all too familiar. But will it be optimum economic game plan? Many of my fellow students of economics have raised the question.

We have argued above that economic inter-dependence of the Asia-Pacific economies and the rest of the world economies has grown to be much too significant. Indeed, these fastest growing economies in Asia have been contributors to the economic dynamism of the rest of the world. We suggest that the global institutional arrangements under the IMF and the WB must be allowed to do their chores under their respective charters. We argue for supplementing the global arrangements by a regional Asia-Pacific institutional arrangement. The rest of the world must be ready to learn form the regional institutional arrangements in Western Europe (Hesse, 1993; Letiche, 1993). We do not propose to replicate EU with its EURO. We do propose that Asia-Pacific regionalism in the context of globalism must be the new order.

VII. ASIA-PACIFIC REGIONAL APPROACH

Indeed, Asia-Pacific economic regionalization of eighteen Asia-Pacific economies (twenty-one as of November 1997) in the form of APEC in 1989 remains a fact. Annual APEC Summits since 1993 have been eventful occasions. Heads of Treasury or Finance Ministries, and of Central Banks of APEC member-economies, joined by their Respective Deputy Heads, have been regularly convening at annual meetings of their own. They last met in Hong Kong October 1997. Indeed, here is a core platform of an Asia-Pacific regional monetary institute. They perhaps elected not to give any warning to their membership, or to the rest of the world, about the financial meltdown that overtook several of their member-economies immediately after their Hong Kong meeting. Of course, they had a good deal to focus on the Thai financial crisis.

The above forum could indeed function effectively as a regional central banking institution for the coordination of monetary policies of the region. They could jointly study and produce the annual report of the member-economies with respect to monetary policy management and ensure their joint accountability by way of pointing out to the phenomenon of "irrational exuberance" for any member-economy. They could play the role of a regional central bank by way of initiating regional market control game. Indeed, the central bank in a given sovereign economy has the responsibility to manage its free market economy by initiating counter-speculative, anti-inflationary monetary policies and/or pro-growth game plans based on rational investment decisions.

Are the APEC member-economies ready to accept the challenge? In 1996–1998 I had a series of personal conferences with members of APEC's Eminent Persons Group (EPG), and with several academic economists engaged in long-term research on APEC in Australia and Malaysia plus the President of Asian development Bank. I reviewed the issue of a possible/feasible APEC macro-economic core, especially the case for an APEC regional monetary institute (Dutta, 1998; Dutta, 1995; Klein, 1994; Yamazawa, 1994; Yoo, 1996). I have reported that there is a sense of awareness of the importance of the issues involved but there is a pragmatic judgement to wait for the things to evolve in its due course. Maybe, the financial meltdown in several Asia-Pacific economies will be an occasion to prompt a sense of immediacy. Maybe, the inauguration of EURO by the EU January 1999 will be an occasion to focus on the issue. Or, there will be an organized effort not to learn the lessons of the experiment in Western Europe and advocate a moratorium on economic regionalization elsewhere in the world – a poor counter-economic argument.

Permit me to add that given the relatively small magnitude of most the APEC member-economies, as measured by their respective shares of world GDP and world export (see Table 1), the ith Asia-Pacific economy, even if it has been successful in restructuring its indigenous macroeconomic core with well-defined monetary and fiscal parameters, will still remain too small and too weak to be exposed to the dynamics of accelerated industrialization and economic development. An intra-regional group macroeconomic core is in order. Or, the wisdom of the economic traditionalists will be that the Asia-pacific economies may give up the accelerated economic growth agenda. Let their choice be sustainable growth defined by predetermined parameters of the global institutions. An optimal option will be a regional Asia-Pacific macroeconomic institutionalization within the overall global framework.

VIII. THE NEW MILLENNIUM AND THE NEW INTERNATIONAL ECONOMIC ORDER

The challenges of the new millennium will, of course, be real. Our process of learning what we have not yet learned must continue to progress. However, the real challenge, I believe, will be to unlearn a great deal of what we had learned in the past. I have argued that absence of a well-structured macroeconomic core of an individual sovereign nation-state based economy will prove a serious bottleneck for its competitive functioning as an open economy. The macroeconomic core refers to a system of monetary and fiscal agenda. In the present context focus is more on the monetary agenda. I have further argued that an intra-regional macroeconomic core within a global macroeconomic system is in order.

The proposal for a global convention of the Heads of the Central Banks of the world is receiving serious consideration. May the IMF be restructured as the new World Central Bank. May the World Bank and the IMF be merged into a new institution to serve as the World Central Bank. Recently, a leader of one of the largest international hedge fund has authored a monograph advocating the institution of the World Central Bank. If the 182-member IMF is reconstituted as the new World Bank, it will continue to suffer from its present limitation, inability to take prompt, *preemptive* decision and action.

Economic regionalization in its many and varied form within the global order has indeed been a part of our economic history. The imperial order became outdated. The post-WWII order of globalization based on the concept of one world, reinforced by institutions, (IMF and WB) struggled to do its job, and until disintegration came following severance of the fixed gold value of the U.S. dollar in August 1971. Another form of polarization we witnessed when the national economies broke into two groups – the North of the rich nations and the South of the poor nations. Our continued reference to the three worlds – the first world and the third world even when the second world experienced its liquidation is a bad mistake. A global economic order continues to elude us.

A map of the worldview of a region has become the focus of the new economic regionalization. The EU illustrates this. The EU has developed its own independent macroeconomic core, allowing each member-economy of a sovereign nation state maintain its own macroeconomic core. On the other hand, EU is consistent with global institutions – IMF, WB and WTO. The challenge of the new millennium may be to accept a new international economic macroeconomic core with regional macroeconomic institutionalization. Let the sovereign nation-state based individual economies in any and all regions of the world welcome this new international economic order (see Dutta, 1999).

REFERENCES

Dutta, M. (1999). *Economic Regionalization in the Asia-Pacific: Challenges to Economic Cooperation* (p. 325). London, U.K.: Edward Elgar Publishing.

Dutta, M. (1998). People's Republic of China and Asia-Pacific Economic Cooperation an invited lecture at the international symposium, sponsored by Shaghai Academy of Social Sciences, Shanghai, China, September 21.

Dutta, M. (1996). The New Industrial Revolution in Asian Economies: Has It Reached Its Ceiling? *Journal of Asian Economics*, *7*(3), 357–364.

Dutta, M. (1996). Asia-Pacific Economic Cooperation: A Conversation with Jang-Hee Yoo. *Journal of Asian Economics*, *7*(3), 537–549.

Dutta, M. (1995). Macroeconomics – Supra-national Macroeconomic Core. In: M. Dutta (Ed.), *Economics, Econometrics and the LINK: Essays in Honor of Lawrence R. Klein* (pp. 59–73). Elsevier: North-Holland, Amsterdam.

Dutta, M. (1995). Symposium on Asia Pacific Economic Cooperation. *Journal of Asian Economics*, *6*(4), 511–588.

Dutta, M. (1995). The Macroeconomic Core of China's Socialist Market Economy. *Journal of Asian Economics*, *6*(2), 291–296.

Dutta, M. (1987). The Asia-Pacific Economic Community: Some Comments. In: M. Dutta (Ed.), *Asia-Pacific Economies: Promises and Challenges*. In: H. Peter Gray (Ed.), *Series on Research in International Business and Finance, Part A* (pp. 93–100). CT. Greenwich: JAI Press.

Dutta, M. (1985). Asian Economic Studies – A New Framework in Studies. In: M. Dutta (Ed.), *United States-Asia Economic Relations* (pp. 1–5). Acorn Economic Communication Series 2, NC, Durham: The Acorn Press.

Dutta, M., & Tantum, D. (1988). Industrialization of Japan and Selected East Asian Countries: The Structural Divergence Hypothesis. In: M. Dutta (Ed.), *Asian Industrialization: Changing Economic Structures, Part A* (pp. 17–46). Greenwich, CT: JAI Press.

Fry, M. J. (1996). How Foreign Direct Investment in Pacific Asia Improves the Current Account. *Journal of Asian Economics*, *7*(3), 459–486.

Harberger, A. C. (1996). Reflections on Economic Growth in Asia and the Pacific. *Journal of Asian Economics*, *7*(3), 365–392. The keynote address given by the author at the Conference on The New Industrial Revolution in Asian Economies, jointly sponsored by American Committee on Asian Economics, The Malaysian Economic Association and the University of Malaya, May 14–17, 1996.

Hesse, H. (1993). Ecu – Now and Later, Some Considerations after Maastricht. *Journal of Asian Economics*, *4*(1).

Klein, L. R. (1994). The Treatment of Pacific Economies in the LINK System. *Journal of Asian Economics*, *5*(3), 313–334.

Klein, L. R. (1990). Can Export-led Growth Continue Indefinitely? An Asia-Pacific Perspective. *Journal of Asian Economics*, *1*(1), 1–12.

Krugman, P. (1994). The Myth of Asian Miracles. *Foreign Affairs*, *73* (November-December), 62–78.

Lau, L. J., & Kim, J. I. (1994). Economic Growth of the East Asian Newly Industrializing Countries. *Journal of the Japanese and International Economies* (in Japanese), 235–271.

Letiche, J. M. (1993). Maastricht: Prospect and Retrospect. *Journal of Asian Economics*, *4*(1), 191–206.

Lloyd P. J. (1996). The Role of Foreign Investment in the Success of Asian Industrialization. *Journal of Asian Economics*, *7*(3), 407–433.

Petri, P. (1993). Is the United States Bowing Out of Asia? *Journal of Asian Economics*, *4*(2), 283–300.

Yamazawa, I. (1994). Asia-Pacific Economic Community: New Paradigm and Challenges. *Journal of Asian Economics*, *5*(3), 301–312.

Yoo, J.-H. (1996). Asia's Economic Growth in the Next Decade: Opportunities and Constraints. *Journal of Asian Economics*, *7*(3), 393–406.

3. THE EURO, DOLLAR, AND INTERNATIONAL MONETARY SYSTEM

Jürgen Schröder

I. INTRODUCTION

At the beginning of January 1999 the Euro will be introduced in 11 of the 15 European Union (EU) member countries. Right from the start Great Britain, Denmark, and Sweden decided not to take part in the European Monetary Union (EMU). Greece didn't qualify because it didn't meet the Maastricht criteria. Monetary policy in the 11 countries, called Euroland, will be conducted by the European Central Bank (ECB). There are no longer any national monetary policies in Euroland. The national central banks will still exist as part of the European System of Central Banks (ESCB), but their only task will be to execute the monetary policy of the ECB. In Euroland the Euro will be legal tender. In each country the national currency will be legal tender for the first three and a half years alongside the Euro. The rate between each national currency and the Euro will be fixed on December 31, 1998, and will be irrevocable. For the first three years, the Euro will be in circulation only as deposit money. Euro coins and notes will be introduced during the first half of 2002, when national coins and notes have to be converted into Euro money. Beginning August 2002 the Euro will be the only legal tender in Euroland.

The fact that 11 sovereign countries decided to give up their national monetary policies in favor of a common currency, and with it a common monetary policy, is unique in monetary history. This quantum leap has

Restructuring Asian Economics for the New Millennium, Volume 9A, pages 25–35.

ISBN: 0-7623-0709-9

considerable economic benefits but also economic costs for the countries involved. The economic benefits are mainly at the microeconomic level, such as savings in transaction costs. The economic costs are mainly at the macroeconomic level, such as losing the exchange rate as an instrument to protect a country against external monetary and real shocks. It has been well known since McKinnon's[1] and Mundell's[2] classical contributions to the theory of optimum currency areas that the more open a country, the more it enjoys the benefits of a common currency and the less open a country, the higher the cost of giving up the exchange rate as a substitute for rigid labor markets when there are external, real, asymmetric shocks. Because the small countries in the EU, like Luxembourg, Belgium and The Netherlands are very open, they are the ones who will greatly enjoy the benefits. Conversely, the larger countries, like France and Germany, are less open and face quite rigid labor markets. From this point of view, they are the ones who will carry the costs of EMU, to say nothing of the fact that Germany has to give up its stable currency, the deutsche mark which is admired all over the world. Nevertheless, it has been France and Germany and their politicians who were the driving forces for European economic and monetary integration since World War II. This suggests that there are benefits other than the purely economic ones connected with economic integration processes. One of these benefits is the expectation of a more stable and peace-oriented political environment in Europe as a result of economic and monetary integration. If this turns out to be true, the integration process certainly would pay off, especially with Europe's far from peaceful history.

In this paper I will discuss the question of whether the Euro will become an international currency and successfully compete with the dollar in international markets. This is an interesting and important question, especially for the future of the international monetary system. I will proceed in the following way.

First, a brief discussion of the functions of an international currency will be given. Second, it will be shown that currently it is the U.S. dollar that mainly fulfills these functions. Third, the criteria for a national currency becoming an international currency will be discussed. Finally, it will be investigated whether the Euro and Euroland will meet these criteria.

2. FUNCTIONS OF AN INTERNATIONAL CURRENCY AND THE IMPORTANCE OF THE U.S. DOLLAR

Under normal monetary and economic conditions, money has the following three functions at the national level: it serves as a "unit of account", "medium of exchange" and "store of value". In open economics, in which the real as well

Table 1. Functions of an International Currency.

	Private	Official
International Unit of Account	Invoice currency	Parity fixing
International Medium of Exchange	Vehicle currency	Intervention currency
International Store of Value	Private reserves	Official reserves

as monetary sectors of the national economies are internationally integrated, a medium is needed that fulfills these functions on the international level.

Table 1 shows the functions of an international currency in both the private and official sectors.[3] The national currency, which is used for invoicing international trade in goods and services, serves as the international unit of account in the private sector. In the official sector, the national currency to which countries peg their currency serves as the international unit of account.

Table 2 shows that the U.S. dollar is the national currency that predominantly serves as the international unit of account in the private sector. Although the U.S. share in world trade is 13.5%, 48% of world exports are invoiced in U.S. dollars. The deutsche mark plays a positive but much smaller role as an international unit of account. Germany's share in world trade is 9% and 15% of world exports are invoiced in deutsche mark. The share of the other EU countries in world trade is larger than the use of their currencies for invoicing world exports. This is also the case for Japan.

In the official sector, the U.S. dollar also serves predominantly as the international unit of account. With n currencies there are n − 1 exchange rates. In a system of fixed exchange rates there is the well known n − 1 problem, i.e. one country can neglect the exchange rate as long as the others peg their currency to it and take care of the external equilibrium. There are about 160

Table 2.

Share in world trade (1992, percent)		Invoice of exports (1992, percent)	
USA	13.5	U.S.$	48.0
Germany	9.5	DM	15.0
EU (excl. Germany)	28.5	EMS currencies (excl. DM)	18.0
Japan	7.1	Yen	5.0

Source: ECU Institute.

countries in the world with national currencies. Roughly two third of these countries have decided on a flexible exchange rate or controlled floating, and one third have their currencies pegged. Again, about one third of the latter are pegged to the U.S. dollar and the other two thirds to currency baskets in which the U.S. dollar is usually an important part, or they are pegged to other currencies.[4]

The second function of an international currency, the international medium of exchange, usually goes along with the first function, i.e. the national currency that serves as the international unit of account also serves as the international medium of exchange. In the private sector this is known as the vehicle currency; i.e. the private transactions in the foreign exchange market take place in this currency. For instance, if an importer in Finland buys goods in India and has to pay in Indian rupees, he does not buy the Indian rupees directly on the foreign exchange market against the Fin-Mark. Instead, he first buys U.S. dollars and then buys the Indian rupees with these in order to pay his bill. Using The U.S. dollar as the vehicle currency is cheaper than going directly from, the Fin-Mark to Indian rupees. This is so because of the huge U.S. dollar foreign exchange market the bid-ask-spread on the dollar exchange rate is much smaller than the bid-ask-spread on the Fin-Mark Indian rupee exchange rate.

Table 3 shows that the U.S. dollar is the predominant vehicle currency, although its importance has decreased. Whereas in 1989 the U.S. dollar's share in two-way foreign exchange transactions was 90% in 1995 it was only 83%. In this period the share of the deutsche mark us a vehicle currency increased from 27% to 37%, again in two-way transactions. In the official sector the U.S. dollar is by far the most important medium of exchange, because foreign exchange market interventions are mainly undertaken in U.S. dollars regardless of whether there is dirty floating or a fixed exchange rate to defend. Even in the

Table 3. Share in Foreign Transactions (Two-way, Percent).

	1989	1995
U.S.$	90	83
DM	27	37
Yen	27	24
Pound Sterling	15	10
Other EMS currencies	3	21

Source: BIS.

Table 4. Share (Percent) in

	Intl. securites markets	Intl. bank assests	Official reserves	
	1997	1997	1975	1997
U.S.$	39.6	44.6	78.2	65.1
DM	10.1	14.3	8.8	13.1
Yen	16.9	10.9	1.8	6.1
Pound Sterling	7.9	4.5	2.8	3.5

Source: BIS.

European Exchange Rate Mechanism (ERM), in which the U.S. dollar is not involved, the U.S. dollar was used quite often as the intervention currency.

The third function of an international currency, the international store of value, is also taken over predominantly by the U.S. dollar in both the private and official sectors. This is shown in Table 4.

If one takes both the international securities market and the international bank assets as a measure for the international store of value, which is a common indicator, the U.S. dollar, with 39.6% and 44.6%, is the predominant national currency that fulfills this function in the private sector. In the official sector, the official reserves of central banks are the international store of value. Again, as Table 4 shows, the U.S. dollar dominates, followed by the deutsche mark and the yen. However, the dominance of the U.S. dollar has decreased. Whereas in 1975 78.2% of all foreign exchange reserves of the central banks were U.S. dollars, in 1997 its share was only 65.1%. During this period, the shares of the deutsche mark and yen have increased from 8.8% to 13.1% and from 1.8% to 1.1%.

3. CRITERIA FOR A NATIONAL CURRENCY BECOMING AN INTERNATIONAL CURRENCY

National money cannot become international currency by law. It must he attractive for people around the world to voluntarily use this particular currency as international currency. One important criterion for a national currency to be used at the international level is its stability and free convertibility; more importantly, people must expect that the currency will be stable and freely convertible in the future. For this to be the case, certain systemic, institutional, and economic policy fundamentals, and some pure economic fundamentals have to he achieved. A necessary but not sufficient condition for a currency to be accepted at the world level as an international currency is a free market economy in that currency's country. If a country is a planned economy, the governments promise of a free currency convertibility is not credible because people expect that sooner or later the exchange rate and the foreign exchange will also be controlled. One extremely important institutional condition for a national currency to be used as international currency is the political and instrumental independence of the central bank. History shown that this is a necessary but again not sufficient condition for the price stability that is necessary for the money to be accepted as international currency. The possibility that politicians can misuse the central bank by financing government deficits with new printed money undermines the confidence in a sound monetary policy and makes the use of this currency unattractive to foreigners. Moreover, it is not only monetary policy but also economic policy that has to be consistent and stable in a stable political environment, because a bad economic policy usually has a negative impact on the currency's external value and its attractiveness to foreigners.

Next to these systemic, institutional, and economic policy fundamentals there are some pure economic fundamentals that have to be achieved if a national currency is to be voluntarily used as an international currency. These fundamentals are described as follows. A country whose currency fills the role of an international currency has to he quite immune to world market instabilities. This is the more likely the less open the economy, i.e. the smaller the export and import shares in the gross national product (GNP). On the other hand, this economy must be large enough to satisfy the other countries' wishes to use the money and capital market of this country without causing a negative impact on key economic variables. To fulfill these two economic conditions simultaneously, the country must have both a large share in the world GNP and small export and import shares in the domestic GNP. Although such a country is not very open, exports and imports are relatively large compared with world exports and imports because of its sheer size. The U.S. was, and still is, the country that best fulfills these conditions, although this advantage is diminishing. After the Second World War the U.S. GNP was about 50% of the world GNP, and both exports and imports over GNP were roughly 4%. At present its share of the world GNP is about 30%, while export and import shares have increased to about 10%. This means that the objective conditions for the U.S. dollar to be used as international currency have deteriorated because the U.S. and the dollar have become more vulnerable to economic disturbances caused by world markets.[5]

Besides these fundamental conditions, which have to he fulfilled for a national currency to be voluntarily used as an international currency, there are some special circumstances that helped the U.S. dollar become an international currency. The U.S. dollar had a special role in the Bretton Woods system. Article IV (1a) of the IMF treaty allowed members to fix their currency against either gold or the U.S. dollar.[6] The possibility of using the U.S. dollar as a numeraire was used by all countries. This gave the U.S. dollar a tremendous starting advantage as an international currency in the official sector. Furthermore, within the Marshall Plan the huge monetary transfer from the U.S. to Western Europe was executed in U.S. dollars, and the European Payments Union (EPU) used the U.S. dollar as a vehicle currency. This gave the U.S. dollar a considerable starting advantage as international currency in the private sector.

4. THE EURO'S CHANCES OF BECOMING AN INTERNATIONAL CURRENCY

The special circumstances after the Second World War that helped the U.S. dollar become a hegemonic currency are at present only to a very limited extent available to help the Euro. Within the framework of the Exchange Rate Mechanism Two (ERM II), the EU countries that did not introduce the Euro can peg their currencies against the Euro before introducing it as legal tender. This certainly gives the Euro an advantage as an international currency in the official sector in these countries. However, in the long run, the fundamentals discussed above determine whether a given currency will he voluntarily accepted as international currency.

Let us first discuss the economic fundamentals. Tables 5 and 6 compare the economic fundamentals of the EU and Euroland with those of the U.S. Both tables show that Euroland performs quite well.

Table 5 shows that Euroland (column 2) has roughly the same population, GDP, and GDP per capita as the U.S. The whole EU, i.e. Euroland plus the four other EU countries (column 1) has a larger population and larger GDP than the U.S. Concerning openness, the EU, with 23.2% of foreign trade to the GDP, is much more open than the U.S., with only 9.4%. This means that the EU is much more vulnerable to what happens in the rest of the world than the U.S. However, half the EU trade is intra Euroland trade, which in the future will be invoiced with domestic money, i.e. with Euro. This implies that the openness will be only about half of what is shown in column 1, i.e. the EU will be as open as the U.S. The same holds for the percentage share in world trade, which is 37.5% for the EU and 13.5% for the U.S. at present. Allowing for the trade

Table 5. Economic Fundamentals (1996).

	EU (15)	EU (Exl. DK, GR, U.K.)	USA	Japan	Germany
Population (mill.)	373	298	270	126	82
GDP (bill. U.S.$)	8608	7153	7636	4604	2353
GDP per capital (U.S.$)	23078	24003	28281	36540	28695
Openness, percentage share of foreign trade(goods) in GDP	23.2		9.4	8.3	20.4
Percentage share in world trade in goods (in brackets after introduction of the Euro)	37.5(20)	31.3	13.5(20)	7.1(10)	9
Growth rate of real GDP 1980–1996	2.1		2.4	3.1	2

Source: BIS, IMF.

within Euroland, the percentage share in world trade will be 20%, the same in the EU and the U.S.[7] Just for comparison, Table 5 also shows the economic figures for Japan and Germany. From these figures it is clear that the economic fundamentals of the Euro are much better than those of the deutsche mark or yen for challenging the U.S. dollar as international currency.

For a national currency to take over the functions of an international currency, a large and well developed domestic capital market is necessary. This is another important economic fundamental.

Table 6 compares the capital markets of the EU, Euroland, U.S., Japan, and Germany and clearly shows the dominance of the U.S. in this area. The capital

Table 6. Capital Markets, 1996 Bill. U.S.$.

	EU(15)	EU (excl. DK, GR, U.K.)	USA	Japan	Germany
National and international securities	9040.4	7599.0	12018.2	5128.0	2221.7
Private	4052.5	3349.2	4879.6	1801.6	1360.4
Government	4987.9	4249.8	7138.6	3326.4	861.3
Stock market capitalization	4499.0	2689.3	8817.5	2996.4	666.0
Capital markets (stocks and securities)	13539.4	10288.3	20835.7	8124.4	2887.7

Source: BIS.

market volumes of national and international securities is 12 trillion U.S. dollars in the U.S. and only 7.6 trillion U.S. dollars in Euroland. The disadvantage of Euroland in the stock market is even greater. The volume of this market is about 8.8 trillion U.S. dollars in the U.S., whereas in Euroland it is only about 2.7 trillion U.S. dollars. Because of this relatively small stock market in particular, the total capital market in Euroland of 10.2 trillion U.S. dollars is only half the size of the U.S. capital market of 20.8 trillion U.S. dollars. Considering that the GDP is about the same in Euroland and the U.S., it is clear that some catching up is needed in the area of the capital market.

What about the systemic, institutional, end economic fundamentals of Euroland? This is a broad field and predictions are quite hard to make. Nevertheless, some comments are worth discussing. Concerning the systemic fundamentals, the eleven countries of Euroland are free market economies like the U.S., but their priorities are different. Compared with the U.S., Euroland is more of a welfare state. The more private economic initiative is handicapped by this welfare state environment, the greater the long-term disadvantage of the Euro vis-à-vis the U.S. dollar as an international currency. In 10 of the 11 Euroland countries, there are now social democratic or socialist parties in the government. Historically, these parties especially have given high priority to welfare-state considerations, but this can change. "New Labor" in Great Britain seems to indicate this.

The important institutional fundamental for a stable and convertible currency, i.e. the political and instrumental independence of the central bank, is very well taken care of in Euroland. The treaty of the ESCB makes the ECB the most independent central bank in the world, with price stability as its primary objective. It says very clearly in Article 105 (1) of the Treaty of the European Union, "The primary objective of the ESCB shall be to maintain price stability. Without prejudice to the objective of price stability, the ESCB shall support the general economic policies in the Community . . .". The election of the first Executive Board is also very promising. The first ECB governor, the Dutchman Wim Duisenberg, and the other five members of the Board have good reputation and are very experienced. In Kenneth Rogoff's[8] words, they are "conservative central bankers", i.e. their objective is price stability. This is promising for the Euro becoming a stable currency. However, the ECB is a new institution, and new institutions are not credible per se; they have to earn credibility by carrying out a convincing policy. Monetary policy will be carried out by the ECB Council, which consists of the Executive Board and the central bank governors of the 11 countries in Euroland. Decisions are made by simple majority vote, i.e. the members of the Executive Board can be overruled in the ECB Council by the 11 national central bank governors.

Last but not least, there are economic policy fundamentals that have to be met in order for the Euro to become stable and challenge the U.S. dollar at the international level in the long run. There are some promising facts in thin area. First, all eleven Euroland countries met the "Maastricht Criteria". Secondly, all 11 Euroland countries agreed to and voted for the "Stability and Growth Pact" which prescribes a ceiling of 3% for the government deficit/GDP ratio in the 11 countries for each future year.[9] A decrease in public sector economic activity in Euroland is essential for a strong Euro. Therefore, it is hoped that Euroland will stick to and carry out this pact in the future. However, there are some economic policy problems in Euroland that have to be solved if the Euro is to be credibly stable in the long run. Unemployment is Euroland's biggest problem. The unemployment rate in Euroland is over 10% while in the U.S. it is only a little over 4%. The main reason for this high unemployment is the rigid labor markets in most of the Euroland countries. Labor mobility within both the countries and Euroland are too low, and – what is even worse – real wage flexibility is much too low in almost all Euroland countries. How damaging this inflexibility in labor markets will be in the future depends very much on the structural economic developments in Euroland. If further economic integration leads to less real divergence in Euroland the possibilities of negative real asymmetric shocks should diminish.[10] In this case flexible labor markets are not that badly needed. However, if economic integration leads to more real divergence in Euroland, the possibility of negative, real, asymmetric shocks increases.[11] In this case flexible labor markets are badly needed, because exchange rate flexibility as an (at least temporary) adjustment instrument is no longer available once the Euro is introduced. In any case, more flexibility in Euroland labor markets is essential to get rid of the high unemployment. Otherwise, the pressure on the ECB to use monetary policy to help overcome unemployment will increase.[12] This pressure on the ECB certainly does not add to the stability of the Euro and the long-term attractiveness of the Euro as international currency.

In conclusion, the economic fundamentals for the Euro becoming an international currency are good. Improvements are still necessary in the area of systemic, institutional, and economic policy fundamentals. Less emphasis on welfare state considerations and more confidence in private initiative will certainly increase the economic vitality of Euroland and thus the international attractiveness of the Euro. The status of the ESCB as a very independent monetary institution is a very important positive institutional fundamental. However, the ECB has to earn credibility in the markets, which takes time. In the area of economic policy fundamentals, more flexibility in the labor markets of the Euroland countries is needed to get rid of the unemployment. For a national currency to be attractive as an international currency. There must not only be a large economy behind that currency but this economy must he strong and people must have confidence in its future strength.

Finally, a word on the time horizon: In the period of the gold standard the British pound served as the main international currency. It took the U.S. dollar almost half a century to replace the British pound. There is no question that it will also take time for a stable Euro to be accepted as international currency.

NOTES

1. McKinnon, R. I., Optimum Currency Areas, American Economic Review, 1963, pp. 717–725.
2. Mundell. R. A. (1961). A Theory of Optimum Currency Areas, American Economic Review, 1961, pp. 657–665.
3. Cooper, R. N., The Gold Standard: Historical Facts and Future Prospects, Brookings Papers on Economic Activity, 1982, pp. 1–45.
4. The current exchange rate arrangements of the various countries are published by the IMF.
5. It might be that these world economic developments have been one reason – among others, for instance, the loose monetary and economic policy in the Johnson and Carter eras – that the Deutsche Mark and the Yen were partly substituted for the U.S. dollar as international currency.
6. The main reason for this was that the U.S.-central bank stood ready to convert dollar reserves of other central banks into gold at a constant price, 35 U.S.$ = 1 ounce gold.
7. With the birth of Euroland, the percentage share of the U.S. trade in world trade will increase, because world trade will decrease by the amount of the intra Euroland trade.
8. Rogoff, K., "The Optimal Degree of Commitment to an Intermediate Monetary Target", Quarterly Journal of Economics, 1985, pp. 1169–1190.
9. Only when a country experiences a severe economic downturn, which is defined in an annual fall of real GDP of at least 2%, is the 3% ceiling not binding.
10. This is the "optimistic" European Commission view. EC Commission, "One Market, One Money", European Economy, 1990.
11. This is Paul Krugman's view. Krugman, P., "Geography and Trade", Cambridge. Mass. 1991.
12. It is well known at least since Barro's and Gordon's contribution that labor market problems cannot be solved through monetary policy. Barro, R. and Gordon, D., "Rules, Discretion and Reputation in a Model of Monetary Policy", Journal of Monetary Economics, 1983, pp. 101–121.

4. THE ARCHITECTURE OF THE MULTILATERAL ORGANIZATIONS

Peter J. Lloyd

1. INTRODUCTION

After the onset of the Asian crisis in July 1998, there has been much discussion about changes in the world economy and especially those in world financial markets. One part of this discussion is the question of rebuilding what has come to be known as the "International financial architecture". In April 1998 the Interim Committee of the International Monetary Fund called on the Executive Board of the IMF to consider steps to strengthen this architecture (see IMF, 1998 and other material on the IMF website www.rmf.org). The organization for Economic Co-operation and Development (OECD), G-7, and other fora are also examining these issues. There has been discussion of the possible creation of a new multilateral institution to oversee the financial sectors of national economies.

In addition to the IMF, the second multilateral organization that plays a role in capital markets is the World Bank, in this case the provision of development capital to developing and transitional economies. This current debate regarding the roles of the IMF and the World Bank is the latest stage in a much longer debate about their roles.[1] (For a view from the IMF and World Bank, see Boughton and Sarwar, 1995.)

The third major multilateral organization is the World Trade Organization (WTO). There has been a vigorous debate about the role of the WTO since it began operation in 1995. In contrast to the debates about the IMF and the World Bank, this one has been about possibilities of extending its powers to the

Restructuring Asian Economics for the New Millennium, Volume 9A, pages 37–68.

ISBN: 0-7623-0709-9

so-called new issues rather than restricting or redirecting its basic powers. These are the issues of trade and the environment, trade and labor standards, trade and competition, and trade and investment, all of which surfaced during the negotiations of the Uruguay Round.

Thus, the role of each of the three "Bretton Woods sisters"[2] is being seriously questioned. The basic role of each of these three institutions was laid down more than 50 years ago at the Bretton Woods Conference in 1944 and, in the case of the General Agreement on Tariffs and Trade (GATT), the Havana Conference. These institutions are now more than 50 years old. There have been major changes in the structure of the world economy during this time and especially in the 1990s.

This paper differs from most of the current discussion in that it looks at the three organizations. As the functions and activities of the three organizations overlap, it is desirable to consider them together. In essence, questions of architecture can be regarded as questions concerning the optimal assignment of functions to each multilateral organization. This may involve other existing organizations or a new one.

As a prelude to consideration of possible reforms of the multilateral organizations, section 1 discusses the changes in the world economy. Section 2 outlines the basic role of each of three main organizations. In outlining the activities of each organization, I concentrate on their objectives and responsibilities, membership, internal architecture, and also their surveillance activities and treatment of developing countries as these are central to the Asian crisis. Section 3 identifies three problems within the architecture of these organizations: a lack of coherence within the organizations, a lack of coherence between them, and gaps in their coverage of financial markets. Section 4 considers possible ways in which the international architecture could be redesigned. This emphasizes the interdependence between markets for goods, services, and capital as well as the interdependence among national markets that is a feature of the new global economy.

1.1. The Formation of a New Global Economy

First, we need to distinguish the "global economy" from the "world economy". The "world economy" is simply the set of all national economies. The term "global economy" means something more. It is a qualitative property of the world economy. By the term "global economy" I mean that the national economies of the world are integrated to such an extent that the interdependencies among them are now so large that one cannot consider events and policies in one national economy separate from those in others. There is a global economy in the economically meaningful sense of a set of markets that span the world economy.

The integration of markets across national borders is a consequence of the lowering of barriers to cross-border trade. Regrettably, it is not easy to document the lowering of these barriers because of the absence of systematic global measures of trade barriers.

For international trade in *goods*, there are estimates of time series of levels of border barriers to trade. The IMF (1997, p. 112) graphs a long-term time series of average tariff rates for 12 advanced countries, weighting the average tariff series of each advanced country by its Gross Domestic Product (GDP). This shows that the average tariffs have fallen from 12% in the early 1960s to 3% in 1990. However, the series is limited in its country coverage and does not include non-tariff measures.

Sachs and Warner (1995) adopt an approach to the analysis of global integration that is useful in the present context. They divide the world into "open" and "closed" economies. They construct a dataset for some 135 countries and then partition these national economies into those that are sufficiently open to be classified as "open" and those that are not. To make this partition, they use various measures. They regard an economy as closed if average tariff rates are 40% or more, non-tariff barriers cover 40% or more of trade, there is a premium in the foreign exchange black market of 20% or more, or there is a state monopoly of major exports or socialist economic system. Their data show that most of the developed countries had opened their economies after the Second World War by the early 1960s, but most of the Developing Countries had not. More of the developing countries opened their economies to international trade in goods during the 1980s and 1990s, although 35 were still closed in 1994 according to their criteria. This dating of the year of opening is a crude but useful way of documenting the liberalization of barriers to trade in goods. It highlights the almost universal nature of the trend in terms of the countries participating in it.

According to their estimates, the open economies accounted for less than 50% of aggregate world production in 1960, but this share rose to almost 70% by 1995. If one uses population weights rather than GDP weights for countries, the share of the open economies is much lower and passes 50% only in the 1990s, "It was not until 1993 that more than 60% of the world's GDP, and more than 50% of the world's population, was located in open economies" (Sachs & Warner, 1995, p. 2). By this time one can certainly regard the existence of a true global economy as a reality, even though it still excludes some large economies such as China and Russia.

This process of integration has been greatly extended by the collapse of the former Soviet Union and the lurch of the former communist or socialist states toward a greater market orientation. This is the movement that Francis Fukuyama (1989) entitled rather grandly the "end of history". By this he meant the emergence of a single dominant ideology. On the political front, this ideology is based on democracy and political liberalism. On the economic front, it is based on capitalism and market liberalization. These features led to the formation of the global economy.

Liberalization of *foreign direct capital* movements has resulted principally from systematic but uncoordinated actions by national governments. Almost every country in the world has liberalized inflows (and in many cases outflows too) of foreign direct capital over the last 15 years, many of them substantially, by allowing rights of establishment and more national treatment. This liberalization came about because there has been a huge change in attitudes of governments to these movements. In the 1960s and 1970s there was concern in many countries over foreign investors because of fears of foreign control of the economy and specific issues such as transfer pricing to avoid profit taxes. However, in the 1980s and 1990s these concerns diminished, and the view that foreign investors were important agents of change who introduced new products and technologies became widespread.

More than 50% of the global foreign direct investment (FDI) flows in recent years have gone into *service* industries. Much of the liberalization of trade in services has followed privatization of key service industries such as telecommunications and power that preceded or accompanied the relaxation of controls that restricted FDI in these industries.

These policy actions have profoundly changed the nature of markets. In its annual report the WTO each year compares the rate of growth of real world merchandise exports with that of real world merchandise output. (See WTO, 1998, Chart II. 1.) These series show clearly that in every year since 1982 the growth in the volume of goods traded internationally has exceeded that in the volume of goods produced globally; in fact, world trade in goods has grown in real terms at about 5% per annum, whereas world merchandise output has grown at about 2% over this period. Hence, more of the world output is exported to other countries, and more of the world expenditure on goods is purchased from other countries. In a similar way, United Nations Conference on Trade and Development (UNCTAD) has related the growing inflows of FDI to production in the countries in which they have invested. Between 1980 and 1994 the ratio of world FDI flows to world gross domestic investment doubled from around 2 to 4%. (See UNCTAD, 1996, Fig. 1.8)

Finally, one must consider the formation of *global financial markets,* that is, markets for stocks, securities, and other financial assets. There has been much empirical work to measure the extent of integration of capital markets across borders. These studies have found a steady but incomplete convergence in the prices of assets in the major markets as national assets markets have been deregulated and cross-border trade in these assets has been liberalized (see Lewis, 1995, for a survey). Reidel (1997) provides a recent survey of financial market integration in the developing countries of Asia.

One important development in recent years has been the growth of hedge funds and other highly leveraged investors operating from the major capital markets in the U.S. and Europe. Another is the establishment of the market for the bonds issued by developing countries for sale there, chiefly in the U.S. These have been called the emerging market bonds. Emerging markets are net borrowers and the volume of trading in emerging market debt grew from only U.S.$95 billion in 1990 to U.S.$5,296 billion in 1996 (Eng, Lees & Mauer, 1998). These two developments played an important part in the Asian crisis. Moreover, the reduction in barriers to the international movement of short-term capital and in the costs of transactions has meant that speculators can shift enormous volumes of funds from one currency to another instantaneously by electronic transfers.

As a result of the liberalization of markets for goods, services, and capital, a more integrated global economy has evolved in the last 10 or 20 years, especially in the 1990s. The changes are so extensive that we may refer to the new global economy.

Figure 1 plots the growth of trade in goods, FDI and foreign exchange, from 1989 to 1997/1998, roughly the decade of the 1990s. To make the series comparable, it is necessary to express all of the series in current prices; there is no quantity series in the case of foreign exchange transactions as they convert one currency into another, and there is no available price deflator for the FDI series. The series are taken from WTO, UNCTAD, and Bank for International Settlements (BIS), respectively. The value of foreign exchange transactions includes both spot transactions and outright forward and Forex swaps but excludes interest rate derivative instruments. The trade in securities is the net purchase of securities in the major industrial countries of Europe, Japan, and North America by nonresidents.

These series show that over the period, FDI has grown more rapidly though less evenly than trade in goods, and the value of foreign exchange transactions has grown more rapidly than the value of trade in goods or FDI. The value of trade in securities has grown more rapidly than the average growth of foreign exchange transactions.

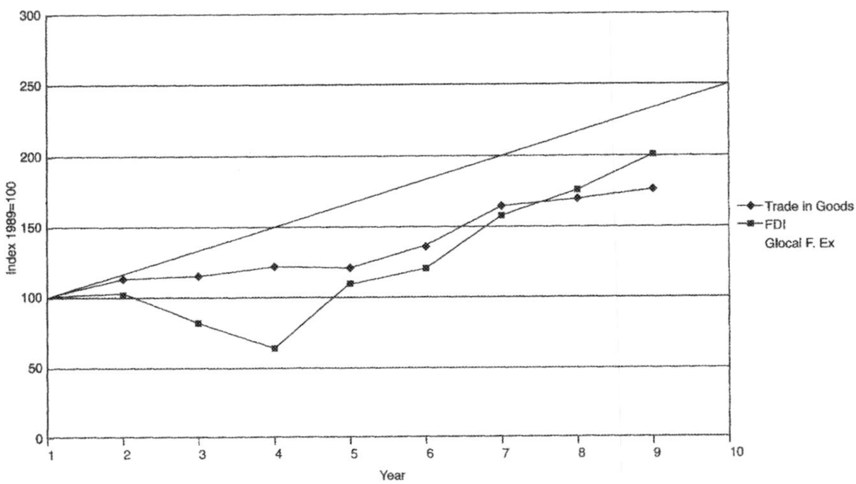

Fig. 1. Growth in Global Trade: Goods, FDI, and Foreign Exchange Transactions.

Perhaps the best way of viewing these series is to focus on foreign exchange transactions. International transactions for the sale of goods, services, and capital all involve foreign exchange transactions.[3] Hence, the foreign exchange market provides the most comprehensive view of cross-border transactions and openness. The rate of growth of foreign exchange market transactions is the average of the rate of growth of the various transactions in the goods, services, and capital markets that result in foreign exchange transactions. As noted, the growth in foreign exchange transactions has exceeded that in the market for goods and services and FDI. Consequently, the demand for foreign exchange for purposes other than goods and FDI must have had more rapid growth. However, statistics of non-FDI capital transactions are less comprehensive.

One part of the growth in non-FDI capital asset sales is growth in portfolio capital and securities. Another part is the establishment after 1992 of foreign exchange derivative markets for currency swaps, options, and other instruments. Non-FDI capital transactions must have grown more rapidly than the growth in goods markets. It is these capital market transactions that have largely driven the growth in the foreign exchange markets. An example of capital assets traded across national borders is foreign bonds. After the emergence of the Eurobond market in 1963, foreign bonds were issued to Japan in the 1970s and other types of foreign bonds, such as Brady and global bonds, emerged in the 1990s. The total net issues of international securities grew from

U.S.$149.9 billion in 1992 to U.S.$595.9 billion in 1997 (BIS, 1998, Table VIII 4).

One consequence of the differentials in the rate of growth of different transactions is that the ratio of the value of annual world exports to annual global foreign exchange turnover fell from 28.5% in 1977 to a mere 1.6% in 1995. (See ul Haq et al., 1996, Statistical Appendix.) Foreign exchange transactions are not now made mainly as a result of sales of goods between countries. Capital account transactions, including derivative transactions (futures and options), now dominate foreign exchange markets.

The concept of the global economy puts the emphasis on the interdependencies among the economics and the way in which the global economy functions rather than on the consequences of opening for individual economics or agents that are more commonly analyzed in the literature on trade liberalization and integration. The formation of the global economy has led to qualitative changes in the behavior of national economics.

The formation of the global economy has generally increased the efficiency of national economies by allowing them to specialize in the production of goods and services in which they have a comparative advantage. Similarly, international trade in capital assets improves the efficiency of national economics. This argument has been put by Stanley Fischer, the First Deputy Managing Director of the IMF, in the following terms:

> Put abstractly, free capital movements facilitate an efficient global allocation of savings and help channel resources into their most productive uses. From the individual country's perspective, the benefits take the form of increases in the pool of investible funds and in the access of domestic residents to foreign capital markets. From the point of view of the international economy, open capital accounts support the multilateral trading system by broadening the channels through which countries can finance trade and investment and attain higher levels of income. International capital flows expand the opportunities for portfolio diversification and thereby provide investors in both the industrial and developing countries with the potential to achieve higher risk-adjusted rates of return (Fischer, 1998, pp. 2–23).

Recently, another side to capital movements has been highlighted by the Asian crisis. This is the increased volatility of asset prices and exchange rates in the markets of some developing countries. This extends to Russian and Latin American markets as well as those in East Asia.

In reviewing the roles of the three main multilateral institutions, I shall start with the WTO. The WTO is especially important, because the liberalization of trade in goods was the original driver of the process of integration and international trade in goods is closely related to the international movement of capital.[4]

2. FROM GATT TO WTO

The GATT was created in 1948 as a temporary substitute for the International Trade Organization (ITO). The ITO was a more ambitious organization intended to accompany the IMF and the World Bank as one of the three pillars of the governance of the world trading system after the Second World War. However, the ITO never came into being because it was not ratified by the government of the United States. Consequently, the scope of the GATT was more limited than that of the planned ITO, which had important chapters dealing with foreign investments, international competition, and other matters not covered by the GATT. The GATT was an organization confined to the regulation of international trade in goods among its members (called contracting parties because of the temporary arrangement). (For a lucid account of the origins and evolution of the GATT, see Jackson, 1989.)

The preamble to the General Agreement on Tariffs and Trade set out the objectives in the following terms:

> Recognizing that their relations in the field of trade and economic endeavor should be conducted with a view to raising standards of living, ensuring full employment and a large and steadily growing volume of real income and effective demand, developing the full use of the resources of the world and expanding the production and exchange of goods.

It then states its responsibilities in the following terms:

> Being desirous of contributing to these objectives by entering into reciprocal and mutually advantageous arrangements directed to the substantial reduction of tariffs and other barriers to trade and to the elimination of discriminatory treatment in international commerce.

The GATT was an organization that regulated the behavior of national governments with respect to government border treatment of imports and exports of goods. With only very minor exceptions relating to unfair trading, state trading, and monopolies, the GATT did not regulate the behavior of private or state-owned producers. The behavior of governments was to be based on four principles:

- Nondiscrimination (most favored nation treatment (MFN)) among members
- Reliance on tariffs as the principal instrument of protection of national producers
- Transparency of government border interventions
- Reciprocity in negotiations relating to market access.

These principles were honored in the breach as more and more exceptions to each of the principles occurred. In relation to MFN, the major concerns were the growth of regional trading arrangements and discriminatory controls on imports such as voluntary export restraints and orderly marketing arrangements. Regional trading arrangements have continued to proliferate in the 1990s, even after the formation of the WTO, and some are now very large in terms of the number of member countries and trade involved. Prior to the Uruguay Round, tariffs had ceased to be the main instrument of protection for many industries in many countries. In particular, the most heavily protected industries, such as agriculture, clothing textiles, and automobiles, were protected in many countries chiefly by non-tariff barriers (NTBS). Some of these were approved by the GATT under waivers or other articles, but many were GATT illegal. NTBS are much less transparent than tariff rates.

The growing dissatisfaction with the limitations of the GATT led to the Punta del Este Declaration in the early 1980s that established the Uruguay Round. After a difficult seven years of negotiations that almost collapsed, the Uruguay Round was completed in December 1993 and signed in 1994. The Uruguay Round led to the establishment of the WTO, which began operation in 1995.

The Agreement establishing the WTO repeats the language of the preamble of the GATT, but the set of objectives of economic policy is expanded by the addition of the objective of sustainable development.

The WTO is an organization that is responsible for administering multilateral trade agreements negotiated by its members. The main agreements are the GATT itself (which was incorporated into the WTO), the General Agreement on Trade in Services (GATS) and the Agreement on Trade-Related Intellectual Property Rights (TRIPS), but there are several dozen other agreements, ministerial decisions, and declarations. The addition of GATS and TRIPS extended the responsibilities of the new organization to trade in services and intellectual property rights. The WTO is now a complex organization whose importance is steadily increasing. Hoekman and Kostecki (1995) provide an excellent introduction to its rules and activities.

All members of the WTO must accept all its laws under the Single Undertaking with the exception of "plurilateral" agreements,[5] which they can choose to opt in. The international law of the WTO is binding on its members. Another major reform accomplished by the Uruguay Round was that of the dispute settlement procedures. The dispute settlement procedures of the WTO, known commonly as DSP, have made the law of the organization enforceable as well as binding. It is largely the enforceability of the WTO international trade law based on the WTO that makes its laws so attractive to Non-Government Organizations (NGOs) and others who want to bring cross-border environmental, labor, and competition law within its compass.

Under Article X of the WTO, its membership is open to any state or customs territory having full autonomy in the conduct of its trade policies. A two thirds majority of the existing members is required to approve accession, but this is preceded by a lengthy process of accession negotiations. As of 31 July 1998, 32 countries have applied to join the WTO (WTO, 1998). A majority of these are transition economies from the former U.S.S.R, including Russia and the Central Asian Republics. China has also applied to accede, but the negotiations are proving especially difficult. The latest country to accede is the Kyrgyz Republic. It became the 133rd member and the first nation of the Commonwealth of Independent States to join the WTO.

The conditions imposed by members for accession have become more stringent over time. Applicant nations are likely to be requested to bind their whole tariff schedule and pressed to apply these bindings to applied rather than ceiling rates. The country will usually have to liberalize access to its markets for other members (such as lowering tariff rates, eliminating subsidies that do not conform to the Subsidies Agreement, and granting commercial presence to foreign service providers). One consequence of the increased stringency is that many of the current accession negotiations have been going on for several years.

Another part of the activities of the WTO that is relevant to the current debates is surveillance of national trade policies of its members. At the center of this work is the Trade Policy Review Mechanism (TPRM). The policies of each member government in relation to trade in goods and services, intellectual property, and related areas are reviewed regularly, every 2 years in the case of the four largest traders – the European Union (EU), U.S., Japan and Canada – and every four or six years for other countries. These country reports are published. Annual reports on trade barriers are also submitted by all members. The objectives of this monitoring are to increase the transparency and understanding of the trade policies of the members' governments and be a part of the assessment of the effects of these policies on the world trading system.

From the point of view of monitoring developments in the world economy, these surveillance powers need to be improved. Apart from the annual reports, between reviews member countries have to report any significant changes in their trade policies, but this requirement is rather vague. A requirement that all countries should immediately report any changes in trade policies would improve the surveillance of global trade policies by the WTO. Rajapatarina (1998) points out other limitations to the TPRM process, which is designed merely to check that a country's trade policies conform to the rules of and commitments to the WTO and improve transparency. He complains that there is no critical examination of the policies themselves. Moreover, this surveillance does not require members to revise their national policies. (Multilateral trade reform is carried out by means of negotiations in rounds or special negotiations at Ministerial meetings and in some cases during the accession negotiations with prospective members.)

2.1. The GATT/WTO and the Developing Countries

For our purposes, it is necessary to outline the treatment of the developing countries in the WTO, because this group is a central concern in the reform of the architecture of multilateral institutions. The original articles of the GATT did not distinguish between developed and developing countries or any other group of countries. All the contracting parties were bound in the same way by the rules. The one exception was Article XVIII, which allowed additional freedom for the "economies which can only support low standards of living and are in the early stages of development" to be able to grant tariff protection for the establishment of particular industries and apply quantitative restrictions for balance of payment purposes. The former allowed these countries to introduce protection for "infant industries," and the latter allowed them to use quantitative restrictions in times of balance of payments crises.

However, from early in its history, the GATT displayed concern over the particular trading problems of developing countries (Hoekman and Kostecki, 1995, chapter 10, provides a history of the treatment of Developing Countries in the GATT and WTO). A landmark in this debate was the publication of the report of the Panel of Experts (1959) (known as the Haberler Report), which documented the slow growth in the trade of developing countries and the instability of the prices of many of the staple commodities they traded. This led to the establishment of Committee III to identify barriers to exports of the developing countries and recommend ways of removing them. The GATT called for duty-free entry on all tropical products, removal of quantitative restrictions affecting the exports of these countries, and a reduction of all tariffs and taxes on primary products of importance to these countries.

In 1966 Part IV was added to the GATT. This was a major departure from the principles of the GATT in that it recognized the problems of the developing countries (or strictly the "less developed" countries, as they were then called by the GATT) as a group and gave them more favorable treatment under the rules. This formally requested the developed countries in the organization to take measures to improve market access for the exports of the developing countries, and it waived reciprocity for commitments made by the developed countries. The latter meant that the developing countries did not have to participate to the same extent in the reductions in trade barriers under the GATT Rounds of negotiations. In the Tokyo Round an enabling clause was adopted that gave the developing countries "special and differential" treatment in the organization. This continued the waiver of reciprocity in these countries and also gave permanent authority for nonreciprocal preferences for exports from the developing countries such as the Generalized System of Preferences system.

The Uruguay Round changed the situation of the developing countries (as they are now officially called in the WTO) rather substantially. While several Uruguay Round agreements affirm the special and differential status of developing countries (see Hoekman & Kostecki, 1995, Annex 6 for a list), they also repeatedly make a distinction between the "least developed" member countries and other developing member countries. In the Subsidies Agreement, the distinction between these two groups is set at a GDP per capita of U.S.$1,000. In effect, these changes have transferred the special and differential treatment largely to the subgroup of developing countries known as the least developed countries, with other developing countries assuming more of the responsibilities of developed country members.

The special provisions for developing countries in the WTO that now hold can he grouped under five headings:

1. a lower level of obligations
2. more flexible implementation timetables
3. best-endeavor commitments by the developed member countries
4. more favorable treatment for least developed member countries
5. technical assistance and training.

Technical Assistance and training has greatly expanded since the establishment of the WTO in 1995.

Since it began operation in 1995, the WTO has had a standing committee devoted to trading problems of developing countries, the Committee on Trade and Development, who has given priority to the problems of the least developed country sub-group. The first WTO Ministerial Conference held in Singapore in December 1996 adopted a comprehensive and integrated plan of action for the least developed countries. In 1997 a high level meeting on integrated initiatives for least developed countries was set up in conjunction with UNCTAD, the IMF, the World Bank, United Nations Development Program and the ITC (which is a joint subsidiary organ of the WTO and UN). In his report to the high-level meeting in October 1997, the Director-General of the WTO again urged members to consider eliminating trade barriers to LDC exports.

However, in two other respects, the situation of developing countries, both the least developed and other developing countries, is now much less special than it was before the Uruguay Round. The first relates to the bindings of tariff rates. Before the Uruguay Round, the coverage of bindings for most developing countries was zero or very low and usually related to ceiling rates rather than the actual applied rates. However, in the Uruguay Round the developed countries insisted that the developing countries bind more of their tariff rates than they had done in earlier GATT Rounds.[6] Secondly, the general success of the multilateral rounds and unilateral actions by national governments in reducing tariff rates and other non-tariff restrictions has eroded the margins of preference enjoyed by developing countries under the GSP and other country-specific preference schemes targeting the exports of the developing countries.[7]

3. THE IMF

Article I of the IMF's Articles of Agreement lays out the six purposes of the fund. Five deal with aspects of the balance of payments of members, the multilateral payment system, and international monetary cooperation. The sixth purpose (listed as number ii) is to "facilitate the expansion and balanced growth of international trade, and to contribute thereby to the promotion and maintenance of high levels of employment and real income and to the development of the productive resources of all members". Thus, since it began operation in 1946, the IMF has had the responsibility among the multilateral institutions of managing the world exchange rate system and related matters of capital flows between countries.

Membership in the IMF is open to all countries "and in accordance with such terms as may be prescribed by the Board of Governors". These include the terms for subscriptions. Accession criteria are less strict than those of the WTO. There are currently over 180 member countries (including China and most of the other transition economies). The Articles do not distinguish between groups of members. The rules apply uniformly to all members; this is known as "uniformity of access". However, the IMF has developed additional loan facilities especially for developing and least developed country members. With respect to their growing concern with the Developing Countries, the history of the IMF is broadly similar to that of the GATT/WTO.

The IMF-based system originally centered on the establishment of par values of member currencies. Members were to fix the values of their currencies, and these pegged rates were to be only infrequently adjusted. Hence, the system was described as one of "adjustable pegged" exchange rates.

A fundamental part of the system was the acceptance by members of full "current account convertibility", that is, unrestricted access by traders to foreign exchange to conduct trade in goods and services. Article VIII (General

Obligations of Members) prohibits members, except with the approval of the IMF, from imposing restrictions on the making of payments and transfers for current international transactions. However, Article XIV (Transitional Arrangements) allows countries that are in transition to the full current account convertibility to introduce or change restrictions on payments and transfers without approval. At present, current account convertibility has been almost universally achieved, although some 40 members still remain under the transitional provisions of Article XIV.

In order to overcome short-term difficulties in the balance of payments because of fluctuations in the demand and supply of foreign exchange, the IMF was to make short-term loans at concessional rates to members. Article V of the agreement requires that there be a balance of payment need for a loan.

In the requirements of current account convertibility and the notification of the exchange rate regime, the IMF is a rules-based organization. However, the IMF does not have rules relating to the domestic banking, financial sector, or monetary policy of its members. The IMF has no effective control over the domestic monetary policies of its members. However, it does establish conditions on its loans that require borrowing countries to carry out reforms relating to the trade and exchange control regimes and in recent years increasingly other monetary and fiscal policies.

Surveillance by the IMF is carried out under Article IV. This surveillance relates to the compliance of each member with its obligations, surveillance of the exchange rate policies, and general oversight of the international monetary system. Legally, the mandate for surveillance is separate and distinct from that for financing, but they do interact. The two principal means of surveillance are the annual consultations with each member country and the semiannual World Economic Outlook discussions, which consider issues from a global perspective. Since 1997 the fund has given increasing attention to issues that have arisen in the present crisis, such as banking soundness, currency board arrangements. and data collection.

The IMF-based system began to change fundamentally in August 1971 when, in response to continuing high rates of inflation and balance of payment deficits, the U.S. ended the convertibility of the dollar into gold. This marked the end of the *de facto* "dollar standard". Several countries floated their exchange rates in the 1970s. The IMF recognized these changes by amending its articles in 1978. The amendment provided that members can adopt the exchange arrangements of their choice. Thus, the system changed substantially from a rule-based to a discretion-based one. The IMF retained a surveillance over members' exchange rate policies.

Since the amendments to the rules relating to par values in 1978, many different exchange regimes have emerged. The IMF classifies them officially as currencies that are pegged (to the U.S. dollar, SDR, or some other currency), those with limited flexibility, those with cooperative arrangements with other countries, those adjusted according to a set of indicators, managed floating, and independently floating. (The numbers are reported monthly in the IMF's *International Financial Statistics.*)

In moving from a pegged to a predominantly floating exchange rate régime, the volatility of exchange rate movements increased around the world. This was contrary to the original expectations. There has been a further increase in the volatility of exchange rate movements in 1997 and 1998.

Restrictions on capital account transactions were relaxed or removed by many countries after 1973 and particularly in the 1990s. This list includes many developing countries (Fischer et al., 1998, Table I). Dooley (1996) provides a survey of the extent of capital controls and views relating to them. This liberalization was done by the unilateral action of countries independent of the IMF, although in recent years the IMF has encouraged movement toward capital account convertibility (Quirk et al., 1995).

The liberalization of international capital flows had greatly reduced the need of the IMF members for short term capital for stabilization purposes, which was the original purpose of the IMF. This has profoundly changed the nature of the organization. The IMF had become more an organization whose primary activity is giving policy advice to its members and conducting research. Loan activity is the supporting function. However, the Asian crisis also saw renewed demand for loans from the IMF and other organizations. In the last financial year, 1997/1998, members of the IMF drew nearly four times as much from the IMF as in the previous year (IMF, 1998).

In September 1997, at its Hong Kong meeting the Interim Committee of the IMF "agreed that the Funds Articles should be amended to make the promotion of capital account liberalization a specific purpose of the Fund and to give the Fund appropriate jurisdiction over capital movements." That is, it agreed to introduce what is known as "capital account convertibility" as a part of the obligations of members. This proposal was extremely contentious in the light of developments after the Asian crisis. Some of the Asian countries have attributed their crises to capital mobility, particularly the freedom of short-term investors to take their capital out. This view is supported by the analyses of some Western economists (see Sachs and Radelet, 1998; Fischer et al., 1998). Countries such as Malaysia and Pakistan have reintroduced capital controls. Some developed countries are also concerned over the mobility of short-term capital. The Interim Committee has subsequently backed away from the proposal.

From July 1997 the asset and foreign exchange markets of several East Asian countries have been falling suddenly and sharply. This is known as the Asian crisis. The five countries most affected are Korea, Indonesia, Malaysia, the Philippines, and Thailand. The crisis has spread to other emerging capital markets in Russia and Latin America. It is now recognized as a crisis of the world financial system rather than a purely Asian crisis (see Sachs & Radelet, 1998). Therefore, it is something of a misnomer to refer to it as the "Asian" crisis.

After the crises in foreign exchange and asset markets in 1997/1998, some Asian countries have changed their exchange rate regimes. Thailand, Indonesia, and Korea abandoned their pegged or controlled rates by floating their exchange rates. There has also been a new demand for loans in the situation of more volatile foreign exchange and asset markets. The IMF has been criticized for its role in the reform packages provided to Korea, Indonesia, and Thailand (for example, see Sachs & Radelet, 1998, and the response of the Deputy Managing Director reported in *The Economist*, 1998).

It is notable that some of the Asian countries that had retained extensive capital controls have not experienced a crisis in their capital and foreign exchange markets in 1997/1998. This applies to China, India, and Vietnam. By contrast, the Asian countries that have experienced major crises all had a much higher degree of capital mobility. Two – Malaysia and Indonesia – had had no restrictions on capital account transactions since the early 1970s.

4. THE WORLD BANK

Article I of the World Bank's Articles of Agreement lays out its purpose in five subparagraphs. Of these, one is "to promote the long-range balanced growth of international trade and the maintenance of equilibrium in the balance of payments by encouraging international investment for the development of the productive resources of members, thereby assisting in raising productivity. the standards of living and conditions of labor in their territories". This resembles the purpose of the IMF dealing with trade, but it is expressed in terms of international investment. Unlike the GATT and the IMF. the World Bank, or the International Bank for Reconstruction and Development to give its full name, was created as an organization to assist the developing countries. Its focus is on long-run growth rather than short-term stabilization as with the IMF.

Membership in the Bank is open to any member of the IMF. It has over 180 members and provides assistance to more than 100 developing countries.

The World Bank provides loans and development assistance to middle income countries, credit-worthy poorer countries, and more recently, also the transition countries. Its funds originally came from member subscriptions but now come mainly from the sale of bonds in international capital markets. Unlike the WTO and to some extent the IMF, the World Bank is not a rule-making or rule-based organization.

World Bank assistance to developing countries has been through a number of fashions reflecting different views on the development process. (For a rather critical insider's view, see Taylor, 1997.) The World Bank sees its mission now as one of reducing poverty. It currently puts emphasis on investing in people, particularly through basic health and education, protecting the environment, supporting and encouraging private sector development, and strengthening the government sector's ability to deliver government services and a stable macroeconomic environment.

In 1996 the World Bank produced a plan called the Strategic Compact to renew and reform the Bank by making it more effective in achieving its mission. However, this plan does not represent a change in the basic position of the Bank among the international organizations.

The World Bank Group now comprises the World Bank itself and its related institutions – the International Development Association (IDA), the International Finance Corporation (IFC), the Multilateral Investment Guarantee Agency (MIGA), and the International Center for Settlement of Investment Disputes (ICSID).

The IDA and IFC complement the World Bank in carrying out its mission. IDA was created in 1960 to provide extra assistance to the group of poorest countries. It provides interest-free loans to these countries. IFC was created in 1956 at a time when greater emphasis was being placed on the role of the private sector in development. It seeks to promote growth in developing countries by providing support to the private sectors of these countries. It invests in commercial enterprises by providing loan and equity capital. It is now the largest multilateral source of loan and equity finance for the private sector in developing countries.

The very rapid growth of FDI to developing countries in the 1970s and more particularly the 1980s reduced the need for multilateral loans. ICSID was created in 1966 to facilitate the settlement of investment disputes between developing country governments and foreign investors. MIGA was created in 1985 to encourage this private capital flow and in particular to issue guarantees against noncommercial risks for flows among member countries. Both of these developments reflect the trend toward globalization, particularly the increasing cross-border flows of FDI.

The Bank is probably the least criticized of the three multilateral institutions in terms of its basic role. Its loan policies and conditions and general policy have all been severely criticized by numerous groups over the years (see Harrigan, 1996 and references therein for a sample), but these criticisms do not go to the main mission of the Bank and its affiliated institutions. Although the customers and policies have changed, the Bank has always been and still is a development promotion agency.

Problems of architecture in the three main multilateral organizations may be considered under the headings of coherence within and between each organization. These are the intra-organization and inter-organization dimensions of the architecture. A third area is that of gaps in their ability to meet stresses in the world trading system.

5. LACK OF COHERENCE WITHIN THE ORGANIZATIONS

Despite the major overhaul of the GATT system in the Uruguay Round, there is general agreement that much work remains to be done to improve the organization. This was recognized in the Round itself by the provision for completion of some of the service agreements and subsequent reviews of many agreements, the so-called built-in agenda. This is unfinished business, left over from the harried timetable of the Uruguay Round. (For a largely WTO view of the state of the organization, see Krueger, 1998.)

In addition to unfinished business, there is a number of shortcomings in the architectural design of the rules of the WTO. The problems derive from the practice of the GATT in early rounds and in the Uruguay Round of dealing with new areas by adding new agreements. The rules are fragmented, with different sectors and instruments being dealt with under separate agreements, Sauvé and Zampetti (forthcoming, p. 19) comment: "Both as an institution and a negotiating forum, the WTO is indeed too segmented (i.e. 'vertical') in design and operation. Its architecture can be viewed as a succession of vertical 'chapels', its whole representing in many instances little more than the incoherent sum of a number of 'partial equilibrium' responses to the 'general equilibrium' challenge of rule-making in a globalizing environment. Such architecture allows for too little of the horizontal cross-linkages and cross-fertilization between policy domains which a broad, generic (i.e. non-goods, services, or IPR-specific), and competition-oriented approach to market access requires in today's global economy".

As examples of architectural incoherence, Sauvé and Zampetti give ". . . the existence of wholly separate rules on goods and services; the existence of establishment-related disciplines for trade in services but not for trade in goods; the imbalance under the GATS between an abundance of provisions relating to investment liberalization and a near absence of provisions on investment protection; the adoption of disciplines on trade-related investment measures for goods-related activities but not for service-related ones . . ." and others mostly related to the separation of the agreements for goods from that for services. One could add others; for example, the separation of the remaining plurilateral agreements from the single undertaking and the absence of negotiations on subsidy rates paralleling those on tariffs and nontariff barriers. These examples call for a unification or codification of the rules.

The third area of perceived shortcoming is the absence of rules relating to aspects of global governance that are closely related to international trade. During and after the Uruguay Round, there was strong agitation by some governments and NGOs for extensions of the WTO rules to cover a number of new areas. These include cross-border policies relating to the environment, labor standards, competition, foreign direct investment, regulation, bribery and corruption, immigration, and other areas. All of these issues have arisen from the greater integration of national economies.

Each of these is contentious and involves difficult problems of architectural design. After the Uruguay Round the WTO set up the Committee on Trade and Environment. At the First Ministerial Meeting in Singapore in December 1996, it set up the Working Group on Trade and Investment and the Working Group on Trade and Competition to hold discussions (but not negotiations) on these issues. I shall not examine them except for some comments below on trade and investment issues.

The internal architectures of the IMF and World Bank Group are less complex than those of the WTO, primarily because they are less rule-based institutions than the WTO. The IMF architecture has had a general coherence because its functions were specifically related to balance of payment issues. The main architectural concern currently relates to its role in the financial crisis and its relationships to other organizations in this area. Similarly, the World Bank Group functions relating to development have given it a general coherence, although there are questions concerning its overlap with the OECD, UNCTAD, and WTO in the area of rules relating to FDI. These concerns are considered below.

6. LACK OF COHERENCE BETWEEN ORGANIZATIONS

The mandates of the three organizations all contain an essentially common set of primary economic objectives that they are meant to pursue. They all derive from a belief in free markets and the consequential pursuit of market liberalization. There is also an overlap in responsibilities for areas of policy; for example, GATT Article XV concerns "Exchange Arrangements". Similarly, the IMF, in regulating exchange controls on current account transactions, exercising surveillance over members' monetary and financial policies, and imposing trade policy conditionality on its loans, has a major impact on the levels of restriction on trade in goods and services. Not surprisingly, this overlap has given rise to some problems in the interfaces between them. This applies chiefly to the IMF and the World Bank on the one hand and the GATT/WTO on the other, because the IMF and the Bank are located in the same city and work closely together. They have both adopted market-oriented policies and advocated opening markets. Their similar policies became known in the 1980s as the "Washington consensus".

The main area of interfaces among the activities of the three organizations is that of policies relating to trade in goods and services. This is the responsibility primarily of the WTO. However, both the Bank, through the conditions attached to its structural adjustment loans, and the IMF, through conditions attached to its loans, have required borrowing countries to undertake comprehensive trade reforms. These conditions have applied to many aspects of trade policy, including tariff reform, abolition of quantitative restrictions, export subsidies, and removal of exchange controls on the trade of goods and greater exchange rate flexibility (see Greenaway & Morrissey, 1996 for an examination of the conditionality of World Bank lending).

The concern here is with coherence. The final act of the Uruguay Round included a ministerial declaration on "The Contribution of the WTO to Achieving Greater Coherence in Global Economic Policymaking". This is known as "global coherence" for short. It recognizes the links between the work of the three institutions and commits them to working together. Concern had been expressed over the lack of consultation by the IMF and the Bank with the GATT in the formulation of the trade reform objectives in their conditional loan programs and the possibility that developing countries might obtain negotiating credit in multilateral GATT negotiations for trade policy reforms introduced under IMF and Bank loan programs. (Sampson, 1998, gives a WTO perspective on global coherence.) The Ministerial Declaration has led to greater cooperation among the three institutions after the Uruguay Round.

However, these problems of coordination are minor. More importantly, the actions of the World Bank and IMF through the trade conditionality of their loans have supported the trade liberalization objective of the GATT/WTO. Indeed, because of the inability of the GATT to recommend changes in the policies of its members outside the formal multilateral negotiations and the "special and differential" status of the developing countries, the GATT did very little to promote trade liberalization among these countries. Even more important, the IMF push for current account convertibility in the developing countries has led to the dismantling of foreign exchange controls in many of these countries. This has greatly liberalized trade in both goods and services and thereby complemented and strengthened the GATT efforts to liberalize trade. These activities of the IMF and the World Bank are in fact an under-appreciated source of trade liberalization.

An important question concerns the resources available to each of the organizations. Sampson (1998, p. 258) notes that the WTO is very under-resourced compared with the IMF: "The travel budget of the IMF is about the same size as the total WTO budget. The IMF has 2,200 staff members: the WTO has 200 professional staff".

Another source of incoherence is the growth in the number of international organizations whose activities overlap with those of the three multilateral institutions. These include the BIS, UNCTAD, United Nations Industrial Development Organization (UNIDO), OECD, Asia Pacific Economic Cooperation (APEC), the regional development banks, G-7 and G-22, and others. All of these, with the exception of the BIS, have arisen since the establishment of the Bretton Woods institutions.

As a single example, consider the OECD. Its members are 29 high-income countries. Its activities cover virtually all aspects of the management of the economies of these countries. Since these countries in the aggregate account for a large part of world production, trade, FDI, and official lending, the OECD is inevitably concerned with the management of the world trading system. Its Trade Committee examines developments in world trade. The OECD compiles the statistics of assistance given to agricultural producers in the OECD countries, of aid given by national governments and multilateral agencies, and of foreign bond issues. The OECD has more responsibilities for the regulation of FDI flows than any other international organization. Hence, the activities of the OECD overlap to a substantial extent with those of all three Bretton Woods organizations.

I shall confine any attention to the two overlaps which bear most directly on the functions and activities of the Bretton Woods institutions. These are the overlaps between the OECD, the World Bank, and others which play a role in regulating world flows of FDI and the overlap between the IMF and BIS activities in the financial system.

FDI flows differ from cross-border flows of goods and services and foreign exchange transactions in that there is no multilateral organization with exclusive or even primary responsibility for regulating them. This is surprising at first sight, but it derives from the fact that such flows were unimportant in the prewar period and the Bretton Woods system designers did not anticipate the growth of FDI.

Instead, a number of organizations have come to share responsibilities in this area. The most important is the OECD. Soon after its formation the OECD completed the code of Liberalization of Capital Movements and the Code of Liberalization of Invisible Operations in 1961 and later the National Treatment for Foreign Controlled Enterprises. These are rule-based agreements, like the GATT/WTO, and are based on the same fundamental principles of non-discrimination and national treatment. However, they apply only to the OECD members, and the national treatment instruments are not binding. APEC drafted the Non-Binding Investment Principles in 1995, but these apply only to the members of this organization and they are nonbinding and rather weak (see Lloyd, 1995). The Trade Related Investment Measures (TRIMS) Agreement of the Uruguay Round has made performance requirements that are trade related subject to the discipline and rules of the WTO, and some aspects of FDI in service industries are covered by the GATS. The International Center on Settlement of Investment Disputes, which is part of the World Bank Group, is the most important of several organizations concerned with settling disputes between foreign investors and governments. The Multilateral Investment Guarantee Agency, another part of the World Bank Group, provides insurance coverage for investors in developing countries against noncommercial (political) risks. There is a host of bilateral investment treaties and provisions in some regional trading arrangements, such as the EU and NAFTA, which also set rules or guidelines for FDI flows (see UNCTAD, 1996, Pact Three and PECC, 1995, Appendix E for outlines of these agreements).

Thus, the rules and guidelines relating to FDI are, to put it frankly, a mess architecturally and organizationally. Since 1995 the OECD has been attempting to negotiate a more comprehensive and truly multilateral set of rules. These are known as the Multilateral Agreement on Investment. (See OECD, 1996 for an early outline. The 144-page text of the current draft agreement is available on the OECD website.) But the negotiations were suspended in April 1998 for a "period of assessment", and they have not been resumed. There are major reservations over sovereignty and divisions over the protection of labor, environmental, and cultural issues. The non-member developing countries (India, Pakistan, Malaysia, and others) and a variety of NGOs object to many provisions and want to see more responsibilities imposed on foreign direct investors to balance the protection of their rights. There is an avowed consensus among delegates of the need for a multilateral framework for investment, but agreement is proving extremely difficult.

In relation to cross-border financial flows, multilateral organizations other than the IMF play significant roles. One is the BIS, located in Basel. The BIS was created at the Hague Conference in 1930 to manage the payments of German reparations after the First World War. It is owned and controlled by central banks. Article 3 of the original statutes laid down its objectives. They are ". . . to promote the cooperation of central banks and to provide facilities for international financial operations. . . ." On its website the BIS states: "One of the major aims of greater international bank cooperation has always been to foster international financial stability. Nowadays with rapidly integrating financial markets in the world, such cooperation is essential. The BIS is thus an important forum for international monetary and financial cooperation between central bankers, and increasingly, other regulators and supervisors". For most of its history the membership of the BIS was restricted to the central banks of the industrialized countries and Eastern Europe, but in recent years it has admitted central banks from Asia (China, Hong Kong, Korea, and India), Latin America, and the Middle East. As of 31 March, 1998, 45 central hanks had rights of representation and voting at the BIS. It also increasingly invites central bank officials from emerging market economies to participate in discussions held at the BIS.

The BIS has several important roles. It is a meeting place for central bankers. A significant portion of the world's foreign exchange reserves are held on deposit at the BIS. It provides the secretariat for the Basel Committee on Banking Supervision, but the Basel Committee is not formally a part of the BIS; this committee provides a forum for the discussion of banking supervision and it is the body that has issued the core principles of banking supervision, a blueprint for effective banking supervision. As a product of these functions, the BIS collates the statistics of foreign exchange market transactions and international banking and securities markets that are supplied by its member central banks, and in the future it will collate data on international debt securities markets and traded derivatives. It is a major center of research on monetary and financial questions. The BIS is establishing the Institute for

Financial Stability. In all of these activities the BIS overlaps with the work of the IMF.

In recent years the G-7 has also played a role in coordinating the policies of the G-7 countries and in particular in trying to stabilize exchange rate movements. Like the rules relating to FDI flows, those relating to financial flows have a low degree of multilateral regulation and a significant degree of incoherence.

7. GAPS IN THE COVERAGE OF THE ORGANIZATIONS

There are major gaps in the coverage of the responsibilities of the three institutions relating to the current problems in world asset and foreign exchange markets. As noted above, there is no multilateral organization with responsibility for the regulation of global FDI flows. No multilateral organization has overall responsibility for the regulation of cross-border movements of nondirect financial asset markets and the related aspects of domestic policies that bear on these, such as prudential controls on financial institutions, capital adequacy, and corporate governance. Where capital account convertibility applies there is no national government regulation of these flows. As one particular instance of the difficulties in the current crisis, there are no statistics of the activities of hedge funds. These gaps are the most serious problem.

These gaps have arisen because of rapid changes in world markets without changes in multilateral organizations to accommodate these changes, in particular the opening of world markets extended beyond the market for goods and services and long-term FDI. The IMF has encouraged member countries to adopt capital account convertibility. Greater mobility of short-term capital and much greater flexibility of exchange rates has led to new patterns of trading in assets such as hedge funds.

In the late 1980s and 1990s, the developing countries engaged in widespread liberalization of markets. They were urged on by the IMF, the World Bank, and the GATT, with each of them becoming more strict with the developing countries in recent years. These countries became increasingly involved in financial asset market trading because of the opening of their economies.

There is a consensus that more needs to be done to establish and enforce rules for international markets, but there is no agreement on what should be done or who should do it. These two questions are inseparable.

Focusing first on the Bretton Woods organizations, would they be set up in the same manner today as they were more than 50 years ago? The answer is certainly no.

An organization is certainly needed to oversee the world trading system. The rules of the WTO need changing, as noted above, but there is no desire and no need to change the basic rules and method of operation of the WTO. The most relevant issue here is the possibility of the WTO taking over the negotiation and ultimately the management of the Multilateral Agreement on Investment (MAI). This has been mooted since the collapse of the negotiations in the OECD. There are two obvious advantages. One is that the membership of the WTO is much larger and more representative than that of the OECD, although the exclusion of China and Russia from the WTO is a serious problem. The second is that the MAI, or any other multilateral agreement on investment that might replace it, deals essentially with rules that bind national governments, just as those of the WTO do. Both sets of rules are based on the same underlying principles of nondiscrimination, national treatment, and transparency. Therefore, such investment rules would sit comfortably with the rules of the WTO. However, if this path is pursued, the rules relating to FDI should be consistent with those in the WTO relating to trade in goods and services; otherwise, the same intraorganizalion incoherence that exists in the present WTO would be exacerbated. For example, there is a danger that an MAI embedded in the WTO might contain provisions concerning labor or environmental standards.

The World Bank would not exist in its present form because of the lesser role of official aid and technical assistance, both of which have been largely privatized. It might be an institution devoted to the analysis of the growth process, probably scaled down from its present size and concentrating on policy advice relating to long-term growth. The role of the World Bank as soft loan provider is debatable. Some developing countries "need" World Bank loans because they cannot borrow enough from existing capital markets, but one must ask why they are unable to borrow. In the eyes of potential lenders, these markets are unattractive. Ultimately, if they are to borrow more from world markets, these countries must take actions which increase the expected rates of return and decrease the perceived risks of investment in these countries. It is possible that the availability of loans on terms more favorable than those in private markets lowers the incentive for these countries to reform their economics and creates a moral hazard.

The biggest question mark is over the IMF, because the world monetary system that it was set up to manage has long ceased to exist. The problems now are not balance of payment disequilibria but crises in financial markets, with associated problems of capital flight, debt relief, banking supervision, and corporate governance.

It is more useful to focus on the reforms that are needed rather than the existing organizations. The architecture should follow from the structure of international rules that are desirable to regulate the global economy.

The reforms that are needed are comprehensive. Some of the failures in the present system arc failures of national governments and some are systemic. Consequently, the reforms needed cover both domestic policies and multilateral rules.

The domestic policies cover a multitude of areas relating to both government and private market behavior. The domestic policies include banking supervision and capital adequacy, corporate governance, bankruptcy laws, FDI and competition in the financial markets, monetary and fiscal policies, and exchange rate management. Basically, the countries of the Asian crisis and many other emerging market economics had liberalized trade in goods, FDI, and capital transactions but lagged in developing supervision and prudential control of financial institutions and other mechanisms to regulate the financial sector.

These questions have traditionally been regarded as the province of national governments. In the area of trade policy, the WTO has set up rules for a number of areas that were previously not regarded as international and were not covered by rules in the old GATT, e.g. technical barriers to trade, subsidies, and government procurement. But this process of rule extension has been difficult and is incomplete. It is likely that multilateral rules relating to the behavior of banks, hedge funds, and other private agents who operate in the international financial market and also those relating to the conduct of monetary and fiscal policies by governments will need to be extended to areas previously regarded as matters for regulation by national governments alone. This process will be similarly difficult.

In the wake of the Asian crisis, there are many proposals to try to control short-term capital flows and exchange rate fluctuations. There is a continuing debate in Asian countries about the merits of fixed, flexible, and floating exchange rates. Some have suggested that the IMF take over from the G-7 the responsibilities for coordinating policies to reduce exchange rate volatility and misalignments and perhaps to move to a system of exchange rate bands. There are proposals for improved capital market surveillance and lending arrangements. Others have suggested that the IMF play a central role in the coordination of debt restructuring or debt relief or that it abandon its conditional financing role or enhance "market discipline" exerted by private capital flows financing by rating the policies of countries.

There has been a debate about the creation of a new multilateral institution to oversee the financial sectors of national economies, but at its 1999 meeting the G-7 countries decided instead to give new surveillance powers to the IMF and increase the lending resources of the IMF and the World Bank. The future of the international system for regulating foreign exchange and capital markets is very uncertain at the present time.

Another area being actively pursued concerns codes of behavior for monetary and fiscal policies, corporate governance, and accounting.

The systemic problems need to be appreciated. The market crises are not wholly the failures of domestic policies in the crisis countries. They arise in part from the interconnection of national markets following liberalization of trade in capital assets. The fundamental problems are those of asset bubbles and capital flight in volatile liquid asset markets. Systemic policies include expanded loan facilities and debt relief. Some economists are questioning the desirability of completely open markets for short-term capital and of freely floating exchange rates.

There has also been a renewal of arguments in favor of restricting capital account convertibility (see the papers in Fischer et al., 1998) by capital controls and in other alternatives such as the Tobin Tax on foreign exchange transactions (see the papers in ul Haq et al., 1998). Western economists are generally agreed that the maintenance of pegged rates is just another example of the harm of trying to fix the price in a market in which demand and supply are constantly changing. Pegged exchange rates lead to substantial misalignments of real effective exchange rates. A flexible rate allows continuous adjustment to accommodate changes in relative price movements or commodity markets. There is a second argument in favor of exchange rate flexibility. Advocates of a flexible exchange rate have argued that it provides a clear signal of the effects of government policies and helps to prevent government policies that are regarded as not credible. However, there is little agreement on the precise exchange rate arrangements which provide the optimum degree of flexibility.

These national and multilateral alterations will substantially change the international architecture. They are too complex and uncertain to be resolved easily. I do not attempt a detailed architectural design, but some general features of the changes that need to be made can be noted.

The future world trading system must be based on open trade in goods, services, and capital, and it must maintain a system of stable but flexible exchange rates.

One safe conclusion regarding the architecture is that there are too many multilateral and regional organizations currently involved in discussions and negotiations relating to inter-national financial markets. These include the IMF, OECD, G-7, World Bank, and BIS as major players and other more minor players, such as APEC. The problem of over-servicing by multilateral or international organizations also applies to trade policy. In addition to the WTO and the activities of the IMF and World Bank, UNCTAD, UNIDO, OECD, APEC, and other UN regional organizations all examine trade policies. This calls for rationalization of the activities of these organizations, but this aspect will not be examined further.

The IMF has the advantage of its lending resources, which are critical to debt relief and bailouts. But it has other disadvantages. In its annual report for 1997/1998, the IMF admitted that it did not predict the crisis. Its lending programs have been severely criticized by the lending countries and others for being too heavily oriented toward fiscal and monetary tightening, which is not appropriate for an asset market crisis. Its lending resources are too limited. It has little experience in the kinds of financial sector reform that are needed in the present Asian crisis. In contrast, the BIS does have experience and expertise in banking supervision and financial sector reform, but it has limited lending resources. It also has a good record in anticipating the crisis: the BIS Annual Report for 1997 (BIS, 1997) anticipated many of the problems that emerged very soon after its publication. However, the BIS is regarded as an organization that is Euro-centric and under the control of central banks and therefore independent of governments.

Again, however, architectural design should not start with the existing organizations. One needs to consider the nature of the rules that are needed in the present global economy. Any new systemic responsibilities should be assigned to the organization that is best suited to assume them. If none is well suited, new organizations may need to he created.

One important feature emerges from the consideration of the broader architecture of the multilateral organizations rather than that of the financial sector organizations alone. There is a need for coherence between the WTO and IMF for the fundamental reason that the markets for goods and services are linked to the market for foreign exchange. The levels of trade restriction affect the determination of the exchange rates; conversely, exchange rates may affect the levels of trade restriction. In fact, exchange rates and trade restrictions are alternative instruments, both from the points of view of providing assistance to domestic producers and achieving equilibrium in the foreign exchange market. This has long been recognized in international trade theory, where a nominal exchange rate devaluation is characterized as equivalent to a combination of a uniform import charge on all imports and a uniform tax on all exports. (This ignores capital flows.) Conversely, one may regard commodity-specific tariffs/export taxes as departures from a unified exchange rate régime (see Bhagwati, 1968).

This interdependence between areas of policy has a number of important policy implications. Attempts by a country to restore balance of payment equilibrium under pegged or inflexible exchange rates may be constrained by restrictions on the market access of its exporters. Conversely, fluctuations in the exchange rate of a country that increase the uncertainty of trading conditions may inhibit trade liberalization. Perhaps the most important link in an era of greater exchange rate flexibility is that an appreciation of the real effective exchange rate of a country for whatever reason may encourage protectionist sentiments. This was confirmed by an empirical study by Grilli (1988) (working, incidentally, for the World Bank!). The causation may run the other way. Maintenance of a higher exchange rate (a real exchange rate devaluation) may be deliberately used to provide protection to a nation's producers. Corden (1985, chapter 17) called this "exchange rate protection". However, deliberate manipulation of the exchange rate in this manner requires a pegged currency and/or controls on capital movements to maintain an exchange rate that is higher than a purely market-determined rate.

In this context the WTO has an important role to play, which is to prevent countries, most particularly the crisis countries, from increasing border protection. This would contract world trade and make it more difficult for all countries to make the necessary financial market adjustments. This danger will be greatly increased if there is a world recession. It was noted above that the surveillance activities of the WTO need to be strengthened. The management of a global economy in which markets for goods, services, and capital are interconnected calls for greater vigilance and improved coordination among multilateral organizations.

NOTES

1. It is also a part of a much wider debate about "global governance". This covers the UN, human rights, and other social and political issues arising from globalization as well as the economic issues (see, for example, Desai and Redfern, 1995).

2. The 1944 Bretton Woods Conference, or the United Nations Monetary and Financial Conference to give its full title, was devoted to monetary and banking issues, excluding trade. It established the charters of the IMF and the World Bank. However, it recognized the need for a trading organization. Negotiations that led to the draft charter of the ITO were held in London, New York, and Geneva in 1946 and 1947.

3. Some international transactions in goods or capital markets do not result in foreign exchange market transactions. For example, in the goods markets barter trade and in FDI the reinvestment of profit in the host country do not involve foreign exchange transactions. However, these examples are exceptional.

4. Growth in the markets for goods and services, capital, and foreign exchange are not independent of each other. FDI stimulates trade in goods in several ways: for

example, it leads to greater exchange of components, materials, and capital goods within multinational corporate groups. Therefore, trade in goods is often a complement of trade in capital. Most services are not supplied by cross-border delivery, as with the trade of goods. For those services that are delivered by commercial presence in the country in which the buyers are located, the integration of markets requires freedom of entry for foreign service suppliers who must be able to invest in the local economy and national treatment for them. Consequently, the liberalization of trade in markets for goods and services has stimulated trade in FDI markets and vice versa.

5. These are the Agreement on Government Procurement, the Agreement on Trade in Civil Aircraft, the International Dairy Agreement, and the International Bovine Meat Agreement. The last two Agreements were terminated in 1997.

6. Developing countries agreed to bind all agricultural tariff rates and a significant increase in binding industrial tariff rates. Since the Uruguay Round the developing countries as a group have bound 72% of tariff lines. While this is still lower than the corresponding coverage among the developed countries (99%), it is much higher than the coverage among the developing countries before the Uruguay Round (22%) (GATT, 1994). The significance of binding an applied rate is that a member country can no longer raise the tariff rate if it wishes. Instead, if a tariff rate is bound at a ceiling rate in excess of the applied rate, the actual tariff rate can be increased up to the ceiling rate.

7. The erosion of preferences was a minor bone of contention among the developing countries in the Uruguay Round. The maximum preference on any tariff item is 100% of the applied MFN tariff rate. As applied tariff rates come down under the Uruguay Round Agreement and the unilateral tariff reforms of member countries, there will be much less scope for preferences. In a similar way tariff preferences have become a major bone of contention in the negotiations between the 70 African Caribbean and Pacific (ACP) countries that are currently covered by the Lome Convention Agreement with the EU. The current agreement, which expires in 2000, is by far the largest preference scheme in the world. The EU is pressing the ACP countries to replace the current non-reciprocal preferences with free trade agreements under Article XXIV, which have reciprocal preferences. (McQueen, 1998, reviews these negotiations and the issues.) Whatever the outcome of the negotiations, it is bound to be less favorable in its treatment of the developing countries.

REFERENCES

Bank for International Settlements, 68th Annual Report (1998). Basel: BIS.

Bhagwati, J. N. (1968). The Theory and Practice of Commercial Policy: Departures from Unified Exchange Rates. Princeton: Princeton University, International Finance Section.

Boughton, B. M., & Sarwar, L. R. (Eds) (1995). Fifty Years after Bretton Wood: The Future of the IMF and the World Bank. Washington, DC: IMF and World Bank Group.

Corden, W. M. (1985). Protection, Growth and Trade: Essays in International Economics. Oxford: Oxford University Press.

Desai, M., & Redfern, P. (1995). Global Governance: Ethics, and Economics of the World Order. London: Printer.

Dooley, M. P. (1996). A survey of the literature on contorts on capital transaction. *IMF Staff Papers*, *43* (December), 639–687.

Eng, M. V., Les, F. A., & Mauer, L. J. (1998). *Global Finance*. Reading, MA: Addison-Wesley.

The Results of the Uruguay Round: Market Access for Goods and Services (1994). Geneva: General Agreement on Tariffs and Trade (1994), GATT.

Fischer, S. (1998). Capital-account Liberalization and the Role of the IMF. In: S. Fischer et al. (Eds), *Should the IMF Pursue Capital-account Convertibility, Essays in International Finance, No, 207*. Princeton University.

Fischer, S. et al. (1998). Should the IMF Pursue Capital-account Convertibility, Essays in International Finance, No. 207. Princeton University.

Fukuyama, F. (1989). The End of History. *The National Interest, 16* (Summer), 3–18.

Grilli, E. (1988). Macro-economic Determinants of Trade Protection World Economy, September.

ul Haq, M., Kaul, I., & Grunberg, I. (1998). The Tobin Tax: Coping with Financial Volatility. Oxford: Oxford University Press.

Harrigan, J. (1996). Review Article – the Bretton Woods Institutions in Developing Countries: Betes Noires or Toothless Tigers? *World Economy, 19* (November, May), 765–779.

Hoekman, B., & Kostecki, M. (1995). *The Political Economy of the World Trading System: from GATT to WTO*. Oxford: Oxford University Press.

International Monetary Fund (IMF) (1997). *International Capital Markets: Developments, Prospects and Key Policy Issues*. Washington, D.C.: IMF.

International Monetary Fund (IMF) (1998). *Annual Report 1998*. Washington, D.C.: IMF.

Jackson, J. (1989). *The World Trading System*. Cambridge, MA: MIT Press.

Krueger, A. (Ed.) (1998). *The World Trade Organisation as an International Organisation*. Chicago: University of Chicago Press.

Lewis, K. K. (1995). Puzzles in International Financial Markets. In: G. M. Grossman & K. Rogoff (Eds), *Handbook of International Economics* (Vol. 3). Amsterdam: North-Holland.

Lloyd, P. J. (1995). An APEC or Multilateral Investment Code. *Journal of Asian Economics, 6*(1), 53–70.

McQueen, M. (1998). Lomé Versus Free Trade Agreements: The Dilemma Facing the ACP Countries. *The World Economy, 21* (June), 411–420.

OECD (1996). *Towards Multilateral Rules*. Paris: OECD.

Pacific Economic Cooperation Council (1996). *Survey of Impediments to Trade and Investment in the APEC Region*. Singapore: PECC.

Panel of Experts (1958). *Trends in International Trade: A Report by a Panel of Experts*. Geneva: General Agreement on Tariffs and Trade.

Quirk, P. J., Owens, E., & Staff (1995). *Capital Account Convertibility, Review of Experience and Implications for IMF Policies*. IMF Occasional Paper No. 131. Washington, D.C.: IMF.

Rajapatarina, S. (1998). Columbian Trade Policies and the 1996 WTO Review. *World Economy*, (June), 515–527.

Reidel, J. (1997). Capital Market Integration in Developing Asia. *The World Economy 20*, (January), 11–19.

Sachs, J. D., & Radelet, S. (1998). The East Asian Financial Crisis: Diagnosis, Remedies, Prospects. *Brookings Papers on Economic Activity, I*, 1–118.

Sachs, J. D., & Warner, A. (1995). Economic Reform and the Process of Global Integration. *Brookings Papers on Economic Activity, I*, 1–118.

Sampson, G. P. (1998). Greater Coherence in Global Economic Policy Making: A WTO Perspective. In: A. Krueger (Ed.), *The World Trade Organisation as an International Organisation*. Chicago: University of Chicago Press.

Sauvé, P., & Zampetti, A. B. (1999). Subsidiary and the New Trade Agenda: How Relevant? In: S. Woolcock (Ed.), *Subsidiary in the Governance of the Global Economy.* Forthcoming.

Taylor, L. (1997). The Revival of the Liberal Creed – the IMF and the World Bank in a Globalized Economy. *World Development, 25,* 145–1512.

The Economist (1998). Reforming World Finance: Lessons from a Crisis. October 3–9.

United Nations Conference on Trade and Development (UNCTAD) (1996). *World Investment Report 1996: Investment, Trade and International Policy Arrangements.* Geneva: UNCTAD.

United Nations Conference on Trade and Development (UNCTAD) (1997). *World Investment Report 1997: Transnational Corporations, Market Structure and Competition Policy.* Geneva: UNCTAD.

World Trade Organisation (WTO) (1998). *Annual Report 1998* (Vol. I). Geneva, WTO.

5. CONTAGION IN THE EAST ASIAN CURRENCY CRISIS[1]

Mardi Dungey[2] and Vance L. Martin

1. INTRODUCTION

When a country or region experiences a financial crisis in the form of large exchange rate fluctuations it is of interest to identify whether the movements can be attributed to idiosyncratic shocks or are a result of contagion. In contrast to previous empirical research on modelling contagion (Eichengreen et al. (1996); Lowell et al. (1998)), a latent factor model is used to distinguish between idiosyncratic and contagion effects for individual exchange rates. This has the advantage that the results are not predicated on the specific indicators chosen. As a natural outcome of the model's framework two types of contagion are identified – fixed and variable. The fixed factor effect arises from the use of a numeraire currency throughout the analysis whereby a common shock has the same fixed/constant effect on each of the exchange rates. The variable factor effect reflects the differing responses in each exchange rate to common shocks.

The structure of the factor model consists of modelling each exchange rate as a linear combination of its idiosyncratic factor, and the fixed and variable contagion effects. The form of the factor model adopted differs from standard factor analysis in two ways. First, unlike in standard factor models where the parameters (loadings) vary across exchange rates for each factor, the fixed factor amounts to imposing a cross equation restriction on the loadings to be equal. For the variable factor no restrictions are imposed. Secondly, conditional variances based on GARCH structures are used to model the second moments.

Restructuring Asian Economics for the New Millennium, Volume 9A, pages 69–83.

ISBN: 0-7623-0709-9

This last property is particularly important given strong empirical evidence of time-varying conditional heteroskedasticity in exchange rates. This model is related to the class of factor models discussed by Diebold and Nerlove (1989), Ng, Engle and Rothschild (1992) and Mahieu and Schotman (1994). As with these models maximum likelihood is infeasible as the likelihood involves a multiple integral which cannot be computed using standard numerical algorithms. To circumvent this problem the indirect estimation procedure of Gourieroux, Monfort and Renault (1993) and Gallant and Tauchen (1996) is used to estimate the parameters of the model.

The application in this paper is to the recent East Asian crisis. The currencies examined are the Thai baht, the Malaysian ringgit, the Indonesian rupiah and the South Korean won. These four currencies are chosen in response to Krugman's (1998) observation that contagion between the MIT currencies (Malaysia, Indonesia, Thailand) can be understood due to the similarity of their export competing sectors but the vector to South Korea is not as transparent. To assess the extent of contagion, the analysis is performed pre-float of the Thai baht in July of 1997, the non-crisis period, and from December 1997 to August 1998, the crisis period, when all the exchange rates were at their most flexible.

In analysing the sources and magnitude of contagion over time, the analysis is performed on both the expected and actual movements in the exchange rate. Whilst most of the empirical literature on contagion has concentrated on actual exchange rates (Sachs, Tornell & Velasco, 1996; Eichengreen, Rose & Wyplosz, 1995, 1996; Lowell, Neu & Tong, 1998), performing the analysis on expectations provides further linkages for the transmission of contagion. For example, Radelet and Sachs (1998: p. 12) have recently argued that the currency crisis in East Asia could be understood as the result of changes in expectations. Thus a test of this proposition could be to identify whether the contribution of contagion to exchange rate expectations has changed over time.

The results show that in both the non-crisis and crisis periods expected exchange rate movements are dominated by idiosyncratic factors. It is not clear that contagion is more important in the crisis period than the non-crisis period for expectations, countering Radelet and Sachs' claim. In contrast, for the observed exchange rate changes contagion clearly dominates idiosyncratic factors in the crisis period. This difference in the impact of contagion between expected and observed exchange rate changes may serve to explain why, despite existing literature clearly relating actual exchange rate changes to currency crises and contagion channels, these events are often poorly anticipated.

The rest of the paper proceeds as follows. Section 2 provides a summary of the recent literature on contagion. The latent factor model of exchange rates is set out in Section 3, whilst estimation strategy is discussed in Section 4. The results of applying the method to data from the East Asian crisis are reported in Section 5, followed by a brief conclusion.

2. THEORIES OF CONTAGION

Although there is a well-established literature on both theoretical and empirical evidence for currency crises, the literature on contagion is considerably more sparse. Flood and Marion (1998) and Eichengreen, Rose and Wyplosz (1996) provide a summary of the currency crisis literature. Recent empirical applications may be found in Sachs, Tornell and Velasco (1996) and Frankel and Rose (1996). There is little empirical work on contagion itself. Sachs et al. in fact look for indicators of crises under the hypothesis that if there is no systematic set of indicators then the financial crises are 'just' due to contagion (Sachs et al., 1996: p. 148). Eichengreen, Rose and Wyplosz (1995, 1996) and Lowell, Neu and Tong (1998) examine contagion empirically. Eichengreen et al. define contagion as an increase in the probability of crisis in a domestic country associated with a crisis elsewhere. In the case of the East Asian crisis this implies a series of contemporaneous crises in geographically related countries. Lowell, Neu and Tong (1998) point out that the probability of this occurring coincidentally is very low.

One of the difficulties with quantifying the effects of contagion is in identifying the sources of shocks to the system. As with most of the exchange rate literature, empirical investigations have difficulties identifying observed factors which may be responsible in some general model. Unobserved factors are a distinct limitation to empirical work. Dooley (1997), Edwards (1998) and Eichengreen et al. (1996) consider it desirable to differentiate unobserved common shocks from shocks due to regional crises and shocks associated with country specific factors (Dooley, 1997; Edwards, 1998). In the framework of this paper this can be understood as differentiating idiosyncratic country factors from fixed and variable contagion effects.

The existing empirical work on crises suggests that there are some indicators which seem to predict exchange rate crises consistently; Sachs et al. (1996), Eichengreen et al. (1995) and Lowell et al. (1998) also provide evidence of some predictable avenues for contagion. This raises the issue of why currency crises and contagion are so difficult to anticipate in reality. In the light of this apparent contradiction this paper examines contagion in both expected *and* observed exchange rate changes. The results support that there are differences between the two, which goes some way to resolving the contradiction between the information given by indicators and the observed lack of anticipation of crises and contagion.

3. A MODEL OF CONTAGION

In this section a factor model of contagion is specified. By construction each of the factors are assumed to be independent. Letting $e_{i,t}$ be the change in the expected exchange rate at time t

$$e_{i,t} = E_t(S_{i,t+1}) - E_{t-1}(S_{i,t}) \tag{1}$$

where $S_{i,t}$ is the log of the exchange rate for currency i relative to the numeraire currency. The model assumes a linear relationship between e_i and the set of factors

$$e_{i,t} = \phi_i C_{i,t} + \delta F_t + \lambda_i V_t \qquad i = 1, 2, \ldots, n \tag{2}$$

where n is the number of exchange rates. The first factor, $C_{i,t}$, is the idiosyncratic factor associated with country i, F_t is the fixed contagion factor and V_t is the variable contagion factor. The term δF_t shows that each country experiences the same level of fixed contagion at a point in time. Of course this amount varies over time according to the changes in F_t, but the effect on country exchange rates is the same. Formally, this property arises from imposing a no-artibrage condition on the model (Dungey (1997)). The last term in (2) captures the variable contagion as there is a common factor across countries, V_t, that has a variable impact on individual countries (λ_i). In general the factor specification in (2) allows for international fixed and variable contagion effects, but in a broader set of countries there may be a need for additional individual regional factors. For example, if the set of currencies considered in this study is extended to include European currencies it may be desirable to add an 'Asian' and 'European' factor to the model.

Following the work of Diebold and Nerlove (1989) and Ng, Engle and Rothschild (1992), Mahieu and Schotman (1994) and Dungey, Martin and Pagan (1997), the factors are specified to include GARCH variances. This reflects a commonly observed characteristic that exchange rate data exhibit conditional heteroskedasticity.[3] Letting $G_t = \{C_{1,t}, C_{2,t}, \ldots, C_{n,t}, F_t, V_t\}$ be the set of n + 2 factors, the dynamic specification of the factors is given by[4]

$$G_{k,t} = \sqrt{h_{k,t}}\mu_{k,t} \tag{3}$$

$$h_{k,t} = 1 - \alpha_k - \beta_k - \alpha_k \mu_{k,t}^2 + \beta_k h_{k,t-1} \tag{4}$$

$$\mu_{k,t} \sim N(0, 1) \tag{5}$$

where $k = 1, 2, \ldots, n+2$.

The restriction $1 - \alpha_k - \beta_k$ in equation (4) is used for identification and normalises the unconditional variance of the factors to be unity (see Diebold & Nerlove, 1989)).

4. INDIRECT ESTIMATION

The latent factor model specified above cannot be estimated by maximum likelihood. The problem is that as the $(n+2)$ factors are latent they need to be integrated out of the likelihood to express the likelihood just in terms of the data. As these factors occur for each t, the order of the integration needed is $(n+2)^*T$ which is not feasibly solved using standard numerical methods. This is a common problem with this class of models; see, for example Gourieroux and Monfort (1995) for a general discussion of the point. To circumvent the problems with using maximum likelihood a simulation procedure based on the indirect estimator of Gourieroux, Monfort and Renault (1993) and Gallant and Tauchen (1996), is used. This estimator achieves consistent parameter estimates whilst efficiency is a function of the choice of the auxiliary model adopted, to be specified below. For the case where the factors exhibit no dependence, Dungey (1997) shows how the model can be estimated by Generalised Methods of Moments (GMM). Dungey, Martin and Pagan (1997) show the relationship between GMM and the indirect estimator for this form of the unconditional model.

Implementation of the indirect estimator involves simulating the latent factor model and matching the characteristics of the simulated data with the sample via an auxiliary model over a repeated number of simulations. Letting $\Psi = \{\delta, \phi_i, \lambda_i, \alpha_i, \beta_{i-}, i = 1, 2 \ldots n\}$ be a $5n+1$ vector of parameters of the theoretical model, the indirect estimator is given as the solution of

$$\hat{\Psi} = \arg\min_{\Psi} \left[\hat{\Pi} - \frac{1}{H} \sum_{h=1}^{H} \tilde{\Pi}_h \right]' \Omega \left[\hat{\Pi} - \frac{1}{H} \sum_{h=1}^{H} \tilde{\Pi}_h \right] \tag{6}$$

where $\hat{\Pi}$ are the parameters of the auxiliary model based on the actual data and $\tilde{\Pi}_h$ are the same set of parameters except based on the hth simulated data set. The form of the weighting matrix Ω, is given by Gourieroux, Monfort and Renault (1993). The sampling properties of the indirect estimator for this class of factor models is investigated by Dungey, Martin and Pagan (1997).

As a necessary condition for identification the auxiliary model must contain at least $5n+1$ moment conditions. The choice of the auxiliary model consists of $n(n+1)/2$ univariate autoregressions based on the squares and cross-products of the exchange rates with a constant and 2 lags. This gives a total of $3n(n+1)/2$ moment conditions. For the case of $n=4$, the necessary condition for identification is satisfied.

This choice of the auxiliary model is motivated by noting that from (2), squaring and taking conditional expectations produces

$$E_{t-1}e_{i,t}^2 = \phi_i^2 E_{t-1}C_{i,t}^2 + \delta^2 E_{t-1}F_t^2 + \lambda_i^2 E_{t-1}V_t^2 \tag{7}$$

As the factors have zero mean, the conditional expectations represent the conditional variances. This equation shows that the conditional variances of exchange rates inherit the conditional variance structure of the factors. It also shows that an AR model in the squares of the exchange rate helps identify the parameters of the model. Further identification is achieved by taking the cross-products. This yields after taking conditional expectations

$$E_{t-1}e_{i,t}e_{j,t} = \delta^2 E_{t-1}F_t^2 + \lambda_i \lambda_j E_{t-1}V_t^2 \qquad i \neq j \tag{8}$$

Hence the inclusion of AR models based on cross-products of $e_{i,t}$ in the auxiliary model helps to identify the underlying parameters.

5. ANALYSIS OF CONTAGION IN EAST ASIA

5.1. Data

The set of East Asian countries used in the empirical analysis are Malaysia, Indonesia, Thailand and South Korea, with the U.S. as the numeraire (see Frankel & Rose, 1996). This choice is governed by the observation made by Krugman (1998) who hypothesised that there was a difference in transmission of contagion between Malaysia, Indonesia and Thailand, the MIT countries, and South Korea.

Daily data on spot exchange rates and short-term interest rates for each of these countries are obtained from Datastream. To assess the impact of contagion during a crisis two data periods are examined; a non-crisis and crisis period. The non-crisis period runs from January 1995 to June 1997 inclusive, prior to the float of the Thai baht. The crisis period is defined as the period from December 23 1997 to 30 August 1998 when the rupiah, won and baht were freely floating. (Thailand floated the baht on 2 July 1997, Indonesia floated the rupiah on 14 August 1997 and South Korea floated the won on 22 December 1997. Conversely, the Malaysian ringgit, which had been loosely pegged to the U.S. dollar before the crisis, was fixed on 2 September 1998.) The interest rate spreads and exchange rates relative to the U.S. dollar for each country are given for the non-crisis and crisis periods in Charts 1 to 8. The pre-crisis period in

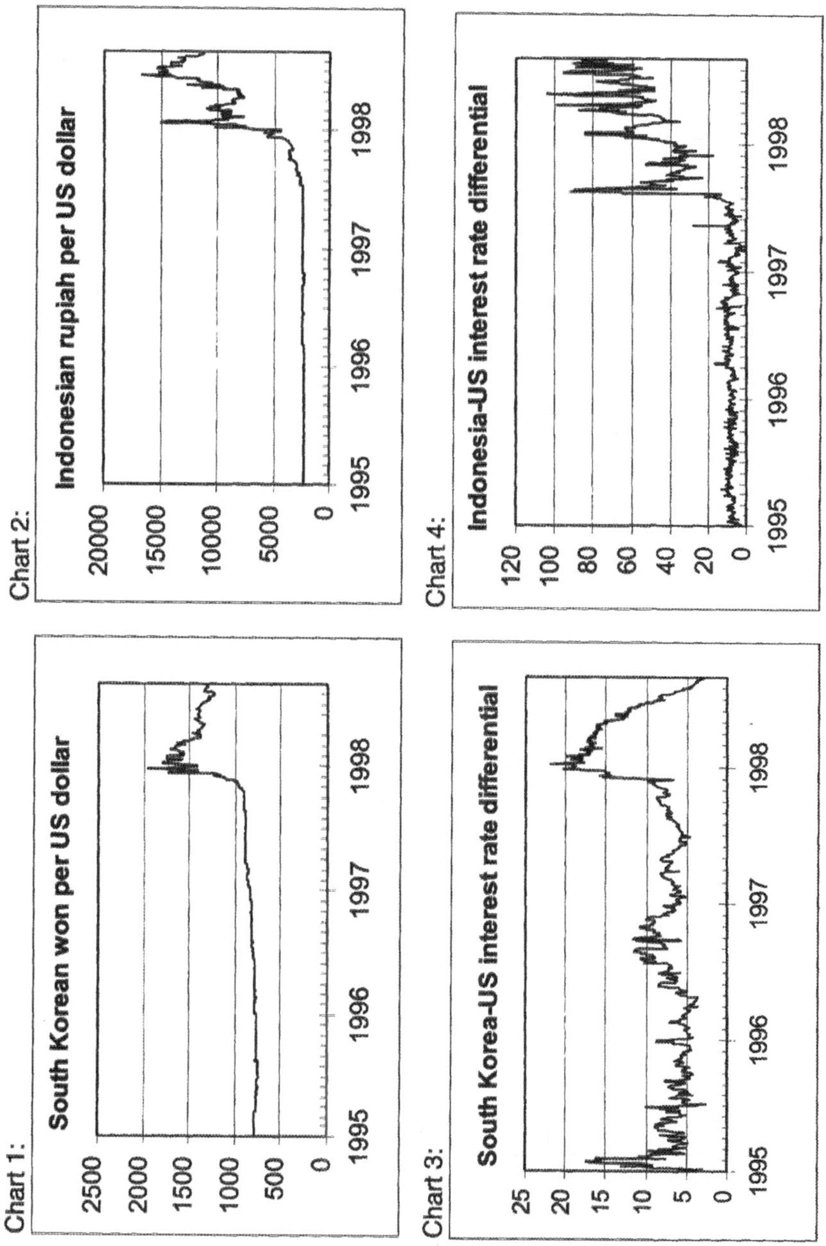
Chart 1:
South Korean won per US dollar
2500
2000
1500
1000
500
0
1995
1996
1997
1998
Chart 2:
Indonesian rupiah per US dollar
20000
15000
10000
5000
0
1995
1996
1997
1998
Chart 3:
South Korea-US interest rate differential
25
20
15
10
5
0
1995
1996
1997
1998
Chart 4:
Indonesia-US interest rate differential
120
100
80
60
40
20
0
1995
1996
1997
1998

Charts 1–4.

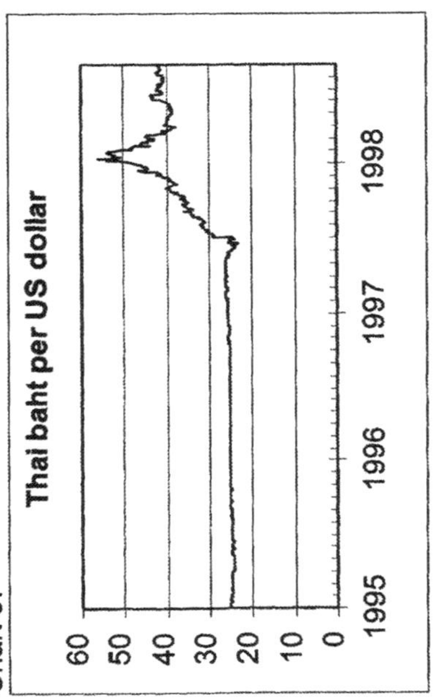
Chart 5:
Thai baht per US dollar
60
50
40
30
20
10
0
1995
1996
1997
1998

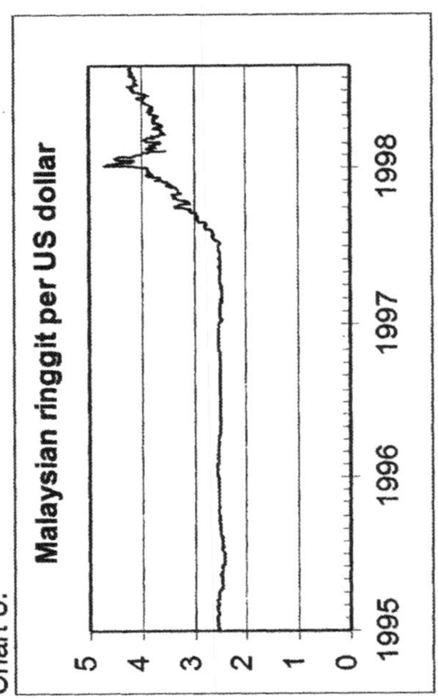
Chart 6:
Malaysian ringgit per US dollar
5
4
3
2
1
0
1995
1996
1997
1998

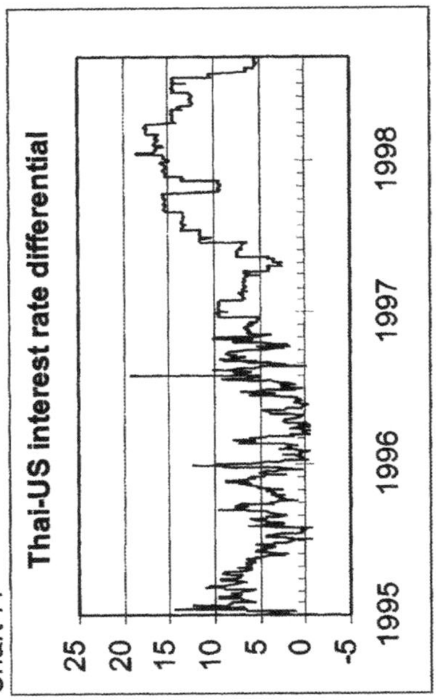
Chart 7:
Thai-US interest rate differential
25
20
15
10
5
0
-5
1995
1996
1997
1998

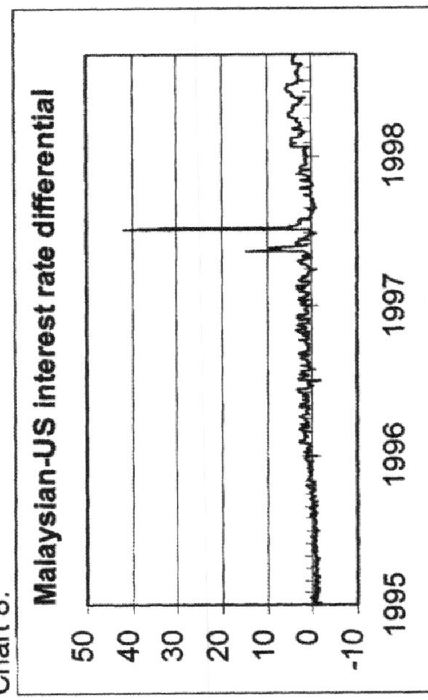
Chart 8:
Malaysian-US interest rate differential
50
40
30
20
10
0
-10
1995
1996
1997
1998

Charts 5–8.

each exchange rate is clearly evident from the charts. For each currency the range of fluctuation against the U.S. dollar was small. In the crisis period, by contrast, the exchange rate changes were substantial. The timing of regime changes can be clearly seen in the graphs. The depreciation in Thailand is followed by Indonesia in the next month, while the won remains relatively stable until its float in late December 1997. The more flexible regime for the Malaysian ringgit means that it reflects the turmoil from the time pressure came onto the Thai baht.

To compute movements in expected exchange rates, uncovered interest parity is invoked

$$E_t(S_{i,t+1}) - S_{i,t} = r_{i,t} - r_{US,t} \qquad (9)$$

where $r_{i,t}$ is the interest rate in country i, and $r_{US,t}$ is the interest rate in the United States. First differencing (9) and rearranging gives an expression for the change in exchange rate expectations

$$e_{i,t} = \Delta(r_{i,t} - r_{US,t}) + \Delta S_{i,t} \qquad (10)$$

The data used in estimation for the non-crisis period are the demeaned sum of changes in the interest rate spread and the exchange rate; the latter is denoted the change in expectations. In the crisis period both expectations data and demeaned changes in interest rate spreads are used.

5.2. Results

Estimates of the loadings from the factor model (2) to (5) are given in Table 1 for expected exchange rate movements.[5] Both the non-crisis period and crisis

Table 1. Coefficient Estimates.

	non-crisis	crisis	
	$e_{i,t}$	$e_{i,t}$	$\Delta S_{i,t}$
δ	0.4486	0.3374	–0.0021
ϕ_K	0.7213	0.3521	–0.0132
ϕ_I	1.6113	6.8449	0.0208
ϕ_T	1.1552	1.4426	0.0016
ϕ_M	–0.4656	–0.5436	–0.0012
λ_K	0.0413	–0.0637	–0.0005
λ_I	–0.6828	3.5623	–0.0355
λ_T	0.2482	0.2190	0.0007
λ_M	0.4569	–0.1523	0.0011

period results are given. The currencies are denoted South Korean won (K), Indonesian rupiah (I), Thai baht (T) and Malaysian ringgit (M).

Also reported in Table 1 are the results for actual exchange rate movements. These results are obtained by the model in (2) to (5) for changes in interest rate spreads, $\Delta(r_{i,t} - r_{U.S.,t})$, and deriving the parameter estimates for $\Delta S_{i,t}$ from the two sets of parameter estimates. As all four currencies were pegged to the U.S. dollar to varying degrees during the non-crisis period, the factor decompositions of actual exchange rate movements are reported just for the crisis period.

Of more interest is to decompose the idiosyncratic and fixed and variable contagion effects. Given the independence of the factors, the unconditional variance of (2) gives the following decompositions:

$$\mathrm{var}(e_{i,t}) = \phi_i^2\,\mathrm{var}(C_{i,t}) + \delta^2\,\mathrm{var}(F_t) + \lambda_i^2\,\mathrm{var}(V_t)$$

As the form of the GARCH specification in equation (4) normalises the unconditional variances of the factors to be unity, the contributions of each factor to the unconditional variance of exchange rate movements can be expressed as proportions.

$$\text{Contribution of idiosyncratic factor: } \frac{\phi_i^2}{\phi_i^2 + \delta^2 + \lambda_i^2} \tag{11}$$

$$\text{Contribution of fixed contagion: } \frac{\delta^2}{\phi_i^2 + \delta^2 + \lambda_i^2} \tag{12}$$

$$\text{Contribution of variable contagion: } \frac{\lambda_i^2}{\phi_i^2 + \delta^2 + \lambda_i^2} \tag{13}$$

Tables 2 and 3 present these decompositions for both actual and expected exchange rate changes. This same information is presented graphically in Charts 9–11.

In both the non-crisis and crisis situations the expectations decompositions are dominated by the country specific factors. That is, expected exchange rate changes are clearly associated with domestic factors.

The impact of contagion on changing expectations in the non-crisis period ranges between 15% and 30% for South Korea, Indonesia and Thailand and up to 65% in Malaysia. In the crisis period total contagion ranges from 7% in Thailand up to 47% in South Korea. The evidence for increased contagion in the crisis period is mixed.

In South Korea contagion was relatively high in the non-crisis period, and was almost entirely fixed contagion. In the crisis period the contribution of

Table 2. Factor Decompositions of Expected Exchange Rate Changes (%).

	country specific	fixed contagion	variable contagion	Total contagion
1. non-crisis				
South Korea	71.95	27.83	0.23	28.05
Indonesia	79.55	6.17	14.29	20.45
Thailand	83.55	12.60	3.86	16.45
Malaysia	34.59	32.11	33.30	65.41
2. crisis				
South Korea	52.81	45.61	1.58	47.19
Indonesia	78.10	0.19	21.71	21.90
Thailand	92.72	5.15	2.13	7.28
Malaysia	68.16	26.44	5.40	31.84

Table 3. Factor Decompositions of Actual Exchange Rate Changes (%).

	country specific	fixed contagion	variable contagion	Total contagion
crisis period				
South Korea	97.41	2.47	0.12	2.59
Indonesia	25.49	0.26	74.25	74.51
Thailand	34.00	59.12	6.88	66.00
Malaysia	21.42	62.03	16.55	78.58

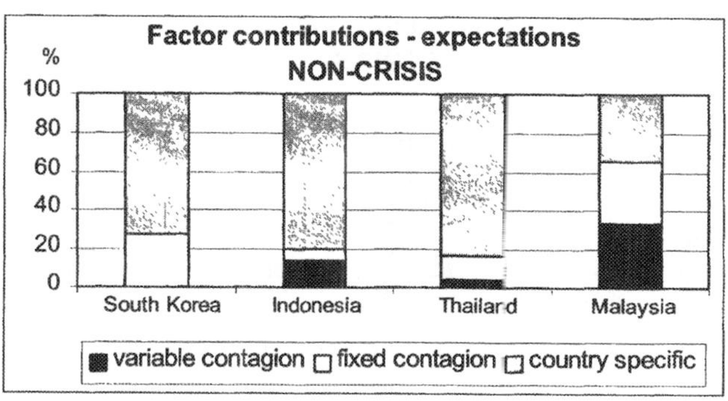

Chart 9.

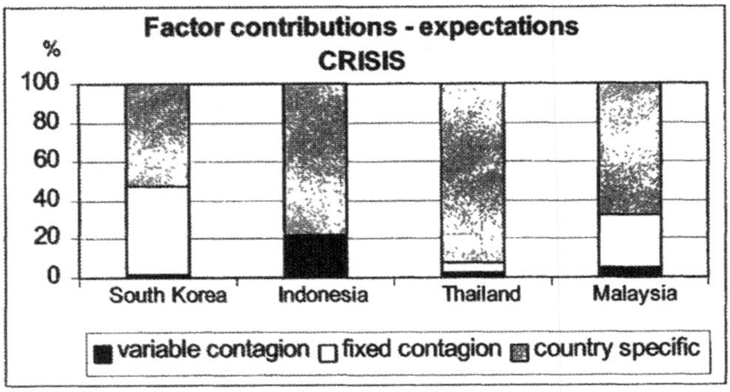

Chart 10.

contagion in expectations increased dramatically, to 47%, and fixed contagion continued to dominate.

Contagion in Thailand was also dominated by fixed contagion in both the crisis and non-crisis periods, but in both cases Thailand experienced the smallest component of contagion of all the countries. The dominance of country specific effects may reflect Thailand's status as a trigger in the East Asian crisis (Lowell et al., 1998).

Malaysia experienced a relatively large contribution from contagion in both the non-crisis and crisis periods, although the contribution in the non-crisis period was over double that in the crisis period. Unlike the other countries, contagion in the non-crisis period was evenly split between fixed and variable contagion in Malaysia. In the crisis period, fixed contagion dominated.

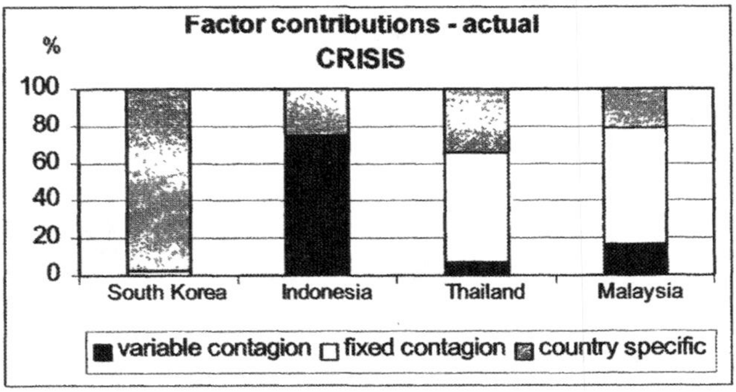

Chart 11.

The contribution of contagion in Indonesia was similar in both periods, of the order of 20% of total movements in expectations. In both periods contagion was dominated by variable contagion, in contrast to the other countries.

The decomposition of the actual exchange rate changes in the crisis reversed the story to changes in expectations. These results are consistent with Krugman's separation of the MIT group from South Korea. The MIT countries have substantial contributions from contagion in the crisis period for the observed exchange rate changes, ranging from 66% for Thailand to 78% for Malaysia, while South Korea has only a 3% contribution from contagion. The majority of the contribution for all countries but Indonesia is from fixed contagion. The high proportion of variable contagion for Indonesia may be responsible for Radelet and Sachs (1998: p. 21) identifying the Indonesian crisis as most clearly originating in contagion.

6. CONCLUSION

A latent factor model was used to quantify the effect of contagion on four exchange rates in the East Asian currency crisis – the South Korean won, the Indonesian rupiah, the Thai baht and the Malaysian ringgit against the U.S. dollar. The factor model distinguished between idiosyncratic and contagion effects for changes in expectations of exchange rates across the non-crisis period, pre-July 1997, and crisis period, after late December 1997 to August 1998, and observed exchange rates in the crisis period. For both sub-periods, changes in expected exchange rates are shown to be dominated by individual country factors. The effect of contagion on expectations for the four countries is also found to be not uniformly increased in the crisis period over the non-crisis period. Perhaps more significant is the finding that the decomposition of expected exchange rate changes is not substantially different over the two periods. This firstly provides contrary evidence to Radelet and Sachs' (1998: p. 12) claim that changes in expectations are an important contributor to the crisis. Secondly, it indicates that empirical research on contagion is correct to focus on actual exchange rate changes rather than being concerned with expectations. Finally, the observation that contagion is important in actual exchange rate movements, but not substantially different in expected changes across non-crisis and crisis periods, supports the notion that crises are not predicted by the market.

For actual exchange rate changes contagion effects clearly dominated idiosyncratic effects in the crisis period for Malaysia, Indonesia and Thailand, while contagion made a relatively small contribution to explaining movements in the South Korean won. This grouping of countries supports Krugman's

(1998) hypothesis that contagion between the MIT countries could be understood as transmitted through trade linkages, but that the links to South Korea are less transparent.

NOTES

1. Paper presented to the ACAES-Chulalongkorn University International Conference on A Macroeconomic Core of an Open Economy for Progressive Industrialization and Development in Asia in the New Millennium.

2. phone: +61 2 6125 0304 fax: +61 2 6125 3700 email: mardi.dungey@anu.edu.au

3. As the factor structure is in terms of expected log exchange rate changes no autocorrelation in the factors is expected. For a model where autocorrelation is included in the factors see Dungey, Martin and Pagan (1997).

4. Given that there is evidence of a need for both GARCH and non-normal errors to model the leptokurtosis observed in unconditional exchange rate distributions it would be of interest to replace the normal specification with a non-normal error structure, such as Student-t (Bollerslev et al. (1992)). This is left for future research.

5. The estimates of α_k and β_k are not included here. They are available from the authors on request.

REFERENCES

Bollerslev, T., Chou, R. Y., & Kroner, K. F. (1992). ARCH Modelling in Finance: A Review of the Theory and Empirical Evidence. *Journal of Econometrics, 52*(1–2), 5–59.

Diebold, F. X., & Nerlove, M. (1989). The Dynamics of Exchange Rate Volatility: A Multivariate Latent-Factor ARCH Model. *Journal of Applied Econometrics, 4*, 1–22.

Dooley, M. P. (1997). A model of crises in emerging markets. NBER Working Paper, No. 6300.

Dungey, M. (1997). A Multilateral Approach to Decomposing Volatility in Bilateral Exchange Rates, Australian National University Working Paper, No. 319.

Dungey, M., Martin, V. L., & Pagan, A. R. (1997). International Bond Spreads and Dynamic Latent Factor Volatility Decompositions, mimeo, Melbourne University.

Edwards, S. (1998). Interest Rate Volatility, Capital Controls and Contagion, NBER Working Paper, No. 6756.

Eichengreen, B., Rose, A. K., & Wyplosz, C. (1995). Exchange Market Mayhem: The Antecedents and Aftermath of Speculative Attacks. *Economic Policy, 0*(21), 249–296.

Eichengreen, B., Rose, A. K., & Wyplosz, C. (1996). Contagious Currency Crises, NBER Working Paper, No. 5681.

Flood, R., & Marion, N. (1998). Perspectives on the Recent Currency Crisis Literature, NBER Working Paper, No. 6380.

Gallant A. R., & Tauchen, G. (1996). Which Moments to Match? *Econometric Theory, 12*(4), 657–681.

Frankel, J. A., & Rose, A. K. (1996). Currency Crashes in Emerging Markets: An Empirical treatment. *Journal of International Economics, 41*, 351–366.

Gourieroux, C., & Monfort, A. (1995). *Statistics and Econometric Models*. Cambridge: Cambridge University Press.

Gourieroux, C., Monfort, A., & Renault, E. (1993). Indirect Inference. *Journal of Applied Econometrics, 8*, S85–S118.

Krugman, P. (1998). Bubble, Boom, Crash: Theoretical Notes on Asia's Crisis, mimeo, MIT.

Lowell J., Neu, C. R., & Tong, D. (1998). Financial Crises and Contagion in Emerging Market Countries, RAND, MR-962.

Mahieu, R., & Schotman, P. (1994). Neglected Common Factors in Exchange Rate Volatility. *Journal of Empirical Finance, 1*, 279–311.

Ng, V. K., Engle, R. F., & Rothschild, M. (1992). A Multi-Dynamic Factor Model for Stock Returns. *Journal of Econometrics, 52*, 245–266.

Radelet, S., & Sachs, J. (1998). The Onset of the East Asian Financial Crisis, mimeo, Harvard Institute for International Development.

Sachs, J., Tornell, A., & Velasco, A. (1996). Financial Crises in Emerging Markets: The Lessons from 1995. *Brookings Papers on Economic Activity, 1*, 147–215.

APPENDIX A

Data Sources

All data are obtained from Datastream. The codes are given in [ ] below. Exchange rates are spot rates against the U.S. dollar and interest rates are call rates.

	Exchange Rates	**Interest Rates**
South Korea	[KOUSDSP]	[KOCALLO]
Indonesia	[USINDON]	[IDIBCAL]
Thailand	[USTHAIB]	[THBBIBN]
Malaysia	[MYUSDSP] Dec-97 to Aug-98 [BBMYRSP] Jan-95 to Jul-97	[MYIBKCL]
USA		[USCALLM]

6. LONG RUN EQUILIBRIUM EXCHANGE RATE BEHAVIOR: ASIAN/YEN EXCHANGE RATES BEFORE THE CRASH

Steven Husted

1. INTRODUCTION

It has now been more than four years since the onset of the Asian financial crisis, and much research has been devoted to its causes and consequences.[1] Certainly, one of the most dramatic features of this crisis has been the sharp deterioration in the exchange rates of many Asian countries. Even countries, such as Australia, India, and New Zealand, that have not usually been identified with the crisis saw sharp depreciations in their exchange rates vis à vis the U.S. dollar during 1997. Many have cited the rapid deterioration in regional exchange rates as being a principal causative factor in the financial crisis that has followed. Radelet and Sachs (1998b, p. 11) argue that the "combination of real exchange rate depreciation and sharply higher interest rates lead to a rapid rise in non-performing loans in the banking sectors of the Asian economies."

Many commentators have focussed on the behavior of nominal (and real) exchange rates prior to the collapse. Bergsten (1998) argues that the "sharp swings in the yen-dollar rate contributed to the outbreak of the Asian crisis in the first place". The Reserve Bank of Australia (1998) also argues that weakness in the yen-dollar rate has led to deterioration in the exchange rates of other Asian economies. The logic behind this reasoning is hard to dispute. Prior

Restructuring Asian Economics for the New Millennium, Volume 9A, pages 85–102.

ISBN: 0-7623-0709-9

to 1997, many countries in Asia linked the value of their currencies to the U.S. dollar, either by nominal or real exchange rate pegging. As the dollar rose in value against the yen, these countries experienced a loss of competitiveness in world markets.

Given these movements in exchange rates and the dire implications these movements have had for the countries involved, a natural question to ask is whether or not the exchange rates of these countries were out of line with economic fundamentals prior to the crash. Surprisingly, only limited research has been done on this issue.[2] This is probably due to two reasons. First, it is now accepted that the relationship between exchange rates and fundamentals is long run. Given the short spans of data that exist since the collapse of Bretton Woods, testing these models is difficult. Added to the problem in this case is the fact that most empirical modeling of long-run exchange rate behavior has focussed on U.S. dollar exchange rates. In Asia, where many countries have tied their exchange rate to the dollar, empirical models have little relevance.

This paper seeks to address both problems described above. It does so by implementing a new tool for the analysis of exchange rate behavior, panel data estimators, on data sets involving the yen exchange rates for a set of Asian economies. Tests of long run behavior necessarily involve long spans of data. In the context of exchange rate modeling, this requires sampling over various exchange rate regimes, which may cause hypothesized relationships to break down. An alternative way of increasing the span of the data in testing equilibrium exchange relationships involves utilizing panel data sets for the recent float (see Chinn & Johnston, 1996; Frankel & Rose, 1996; MacDonald, 1988, 1996). Such tests have re-examined purchasing power parity (PPP) and the time series properties of real exchange rates.[3,4] In terms of panel PPP tests, researchers find that homogeneity restrictions hold and that the implied real exchange rate is strongly mean-reverting with around one-half of a disturbance to PPP being extinguished after 4 years.[5] Most recently, Husted and MacDonald (1998) and (1999) and Mark and Sul (1998) have used panel estimators to test the monetary approach to exchange rate determination. All three papers report strong evidence in favor of the monetary approach, using panels of data from OECD countries and dollar, DM, and yen exchange rates.

This paper follows the lead of these earlier studies. I employ a panel of data from nine Asia/Pacific countries and test both PPP and the monetary approach as determinates of long-run equilibrium exchange rates. The evidence in favor these two hypotheses is strong. The results indicate that the collapse of some of these exchange rates in 1997 was not due to exchange rates being out of line with fundamentals. In other words, the evidence suggests that these currencies were not "over-valued" so that their subsequent falls were not market corrections.

The remainder of the paper proceeds as follows. Section 2 offers a brief derivation of the models that underly this analysis. Section 3 discusses econometric methodology. Section 4 provides detail on data and empirical results. Section 5 concludes the paper.

2. EXCHANGE RATE MODELS

It is possible to identify two models of the nominal exchange rate that would generally be accepted as long-run constructs. The simplest, perhaps, is that of absolute purchasing power parity (PPP) in which the exchange rate is determined by relative prices:

$$s_t = p_t - p_t^* \quad (1)$$

where s_t denotes the spot exchange rate (home currency price of a unit of foreign exchange), p_t denotes a consumer price level and an asterisk denotes a foreign magnitude. Lower case letters in this and future expressions indicate that a variable is written in logs (the exception being interest rates). The standard approach to testing the simple PPP hypothesis is to estimate an empirical version of equation (1), expressed here as:

$$s_t = \gamma_0 + \gamma_1 p_t + \gamma_2 p_t^* + \nu_t \quad (2)$$

For PPP to be a valid long-run, or equilibrium, relationship ν_t should be an I(0) process (weak-form PPP) and $\gamma_1 = \gamma_2 = 1$ (strong-form PPP). An alternative way of expressing this is

$$q_t = \zeta + \delta q_{t-1} + \nu_t \quad (3)$$

where $q_t = s_t - p_t + p_t^*$ and for PPP to hold as a long-run relationship δ should be significantly less than one.

On using standard money market equilibrium conditions, relationship (1) may be transformed into a monetary approach equation. The money market equilibrium conditions for the home and foreign country are given by (4) and (5), respectively:

$$m_t - p_t = \kappa + \alpha y_t - \beta i_t \quad (4)$$

$$m_t^* - p_t^* = \kappa^* + \alpha^* y_t^* - \beta^* i_t^* \quad (5)$$

where m denotes the money supply, y is real income, i is the interest rate, κ is an arbitrary constant, α is the income elasticity of demand for money and β is the interest rate semi-elasticity. On rearranging (4) and (5) for the home and foreign price levels and substituting into (1) a monetary approach equation may be derived which is the long-run solution for the flex-price monetary model of Bilson (1978), Frenkel (1978) and Hodrick (1978) and the sticky price monetary model of Dornbusch (1976) and Frankel (1979). This equation is reported here as (6).[6]

$$s_t = \psi + m_t - m_t^* - \alpha y_t + \alpha^* y_t^* + \beta i_t - \beta^* i_t^* \tag{6}$$

Equation (6) asserts that an exchange rate in equilibrium is driven by relative excess money supplies. This equation may be re-expressed as a reduced form, as:

$$s_t = \omega_0 + \omega_1 m_t + \omega_2 m_t^* + \omega_3 y_t + \omega_4 y_t^* + \omega_5 i_t + \omega_6 i_t^* + u_t \tag{7}$$

where the ωs are the parameters to be estimated and u_t is a disturbance term. The strict form of the monetary model implies $\omega_1 = -\omega_2 = 1$, ω_3 and ω_4 should take on values which are close to estimated income elasticities from money demand functions and ω_5 and ω_6 should take on values close to interest rate semi-elasticities from the demand for money. The hypothesized values of ω_1 and ω_2 would be those expected, in equilibrium at least, by proponents of both the flex-price and fix-price monetary models. However, it is important to note that even within the monetary tradition there is a view that indicates that these coefficients may be less than one in absolute value; this view is attributed to Mussa (1976).

Baillie and Selover (1987), Kearney and MacDonald (1990), Meese (1986), Boothe and Glassman (1987) all estimate (7), or simplified versions thereof, using the Engle-Granger two-step methods and find no evidence of a long-run relationship in the sense that the residual series is an I(1) process. However, subsequent to these studies, MacDonald and Taylor (1993, 1994) have demonstrated that using the multivariate methods of Johansen (1988) a long-run version of the pure monetary model (7) can be defined for the recent float in the sense that the residuals are I(0) and point estimates are close to their a priori values. However, the speed of adjustment in such relationships proves to be painfully slow.

As noted in the discussion of equation (3), one feature of the PPP hypothesis and therefore the monetary approach is that the real exchange rate is assumed to be stationary. Several authors have recently noted that this assumption may be violated in certain countries, especially in Asia (see Isard & Symansky, 1996; Chinn, 1997, 1998). To deal with this problem, Chinn (1998) suggests adding an additional variable to the empirical specification. This variable allows for movements in relative prices of tradables to nontradables within and across countries that may either be due to relative differences in productivity in tradable and nontradable sectors (see Balassa, 1964; Samuelson, 1964) or international differences in preferences for nontradables (see DeGregorio & Wolf, 1994).

Since nontradable prices are not directly observable, it is necessary to construct an empirical proxy for the relative price term. Chinn (1998) suggests using

$$rp_t = \log\left[\frac{(PPI/CPI)}{(PPI^*/CPI^*)}\right] \tag{8}$$

where PPI is the producer price index, assumed to be comprised largely of tradables, and CPI is the consumer price index, assumed to include many nontradable goods prices. Chinn argues that the sign of this term should be positive.

3. ECONOMETRIC METHODS

The approach adopted in this paper to estimate equations (2) and (7) involves exploiting panel data sets defined with respect to the Japanese yen. The use of panel data in the current context has a key advantage over single equation estimates. It is now widely accepted that low frequency data, such as annual data, have a much higher signal-to-noise ratio than monthly or quarterly data in the context of estimating an equilibrium relationship (Shiller & Perron, 1985). However, in trying to estimate, say, a PPP relationship for the recent floating period with annual data, a researcher will have only approximately 23 useable observations. Although this may be sufficient to obtain homogeneity, it is unlikely that it will produce evidence of a cointegrating relationship (i.e. a stationary residual series) or, in the context of the real exchange rate autoregression (3), statistically significant mean-reversion. The way to increase the power of such tests involves increasing the span of the data (see MacDonald, 1995). One way the span may be increased simply involves lengthening the historical time series of the data (see Diebold, Husted & Rush, 1991). However, such an approach is fraught with difficulty – there may be several exchange rate regime changes throughout the period and the definition of the explanatory variables may have changed beyond all recognition from the beginning to the end of the sample. An alternative way to increase the span of the data involves expanding the panel dimensions. In a univariate context, Levin and Lin (1993) have demonstrated that a substantial increase in power may be obtained by testing an equation like (3) in a panel context. Pedroni (1995) has made a similar case for estimating cointegrating relationships in a panel context. I therefore propose using panel methods to estimate the monetary model (7) and also as a baseline relationship PPP, as represented in equation (2).

My testing methods may be illustrated by starting with the following standard panel framework:

$$s_{it} = \alpha + \beta' \mathbf{x}_{it} + \{\Sigma_i \gamma_i D_i\} + \{\Sigma_t \delta_t D_t\} + u_{it} \tag{9}$$

where the i subscript indicates that the data have a cross sectional dimension (running from 1 to N), $\mathbf{x}_{it}$ is a matrix, consisting of any of the sets of explanatory variables from the models noted above, D_i and D_t denote, respectively, country-specific and time-specific fixed effect dummy variables (although not noted here, it is straightforward to incorporate random effects into (9)). In a standard panel setting several modeling strategies are available for the disturbance term: it may be assumed random, heteroscedastic, autoregressive (with either a common autoregressive terms across individual panel members or different autoregressive terms across members), it may be spatially correlated or some combination of these assumptions may be used.

However, the key feature of recent work with panel data sets relates to the potential non-stationarity of the data. If the dependent and explanatory variables in (9) are stationary then (9) would represent an appropriate way to estimate the desired relationship. If, however, the variables in (9) are non-stationary then it has to be determined if they form a cointegrating set or not. If they are I(1) and not cointegrated, an appropriate way to proceed simply involves recasting (9) with the dependent and independent variables in first differences. If, however, s_{it} and $\mathbf{x}_{it}$ are cointegrated then simply specifying (9) in difference form would be inappropriate since information pertaining to the long-run relationship would be lost.

In the single equation setting, testing for stationarity and determining the existence of cointegration amongst a vector of non-stationary variables is relatively straightforward since there are a number of alternative estimators available (see Campbell & Perron, 1991) for a discussion of the relative merits of a range of different estimators). Unfortunately, cointegration in the panel context is not well developed. Several approaches to this problem have been attempted in the recent literature. Husted and MacDonald (1998) employ a two step approach wherein they use a variety of panel estimators to estimate a cointegrating vector and then test for the stationarity of the residuals from these specifications using the Levin-Lin panel unit root test. In so doing, they acknowledge the problem that they are using generated residuals in their second stage so that the conventional critical values for the Levin-Lin procedure do not apply. Consequently, they also supplement their results with alternative specifications such as error correction models that do not employ non-stationary data in the setup. Pedroni (1997) has proposed a test for panel cointegration that is somewhat similar to the procedure described above. However, his test is difficult to implement empirically, especially in the case of a common cointegrating vector. Finally, Mark and Sul (1998) also propose a two step procedure that uses the panel unit root tests of Im, Pesaran and Shin (1997) in the second stage.

The Levin-Lin method involves assuming that the $\mathbf{x}_{it}$ matrix in (9) is simply equal to a scaler, defined as the lagged dependent variable. The stationarity of the exchange rate may be gauged by the estimated t-ratio on the lagged dependent variable. In circumstances where all of the deterministic elements in (9) are excluded apart from the single constant term, α, Levin and Lin demonstrate that the t-statistic converges to a standard normal distribution. Including individual-specific effects (either or both) but excluding time-specific intercepts, Levin and Lin (1992) show that the t-ratio converges to a non-central normal distribution, with substantial impact on the size of the unit root test. However, Levin and Lin (1993) argue that unless there are very strong grounds for exclusion, time specific intercepts should always be included in this kind of panel test. The reason for this is that the inclusion of such dummies is equivalent to subtracting the cross section average in each period. This subtraction may be dispensed with in cases where the units in the panel are independent of each other, however in the case of panels involving bilateral exchange rates this would clearly be inappropriate.[7]

In testing for cointegration. I apply Levin-Lin tests to the residuals from (2) and (7). In particular, I report adjusted (ALL) and unadjusted (LL) Levin-Lin statistics, both constructed on the basis of no deterministic elements and a country-specific constant. The use of Levin-Lin t-ratios, of course, leaves the results reported below open to the generated residual criticism (although it should be noted that Frankel and Rose (1996) in their panel tests of PPP use a more primitive version of the Levin-Lin t-ratio, which does not adjust the data for cross sectional means). To address the generated residual issue, I follow Husted and MacDonald (1998) by using a method proposed by Hansen (1990). In that paper, Hansen describes what he refers to as the curse of dimensionality which is that in a residual-based cointegration test the cointegrating vector is random, imparting randomness into the generated residual series, and that this randomness is a function of the number of variables in the system. The particular implication that this has is to give dramatic reductions in power as the number of variables in the system increases, rendering cointegration tests ineffective in moderate sample sizes with moderately large systems. The reason that the test distributions depend on the dimensionality of the system is because they involve estimation of all n unit roots in the system. Hansen's proposition is to first adjust the non-stationary variables using a Cochrane-Orcutt estimator, and his Theorem 1 demonstrates that under the null hypothesis of no cointegration, the Cochrane-Orcutt estimate of the cointegrating coefficient converges to a constant rather than a random variable as it does using least squares. This suggests, therefore, that a test for cointegration constructed from the Cochrane-Orcutt estimates will not display the curse of dimensionality. In particular, Hansen's test statistics have the advantage that their limiting distributions are always the same as those of univariate unit root tests. Since almost all of the traditional panel estimators implemented in this paper use a Cochrane-Orcutt type correction this should help provide confidence in the results of the panel cointegration tests. More generally, my approach involves using several panel estimators to act as a robustness check on the simplest model. Since one of the estimators involves a correction for contemporaneous correlation, it directly addresses the point made by O'Connell (1996) that a failure to control for cross sectional dependence in a panel unit root test can have a dramatic effect on statistical size.

4. DATA AND EMPIRICAL RESULTS

The data for this study are collected from the International Monetary Fund's *International Financial Statistics* CD-ROM disc and are for the sample period 1974–1996 for nine Asian countries. The exchange rate is defined as the home currency price the Japanese yen (line ae for the individual countries divided by line ae for Japan), money supplies are M1 (line 34), income terms are real GDP (line 99b.p), interest rates are short rates (line 60 or 61), prices are the producer (PPI) and consumer price (CPI) indices (lines 63 and 64).[8] All variables, apart from the interest rate terms, have been converted into natural logarithms. The nine countries that make up the panel are Australia, India, Indonesia, Korea, Malaysia, New Zealand, Philippines, Singapore, and Thailand.[9]

4.1. Single Country Two Step Cointegration Results

As a motivation for the panel results, I first report in Tables 1–3 individual country estimates of cointegrating vectors for the two exchange rate models. PPP results are found in the first two tables. Table 1 uses CPI as a measure of prices; Table 2 uses PPI. As the results indicate, there appears to be substantial support for the strong-form PPP hypothesis no matter which price series is used. Virtually all of the coefficients are correctly signed, and most estimates are near unity in absolute value. The relative price term appears to add little to the explanatory power of the model: in most cases the R^2 value hardly changes and parameter estimates do not exhibit dramatic changes in value when RP is included in the model.

Despite the apparent success of these results, there is only limited evidence against the null of no-cointegration. In Table 1 only 4 of 36 augmented Dickey Fuller (ADF) and Phillips-Perron (PP) t-ratios are below the 5% critical value (of –4.14). These countries are Australia, New Zealand, and the Philippines.[10] Evidence against the no-cointegration null is somewhat higher in Table 2, where 13 of 36 ADF and PP t-ratios are smaller than the 5% critical value (Australia, Indonesia, Korea, Malaysia, and the Philippines).[11] Overall, the evidence in these two tables is similar to that reported by Husted and MacDonald (1999) who test for PPP using yen exchange rates for a panel of 21 OECD countries. They report similar parameter values and find evidence against the no-cointegration null in one third of the cases.

Table 1. Two-Step PPP Cointegration Results: All Countries, CPI Price Measure.

	P	P*	RP	γ_0	R^2	DW	ADF	PP	k
Australia	1.63	–1.37		–5.7	0.93	0.96	–5.17	–2.70	2
	0.74	–0.80	2.33	–4.3	0.96	1.48	–3.43	–3.43	0
India	1.77	–0.73		–6.8	0.97	0.83	–3.06	–2.54	2
	1.38	–1.78	3.00	–0.2	0.99	1.65	–3.72	–3.72	0
Indonesia	2.13	–2.05		2.1	0.94	0.86	–2.18	–2.18	0
	1.44	–2.78	2.14	8.9	0.96	1.52	–3.60	–3.60	0
Korea	1.06	–0.50		–0.9	0.92	1.09	–3.03	–3.03	0
	1.62	–2.41	1.60	5.2	0.92	1.20	–3.34	–3.34	0
Malaysia	1.66	–0.23		–10.7	0.77	0.50	–1.64	–1.64	0
	1.61	0.09	–0.13	–11.9	0.77	0.51	–1.65	–1.65	0
New Zealand	1.00	–0.33		–7.4	0.97	0.99	–4.44	–2.49	3
	1.00	–1.19	1.27	–3.4	0.97	1.10	–2.44	–2.44	0
Philippines	1.17	–0.61		–4.4	0.96	1.03	–2.44	–2.44	0
	0.89	–0.89	1.65	–1.6	0.99	2.70	–6.68	–6.97	1
Singapore	–0.23	1.09		–8.3	0.59	0.63	–2.00	–2.00	0
	1.32	2.21	–0.83	–8.5	0.61	0.70	–2.81	–2.38	1
Thailand	1.11	–0.01		–6.8	0.86	0.54	–2.61	–2.01	1
	0.73	–0.12	1.76	–4.5	0.94	0.93	–2.54	–2.54	0

Table 3 provides estimates of the monetary model, equation (7). Slightly more than half of the coefficients are correctly signed. The variable that performs least well is the home interest rate, with only 2 of 18 coefficients correctly signed. Foreign (i.e. Japanese) income has the most correctly signed coefficients (15 of 18). Money supplies have correctly signed coefficients in half the cases. Evidence against the no-cointegration is limited. ADF and PP t-ratios are large in absolute value. But, few are smaller than the 5% critical value (of − 5.53), which is based on six right hand side variables. Critical values for larger specifications, such as the equations that include rp are not available. Overall the model does not perform as well as PPP. This should not be surprising. The monetary model requires not only that PPP holds, but also that price levels are determined solely by excess supplies of money. Both of these propositions seem sensible as long run constructs, but may not occur with sufficient frequency to be detected in relatively short spans of data. This reasoning provides the motivation for the panel estimates that follow.

Table 2. Two-Step PPP Cointegration Results: All Countries, PPI Price Measure.

	P	P*	RP	γ_0	R^2	DW	ADF	PP	k
Australia	1.24	−1.43		−3.7	0.95	1.27	−4.48	−3.02	2
	0.74	−0.79	1.53	−4.4	0.96	1.47	−3.42	−3.42	0
India	1.67	−1.78		−1.5	0.98	1.26	−3.32	−3.32	0
	1.30	−1.77	1.41	0.2	0.99	1.81	−3.42	−3.42	0
Indonesia	1.40	−2.81		9.1	0.97	1.58	−2.33	−2.33	0
	0.75	−2.57	1.74	11.1	0.98	2.51	−6.36	−6.36	0
Korea	1.36	−2.38		6.4	0.95	1.94	−4.91	−4.91	0
	1.33	−2.40	0.34	6.6	0.95	2.01	−5.28	−5.28	0
Malaysia	1.06	−3.21		5.9	0.93	1.32	−2.29	−2.29	0
	0.08	−2.76	1.16	8.3	0.95	1.73	−4.38	−4.38	0
New Zealand	1.00	−1.05		−4.1	0.97	1.09	−3.35	−2.50	3
	0.96	−1.02	0.14	−4.0	0.97	1.08	−3.07	−2.43	3
Philippines	1.05	−1.17		−1.2	0.98	2.11	−4.97	−4.97	0
	0.89	−1.05	0.78	−0.9	0.99	2.77	−4.47	−9.79	3
Singapore	0.79	−1.08		−3.1	0.02	0.29	−1.95	−1.95	0
	1.49	0.85	1.88	−1.5	0.17	0.27	−1.22	−1.22	0
Thailand	1.49	−1.51		−1.7	0.92	0.76	−1.84	−1.84	0
	1.07	−0.75	1.13	−3.1	0.93	0.85	−2.10	−2.10	0

4.2. Panel Estimates

Panel estimates of equations (2) and (7) were obtained using a variety of a variety of techniques and are reported in Tables 4 and 5. Before discussing these estimates, I provide some detail on the estimators used in this section. They are: OLS with fixed effects, that is individual intercepts for each unit (denoted in the Tables by $\gamma_{0i} \neq \gamma_{0j}$); Maximum Likelihood estimates with fixed effects and a common AR1 process specified for the error term (this is denoted in the Tables as MAX 1 ($\rho_i = \rho_j$)); and Maximum Likelihood with fixed effects and an AR1 process which differs across individual units (represented in the Tables by MAX 2 ($\rho_i \neq \rho_j$)); finally, I report a Maximum Likelihood estimator which allows fixed effects, differing AR1 processes across individual units, contemporaneous residual correlation and a correction for heteroscedasticity (this is labeled 'POOL' ($\rho_i \neq \rho_j$) in the Tables). Each of these estimators is used over the entire panel of data, and slope coefficients are assumed to be equal across countries.

Table 3. Two-Step MAER Cointegration Results: All Countries.

	M	M*	Y	Y*	i	i*	RP	R^2	ADF	PP	k
Australia	−0.89	−1.47	2.98	4.24	−2.52	−5.38		0.96	−3.93	−4.38	3
	0.83	−0.83	2.06	2.41	−1.48	−0.45	3.22	0.97	−5.22	−5.22	0
India	0.92	−0.14	0.83	−1.00	1.33	−0.39		0.97	−2.63	−2.63	0
	0.71	−1.65	2.19	−1.39	2.04	1.55	3.23	0.98	−4.20	−4.20	0
Indonesia	−0.43	−0.70	0.34	6.11	−1.10	−4.34		0.97	−3.76	−3.76	0
	0.78	0.52	1.04	2.75	−2.38	−1.47	2.08	0.98	−5.03	−5.03	0
Korea	0.23	0.39	−0.44	1.28	−2.01	0.94		0.95	−4.38	−4.19	3
	0.36	0.31	1.12	−2.30	−1.28	6.65	2.72	0.97	−4.86	−5.03	2
Malaysia	0.51	1.40	−2.20	1.72	−1.68	1.32		0.91	−2.63	−2.63	0
	0.60	2.01	−2.81	2.86	−6.66	3.03	−0.75	0.96	−5.21	−5.21	0
N. Zealand	−0.19	−2.06	0.73	5.58	−0.33	−5.36		0.98	−5.50	−5.50	0
	0.08	−2.12	1.06	4.60	−0.11	−3.48	1.18	0.98	−5.15	−5.15	0
Philippines	0.49	−1.56	−2.21	5.06	−0.09	−4.82		0.99	−5.56	−5.56	0
	0.53	0.33	−1.62	1.01	−0.25	1.61	1.58	0.99	−4.52	−4.52	0
Singapore	−0.13	0.33	−0.94	2.23	−1.29	−1.69		0.82	−4.20	−3.61	3
	0.11	0.35	−1.00	2.28	−1.14	−1.79	−0.05	0.82	−4.18	−3.63	3
Thailand	0.67	−0.03	−1.32	2.24	−1.90	−1.45		0.96	−4.85	−4.50	1
	0.51	0.16	−1.26	2.14	−1.71	−0.45	0.67	0.97	−5.15	−4.61	1

The above-noted 'traditional' panel models facilitate testing for cointegration in the various nominal models. As an alternative check on the existence of cointegration, I also provide estimates from two error correction models (ECMs) in a panel context. The ECMs are constructed with the current change in the exchange rate on the left had side and two lagged changes of all the variables in the model as well as the lagged equilibrium error on the right. The equilibrium error is the residual taken from a first stage regression based on

Table 4. Panel PPP Results.

Method	P	P*	μ	R^2	DW	LL	AL
CPI price measure							
OLS ($\gamma_{0i} \neq \gamma_{0j}$)	1.18	–0.11		0.99	0.59	–5.28 [–5.23]	–5.06 [–1.49]
MAX 1 ($\rho_i = \rho_j$)	0.91	0.27		0.95	1.75	–12.5 [–13.0]	–11.9 [–9.7]
MAX 2 ($\rho_i \neq \rho_j$)	0.94	0.30		0.95	1.66	–12.4 [–13.7]	–11.8 [–10.3]
POOL	0.88	0.24		0.69	1.54	–12.77 [–13.5]	–12.20 [–9.97]
ECM (OLS)			–0.28 (0.06)	0.19	2.03		
ECM (POOL)			–0.32 (0.06)	0.18	1.91		
PPI price measure							
OLS ($\gamma_{0i} \neq \gamma_{0j}$)	1.23	–1.87		0.99	0.70	–5.56 [–6.28]	–5.33 [–2.52]
MAX 1 ($\rho_i = \rho_j$)	1.13	–1.60		0.96	2.13	–16.0 [–16.8]	–15.3 [–12.9]
MAX 2 ($\rho_i \neq \rho_j$)	1.20	–1.88		0.96	2.05	–11.5 [–12.2]	–11.0 [–8.8]
POOL	1.19	–1.65		0.83	1.80	–13.53 [–15.2]	–12.93 [–10.75]
ECM (OLS)			–0.19 (0.08)	0.09	2.04		
ECM (POOL)			–0.22 (0.08)	0.07	1.87		

OLS fixed-effects estimates of (2) and (7) respectively. The Tables report OLS estimates of the ECM model (labeled 'ECM (OLS)' and full pooling estimates using the Pool estimator referred to above (ECM(POOL)). The focus of interest in ECM models is the coefficient on the lagged equilibrium error, reported in the Tables under the heading μ. The value of μ provides information on whether or not cointegration exists and, if so, the speed at which the exchange rate responds to disturbances in the equilibrium relationship. Consequently, I report both the coefficient estimate and in parentheses below its estimated standard error.

4.3. Panel PPP Estimates

Panel PPP estimates are found in Table 4. There I report results for all of the models described above, using alternately CPIs and PPIs as the price level variable. Based on the single country results discussed above, I do not report panel estimates for models that include the relative price term, rp_t.[12] Several features of the results are worth noting.

First, the various estimates provide some, though less than compelling, evidence in favor of the PPP hypothesis. The strongest evidence is from the models that use the PPI price measure. In every specification, signs are correct and magnitudes are roughly in line with the strong-form hypothesis. These results are consistent with the assumption that PPP holds for tradable goods prices, since PPIs are believed to be heavily weighted with tradable goods. PPP appears to hold less well when the CPI price measures are used. Only one of the four foreign (i.e. Japanese) price coefficients is correctly signed.

One consistent pattern of behavior emerges no matter which price series is used in the setup. In particular, the own price coefficient is remarkable similar throughout and quite close to unity. This suggests a strong, positive, one-to-one relationship between own price levels and the exchange rate across all the countries in the panel. But, as noted above, the degree and direction of the relationship between Japanese prices and the exchange rate appears to depend crucially on the data series used to proxy for price levels.

Whatever the specific relationship, there is strong evidence against the null of no-cointegration. This evidence comes in two forms. First, with two possible exceptions, the LL and AL statistics indicate strong evidence that the residuals are stationary across all of the traditional estimators (the numbers not in square brackets are the t-ratios calculated on the basis of no deterministic, while those in square brackets elements include a constant in the deterministic set). Recall that these statistics are constructed to have Standard Normal distributions, so that if the data being tested were actual rather than generated, one would reject the unit root null for values smaller than –1.96. The fact that we are testing regression residuals means that the appropriate test statistic may be somewhat smaller, although in the case of the maximum likelihood estimators (MAX1, MAX2, and POOL) we believe that this problem is minimal. Nonetheless, the statistics we derive are minimally several orders smaller than the Standard Normal critical values. Second, in each of the ECM specifications, we find statistically significant estimates of μ. These values are similar within and across the two specifications. They suggest that following a shock to the long-term relationship between the exchange rate and prices, the half life of the adjustment back to equilibrium is about 2.5 years. This rate is similar to the adjustment speed reported by Husted and MacDonald (1999) in panels involving the U.S. dollar exchange rates of European OECD countries.

4.4. Panel Monetary Approach Regressions

Table 5 reports estimates of the monetary approach model, equation (7), using several alternative panel estimators. The results in support of the monetary model are striking. First, a very high proportion of the coefficient estimates have the correct sign, and many of these parameters have magnitudes consistent with the standard monetary model. In particular, the coefficients on home money, and home and foreign output are all correctly signed. The fixed effects OLS estimates are all correctly signed, and the fit is very good. However,

Table 5. Panel MAER Results.

Method	M	M*	Y	Y*	i	i*	μ	R^2	LL	AL
OLS ($\gamma_{0i} \neq \gamma_{0j}$)	0.57	–1.36	–0.47	2.58	1.68	–5.16		0.99	–3.13	–3.01
									[–2.6]	[1.2]
MAX 1	0.24	0.42	–0.57	1.30	–0.33	0.18		0.95	–10.2	–9.7
($\rho_i = \rho_j$)									[–11.0]	[–7.2]
MAX 2	0.27	0.01	–0.58	1.94	–0.31	–0.85		0.95	–7.3	–7.0
($\rho_i \neq \rho_j$)									[–7.7]	[–3.7]
POOL	0.33	–0.20	–0.59	1.97	0.26	–1.47		0.84	–11.4	–10.9
									[–10.7]	[–7.0]
ECM							–0.14	0.27		
(OLS)							(0.06)			
ECM							–0.16	0.30		
(POOL)							(0.06)			

simple GLS corrections for residual autocorrelation (i.e. MAX1 and MAX2) lead to sign changes on two variables. When these corrections are accompanied by GLS corrections for heteroskedasticity and contemporaneous correlation with the POOL estimator, the hypothesized signs reemerge.

As before, the LL and AL statistics provide information on whether or not these are long-run relationships in the data. And again, with limited exceptions, the statistics are much lower than the critical value of −1.96. This suggests relatively strong evidence of cointegration, most especially with the preferred POOL estimates. The error correction results reinforce this conclusion. The estimates of μ are statistically significant, although they are smaller than those found in Table 4. They suggest that the half-life of a shock is 4.6 years, more than twice as long as a shock to PPP. This evidence provides further support for the use of a panel of data rather than data from a single country to test the monetary model.

5. SUMMARY AND CONCLUSIONS

This paper has sought to provide information on the long-run behavior of Asia-Pacific nominal exchange rates vis-à-vis the yen over the post Bretton Woods period. In particular, I report tests of PPP and the monetary approach to exchange rates over the period 1974–1996 individually for nine countries and for these countries as a panel. Parameter estimates consistent with the PPP hypothesis emerge both for the individual countries and in the panel, especially when price levels are proxied by the PPI. Tests for cointegration between exchange rates and price levels suggest that a long-term relationship for the countries as a group is consistent with the data. Evidence in favor of the monetary model emerges strongly from the panel estimates, but not in the individual country estimates. Again, a cointegrating relationship between exchange rates and money market fundamentals is evident in the panel tests. To supplement these tests, I also report estimates from error correction models. These results are also consistent with cointegration, however half-lifes to shocks differ across models. The half-life to a PPP shock appears to be about 2.5 years, two years less than the half-life to a monetary approach shock.

The results in this paper suggest that sensible statements can be made about equilibrium values of Asian/yen exchange rates. Is there any evidence from this analysis that actual exchange rates at the end of 1996 were substantially overvalued? To address that question, I compare actual and estimated values from the panel OLS monetary model. For most of the countries, the model does a very good job of replicating actual behavior. In virtually cases, the model fits the data more closely at the end of the period than in the beginning. In particular, for many of the countries most closely associated with the Asian crisis, Korea, Malaysia, the Philippines, and Thailand, the model works especially well. With the exception of Malaysia, the model suggests that exchange rates at the end of 1996 were very close to actual values. In Malaysia, the ringgit appreciated strongly in 1996, especially when compared to long run equilibrium. For Indonesia, the model suggests that the rupiah has been undervalued vis a vis the yen for most of the past ten years. These findings suggest that the currency falls experienced by these countries represent shifts in long run mean values unrelated to underlying fundamentals.

NOTES

1. See, for example, Radelet and Sachs (1998a) and the references therein.
2. An exception is Chinn (1998).
3. See Frankel and Rose (1996) and MacDonald (1988).
4. See Chinn and Johnston (1996), Wu (1996) and MacDonald (1996).
5. These results are identical to those obtained with long runs of historical data.
6. This kind of reduced form may also be derived from the generalized asset pricing model of Lucas (1982).
7. Subtracting the cross section mean reduces the potentially problematic effect that contemporaneous correlations can have in a panel unit root context; see O'Connell (1996).
8. Several interest rate values were missing from the data. Simple interpolations were used to construct these values. Also missing were the first eleven observations on the Malaysian PPI. These were generated using a regression of Malaysian PPI on Indonesian PPI.
9. Data problems for both the People's Republic of China and Taiwan led me to exclude them from the panel.
10. In the case of the Philippines this is only true for the specification that includes the relative price term.
11. In the cases of Indonesia and Malaysia, this is only true for the specification that includes the relative price term.
12. These results are available on request.

REFERENCES

Baillie, R., & Selover, D. (1987). Cointegration and Models of Exchange Rate Determination. *International Journal of Forecasting*, *3*, 43–51.

Bergsten, C. F. (1998). How to Target Exchange Rates. *Financial Times*, November 20.

Bilson, J. F. O. (1978). Rational Expectations and the Exchamge Rate. In: J. Frenkel & H. Johnson (Eds), *The Economics of Exchange Rates*. Reading, MA: Addison-Wesley.

Boothe, P., & Glassman, D. (1987). Off the Mark: Lessons for Exchange Rate Modelling, *Oxford Economic Papers*, *39*, 443–457.

Campbell, J. Y., & Perron, P. (1991). Pitfalls and Opportunities: What Macroeconomists Should Know about Unit Roots. In: O. J. Blanchard & S Fischer (Eds), *NBER Economics Annual 1991*. Cambridge: MIT Press.

Chinn, M. (1997). The Usual Suspects: Productivity and Demand Shocks and Asia-Pacific Real Exchange Rates. *NBER*, Working Paper, 6108.

Chinn, M. (1998). On the Won and Other East Asian Currencies. *NBER*, Working Paper, 6671.

Chinn, M., & Johnston, L. (1996). Real Exchange Rate Levels, Productivity and Demand Shocks: Evidence from a Panel of 14 Countries. *NBER*, Working Paper, 5709.

Diebold, F. X., Husted, S., & Rush, M. (1991). Real Exchange Rates Under the Gold Standard. *Journal of Political Economy*, *99*, 1252–1271.

Dornbusch, R. (1976). Expectations and Exchange Rate Dynamics, *Journal of Political Economy*, *84*, 1161–1176.

Frankel, J. (1979). On The Mark: A Theory of Floating Exchange Rates Based on Real Interest Differentials, *American Economic Review*, *69*, 610–622.

Frankel, J. (1982). The Mystery of the Multiplying Marks: A Modification of the Monetary Model, *Review of Economics and Statistics*, *64*, 515–519.

Frankel, J. (1983). Monetary and Portfolio-Balance Models of Exchange Rate Determination. In: J. S. Bhandari & B Putnam (Eds), *Economic Interdependence and Flexible Exchange Rates*. Cambridge, MA: MIT Press.

Frankel, J., & Rose, A. (1996). A Panel Project on Purchasing Power Parity: Mean Reversion Within and Between Countries, *NBER*, Working Paper, 5006.

Frenkel, J. (1978). A Monetary Approach to the Exchange Rate: Doctrinal Aspects and Empirical Evidence. In: J. Frenkel & H. Johnson (Eds), *The Economics of Exchange Rates*. Reading, MA: Addison-Wesley.

Hansen, B. (1990). A Powerful Simple Test for Cointegration Using Cochrane-Orcutt. *Mimeo*, University of Rochester.

Hodrick, R. (1978). An Empirical Analysis of the Monetary Approach to the Determination of the Exchange Rate. In: J. Frenkel & H. Johnson (Eds), *The Economics of Exchange Rates*. Reading, MA: Addison-Wesley.

Husted, S., & MacDonald, R. (1998). Monetary-Based Models of the Exchange Rate: A Panel Approach. *Journal of International Financial Markets, Institutions & Money*, *8*, 1–19.

Husted, S., & MacDonald, R. (1999). Nominal Equilibrium Exchange Rate Models: A Panel Perspective. In: R. MacDonald & J. Stein (Eds), *Equilibrium Real Exchange Rates*. Kluwer Press.

Im, K., Pesaran, M., & Shin, Y. (1997). Testing for Unit Roots in Heterogeneous Panels, *Mimeo*. Cambridge University.

Isard, P., & Symansky, S. (1996). Capital Flows and Monetary Policy in East Asia. In:Hong Kong Monetary Authority (Ed.), *Monetary and Exchange Rate Management with International Capital Mobility*. Hong Kong: Hong Kong Monetary Authority.

Johansen, S. (1988). Statistical Analysis of Cointegrating Vectors. *Journal of Economic Dynamics and Control*, *12*, 231–254.

Kearney, C., & MacDonald, R. (1990). Rational Expectations, Bubbles and Monetary Models of the Exchange Rate: The Australian/U.S. Dollar Rate During the Recent Float. *Australian Economic Papers*, *29*, 1–20.

Levin A., & Lin, C. F. (1992). Unit Roots in Panel Data: Asymptotic and Finite Sample Properties. *Mimeo*, University of California.

Levin A., & Lin, C. F. (1993). Unit Roots in Panel Data: Asymptotic and Finite Sample Properties [revised]. *Mimeo*, University of California.

Lucas R. (1992). Interest Rates and Currency Prices in a Two-Country World. *Journal of Monetary Economics*, *16*, 335–360.

MacDonald, R. (1988). Purchasing Power Parity: Some Long Run Evidence from the Recent Float. *De Economist*, *136*, 239–252.

MacDonald, R. (1993). Long-Run Purchasing Power Parity: Is it For Real?, *Review of Economics and Statistics*, *75*, 690–695.

MacDonald, R. (1995). Long-Run Exchange Rate Modeling: A Survey of the Recent Evidence. *Staff Papers*, International Monetary Fund, *42*, 437–489.

MacDonald, R. (1996). Panel Unit Root Tests and Real Exchange Rates. *Economics Letters*, *50*, 7–11.

MacDonald, R., & Taylor, M. P. (1993). The Monetary Approach to the Exchange Rate: Rational Expectations, Long-Run Equilibrium and Forecasting. *Staff Papers*, International Monetary Fund, *40*, 89–107.

MacDonald, R. (1994). The Monetary Model of the Exchange Rate: Long-Run Relationships, Short-Run Dynamics, and How to Beat a Random Walk. *Journal of International Money and Finance*, *13*, 276–290.

MacKinnon, J. (1991). Critical Values for Cointegrating Tests. In: R. F. Engle & C. W. J Granger (Eds), *Long-run Economic Relationships: Readings in Cointegration* (pp. 267–276). Oxford: Oxford University Press.

Mark, N., & Sul, D. (1998). Nominal Exchange Rates and Monetary Fundamentals: Evidence from a Seventeen Country Panel. *Mimeo*, Ohio State University.

Meese, R. (1986). Testing for Bubbles in Exchange Markets: A Case of Sparkling Rates? *Journal of Political Economy*, *94*, 345–373.

Mussa, M. (1976). The Exchange Rate, the Balance of Payments and Monetary and Fiscal Policy Under a regime of Controlled Floating, *Scandinavian Journal of Economics*, *2*, 229–248.

O'Connell, P. G. J. (1996). The Overvaluation of Purchasing Power Parity. *Mimeo*, Harvard University.

Pedroni, P. (1995). Panel Cointegration; Asymptotic and Finite Sample Properties of Pooled Time Series Tests with and Application to the PPP Hypothesis. *Mimeo*, Indiana University.

Radelet, S., & Sachs, J. (1998a). The East Asian Financial Crisis: Diagnosis, Remedies, Prospects. *Brookings Papers on Economic Activity*, 1–90.

Radelet, S., & Sachs, J. (1998b). The Onset of the East Asian Financial Crisis. *NBER*, Working Paper, 6680.

Reserve Bank of Australia (1998). *Semi-Annual Statement on Monetary Policy*. Sydney: Reserve Bank of Australia.

Shiller, R., & Perron, P. (1985). Testing the Random Walk Hypothesis: Power Versus Frequency of Observation, *Economics Letters*, *18*, 381–386.

SECTION TWO

7. SCHOOLING IN ASIA: SELECTED MICRO EVIDENCE ON DETERMINANTS AND EFFECTS AND POLICY IMPLICATIONS

Jere R. Behrman[1]

Education, which in empirical work usually is represented by years or levels of schooling completed, is of considerable interest in much of the applied and policy-related development literature. Many hold the view that the empirical evidence for strong public policy support for schooling in developing countries is very strong. Often the extraordinary longer-run transformations in many of the economies of East and Southeast Asia are thought to support strongly substantial public subsidies for schooling in developing countries to order best to attain their goals.

The objective of this paper is to revisit the question of to what extent such claims are warranted by available empirical evidence. To attempt to sort out the empirical effects of schooling it is necessary to consider the tripartite foundations for good empirical analysis in the particular context of the determinants of and the impact of investments in human resources and related policy issues: (1) theory, (2) data, and (3) estimation. These three dimensions of empirical social science analysis are critically interrelated. Theory provides frameworks for exploring systematically various dimensions of the determinants of and the impact of human resource investments, points to what data are needed for such explorations, and to some of the probable estimation issues that should be addressed. Data, of course, are essential for empirical analysis,

Restructuring Asian Economics for the New Millennium, Volume 9A, pages 105–164.

ISBN: 0-7623-0709-9

limit the extent to which analyses can be undertaken, and shape most of the estimation problems. Most data that have been used for explorations of empirical issues are behavioral, not experimental data. The estimation problems, thus, reflect implications of theory and of the data that are available and, often more importantly, of the data that are not available – such as variables that are not observed or measured only imperfectly in the data but that may affect importantly human resource investments as well as subsequent adult productivity and other outcomes of interest. What can be said about the impact of schooling from empirical estimates (or about any other empirical estimates), thus, is conditional on what underlying model of behavior is assumed.

One basic proposition of this paper is that most of the studies that purport to provide micro empirical evidence of a strong impact of schooling on productivity, income, and other outcomes of interest do not persuasively identify the impact of schooling because they are not well-based on explicit plausible underlying models of behavior.

A second basic proposition of the paper is that even many of the studies that do attempt to deal with the estimation problems that arise because schooling, as well as schooling outcomes, is determined by behaviors, are not very illuminating regarding policy justifications for public subsidies for schooling

The paper begins in Section 1 by reviewing what simple but widely accepted economic models of investments in human resources suggest (a) are the determinants of such investments and (b) are the implications for attempting to estimate the impact of such human resources. Section 2 then reviews what this simple framework implies for policy justifications for public subsidies for schooling. Section 3 summarizes the types of empirical relations that frequently are estimated and some estimation problems that result in part because data are generated by behaviors such as those that are summarized in the previous sections. Section 4 then surveys some selected micro studies on both the determinants and the impact of schooling in the Asian context.

1. ANALYTICAL FRAMEWORK FOR CONSIDERING THE MICRO DETERMINANTS OF AND THE IMPACT OF HUMAN RESOURCE INVESTMENTS

Good theories about the determinants of and the impact of human resource investments such as schooling abstract the essence of complex empirical phenomena in ways that lead to testable empirical propositions about behavior and policy choices in the presence of imperfect information. Becker's (1967) Woytinsky Lecture provides a simple but very useful framework with which to think about investments in human capital (Fig. 1). Human capital investment

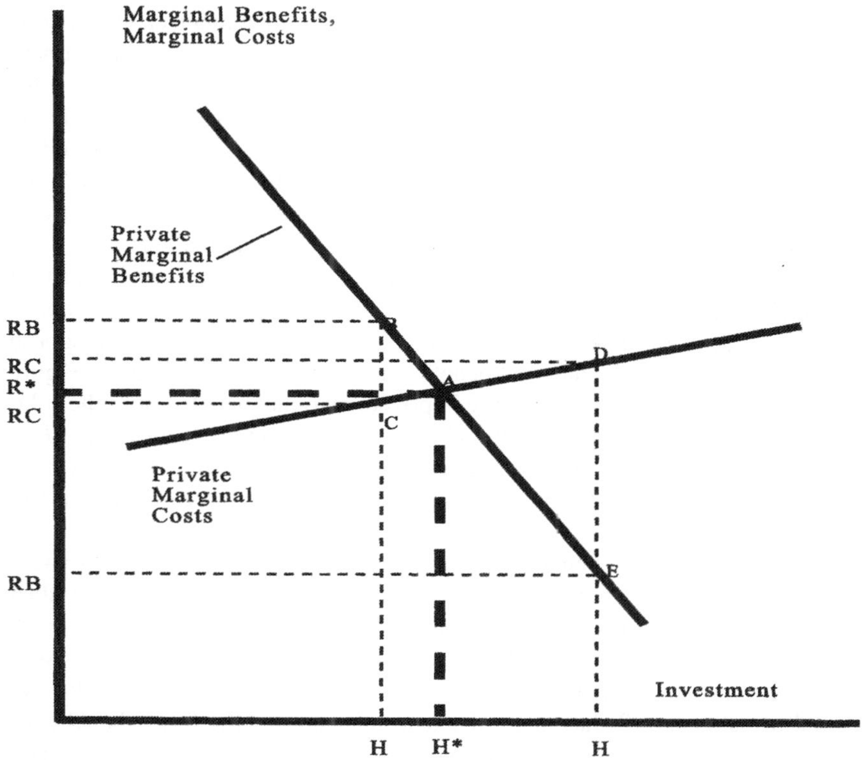

Fig. 1. Private Marginal Benefits and Private Marginal Costs of an Investment for an Individual.

demands reflect the equating of the expected present discounted value of marginal private benefits and expected present discounted value of marginal private costs for human capital investments in a given individual (though for expositional simplicity I will not repeat "expected present discounted value" in what follows).[2] The marginal private benefit curve depends importantly, *inter alia*, on the expected private gains in productivities (or perhaps in wages/salaries that may be related to productivities) due to human capital investments. These benefits depend on the marginal impact of the human capital investment on productivity in production functions for the relevant outcomes and on the marginal rewards that accrue to the investor because of that impact. The marginal private benefit curve is – at least at high enough investments levels –

downward-sloping because of diminishing returns to human capital investments (given genetic and other endowments).[3] If the returns to an investment are not reaped until it is completed, as generally is assumed to be the case for schooling, there also is likely to be eventual diminishing returns because the longer is the investment period, the shorter is the post-investment period in which to reap the returns from higher productivity due to the investment. The marginal private cost may increase with human capital investments because of increasing marginal private costs of borrowing on financial markets.

Private returns net of costs are maximized at level H*, where private marginal benefits are equal to private marginal costs.[4] This framework implies that private decision makers will tend to change their investments in response to changes in the private marginal benefits or the private marginal costs. Figure 2 illustrates the impact of an increase in private marginal benefits from the solid to the dashed line so that the optimal investment increases from H* to H**. Figure 3 illustrates the impact of a decrease in private marginal benefits from the solid to the dashed line so that the optimal investment increases from H* to H***.

The marginal private benefit curve may be higher for one of two otherwise identical individuals except for the difference noted below (i.e. the dashed rather than the solid line in Fig. 2) because one individual (or whomever is investing in that individual, such as the individual's parents): (1) has greater genetic endowments that are complementary with resources devoted to human capital investments; (2) has lower discount rates so that the future benefits of human capital investments have greater value at the time of the investment decision; (3) has human capital investments options of higher quality (e.g. access to higher quality public schools); (4) has better health and a longer expected life due to complementary investments so that the post-investment period in which that individual reaps the returns to the human resource investment is greater; (5) has greater marginal private benefits to a given level of such investments because of schooling or labor market discrimination that favors that individual due to gender, race, language, family, village, or ethnic group; (6) has returns to human resources investments that are obtained more by the investor or the relevant decision maker (e.g. if traditional gender roles dictate that children of one sex, but not the other, provide old-age support for their parents, parental incentives may be greater to invest in children who are likely to provide such support); (7) has greater marginal private benefits to a given level of investment because of being in a more dynamic economy in which the returns to such investments are greater; (8) has greater marginal private benefits to a given level of such investments because of greater externalities from the human capital investments of others in the same

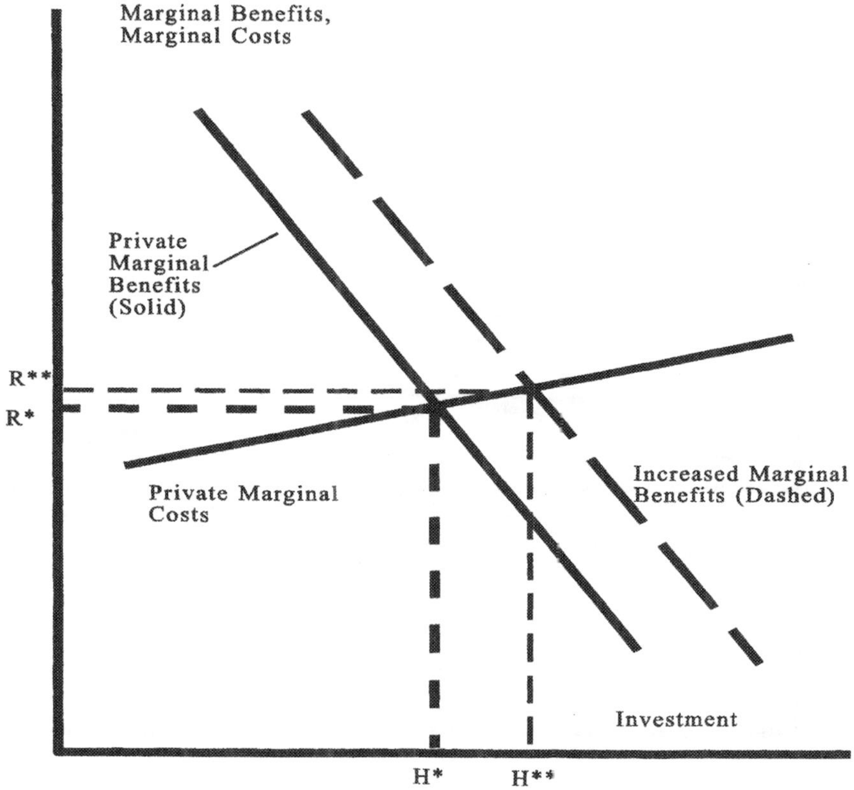

Fig. 2. Private Marginal Benefits and Private Marginal Costs of an Investment for an Individual, with Increased Marginal Benefits (Dashed Line).

economy; or (9) lives in a more stable economy so that the discount rate for future returns is lower (because risk is less) and thus the marginal private benefit of future returns greater.

If the marginal private cost is lower for every level of human capital investment as for the dashed line in comparison with the solid line in Fig. 3, the equilibrium human capital investment is greater, with the marginal private benefit lower at the higher investment level. The marginal private cost might be lower for numerous reasons. Compare two otherwise identical individuals except that one individual: (1) has lower private cost access to programs related to human resource investments because of closer proximity to such services or lesser user charges; (2) has less opportunity costs for time used for such

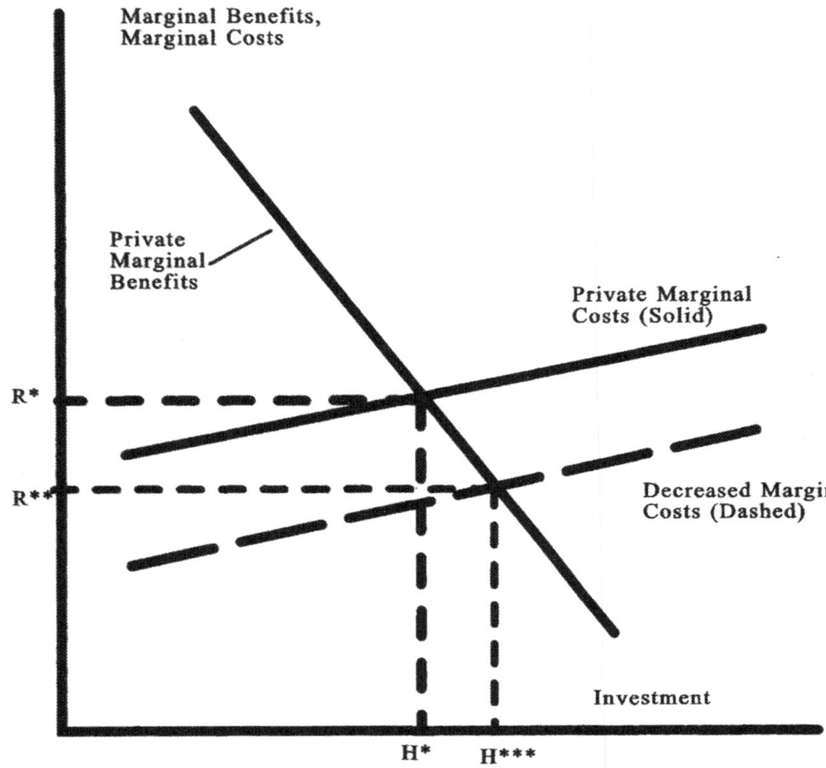

Fig. 3. Private Marginal Benefits and Private Marginal Costs of an Investment for an Individual, with Decreased Marginal Costs (Dashed Line).

investments; or (3) is from a household with greater access to credit (or less need for credit) for financing such investments because of greater wealth or status or better connections.

This maximization process leads to dynamic decision rules or demand relations for human capital investments in individual i that depend on all relevant prices P and on all relevant resources R and on all the parameters of the relevant production functions (including those for the production of human resources) and on preferences:

$$H_i = H(P, R \mid \text{production parameters, preference parameters}). \quad (1)$$

The prices include all prices that enter into the investing household's decision-making process, including the prices paid by the household for schooling, other forms of education, health and nutrition, and other consumption and investment purposes and for transferring resources over time (i.e. the interest rate) and for insuring against uncertainty. At the time that any human resource investment decision is made, these prices include all past and current prices for these goods and services (perhaps embodied in current stocks of human capital), as well as expected future prices (including expected future returns to human capital investments). The resources include all resources of the individual, household (identified by ownership if there is intrahousehold bargaining), educational and health/nutrition service providing institutions, and community that affect any household decisions. These resources include human resources that reflect past investments, financial resources, physical resources, genetic endowments, and general learning environments. If household decisions are made as the result of collective interactions among household members with somewhat divergent interests, moreover, the question of who controls resources may be important. For example, a number of observers claim that mothers on the average care more about human resource human resource investments in children than do fathers so that intrahousehold shifts in control over resources towards mothers tends to increase investment in human resources (e.g. Thomas, 1990). These resources also include the quantity and quality of social services.

This simple framework systematizes five critical, common sense, points for investigating dimensions of the determinants and the effects of human capital investments such as schooling.

First, the impacts of changes in policies may be hard to predict by policymakers and outside analysts. If households face a policy or a market change, they can adjust all of their behaviors in response, with cross-effects on other outcomes including savings, interhousehold transfers and consumption of adult goods, not only on the outcome to which the policy is directed. Provision of subsidized schooling, for example, effectively may be an income subsidy to the household if the household otherwise would have paid school fees, which the household decision makers can divert to whatever use they wish.

Second, the marginal benefits and marginal costs of human capital investments in a particular individual differ depending upon the point of view from which they are evaluated: (i) There may be externalities or capital/insurance market imperfections so that the social returns differ from the private returns and (ii) there may be a difference between who makes the investment decision (e.g. parents) and in whom the investment is made (e.g. children) that may result in gender (or birth-order) differentials in incentives for investments in children given traditional gender (birth-order) roles in old-age care for parents. To evaluate the determinants of human capital investments and their impact it may be important to be clear about from whose perspective the investment decision is being made.

Third, human capital investments are determined by a number of individual, family, community and (actual or potential) employer characteristics, only a subset of which are observed in social science data sets such as are available to analyze schooling determinants and effects. To identify the impact of the observed characteristics on human capital investments, it is important to control for the correlated unobserved characteristics. For example, households with greater income also may live in areas in which expected rates of returns from human capital investments tend to be greater and learning environments are better. If so and if only household incomes and not expected rates of return nor learning environments are observed in the data, the impact of household income on such investments is likely to be overestimated with usual procedures.

Fourth, to identify the impact of human capital investments, it also is important to control for individual, family, and community characteristics that determine the human capital investments and also have direct effects on outcomes of interest. Otherwise the estimated effect includes not only the impact of the human capital investments, but also the effects of individual, family, and community characteristics that directly affect the outcomes of interest and are correlated with the human capital investment because they partly determine those investments. For example, if innate ability affects both investments in education and then wages directly due to the impact of ability on production in relation (1) and if there is not control for innate ability in estimates of the impact of schooling on wages, the estimated effects are likely to be biased because the representation of education is partly proxying for innate ability in these estimates.

Fifth, empirically estimated determinants of and effects of human capital investments are for a given macro economic, market, policy, and regulatory environment in which there may be feedbacks both at the local and at the broader level, depending in part on how integrated local communities are into markets with broader geographical coverage. The process of economic development, for example, is likely to change the marginal benefits and marginal costs for micro human resource investments related to schooling as well as other micro human resource investments such as in human, physical and financial resources. The marginal benefits curve for having children from the perspective of potential parents, for example, is likely to shift downward if there are reduced expected rates of returns (wages, productivities) for unskilled labor and increased alternatives to children for parental old-age support due to improved capital markets and increased formal pension systems. But the marginal benefits curve is likely to shift upwards if, as part of the development process, the life expectancy for children increases or if parents want to use their increased income in part to obtain more "child services." The marginal cost curve for having children from the perspective of parents is likely to shift upwards due to increased time costs of raising children due to higher wages, but there may be tendencies in the other direction because, for instance, of lower financing costs due to improved capital markets and to lower private schooling and health care costs due to expansion of social services related to education, health and nutrition. Whether the tendencies for increasing or decreasing the number of children dominate with economic development within any particular market, policy and institutional context is an empirical question. Widespread experience suggests, however, that the factors inducing parents to choose fewer children predominate, while at the same time human resource investments in the human resources of children (and of adults) and in physical and financial assets tend to increase.

Estimation of the demands for human capital investments in relation (1) that are sensitive to these considerations can be informative for addressing a number of relevant questions. For example, how responsive are human capital investment decisions to the prices of resources or programs for human resource investments? How important are incomplete markets, particularly for capital and insurance, in human resource investments decisions? Do limitations in such markets mean that individuals from poorer backgrounds face relatively severe constraints on human resource investments because their families have very limited resources for self-financing human resource investments and can not readily finance them through capital markets? What role do information imperfections play in the household human resource investments decisions? Are potential household/individual investors in human resource investments well-informed about the potential returns to human resource investments? Are there important interactions between household characteristics and program characteristics?

Embedded in the framework for human resource investments are a number of production functions, such as for cognitive achievement, health and other outcomes of interest. The assessment of the *direct* impact (as opposed to the *total* impact in relation 1) of some important policy program components may be attained by estimating these production relations. For example, consider a production function for cognitive achievement CA_i for the ith child depending on pre-school human resource investments (PS_i), ability (A_i), health (H_i), nutrition (N_i), school quality (Q_i), time in school (S_i), family background characteristics (F_i) such as schooling of other household members and other factors:

$$CA_i = CA(PS_i, A_i, H_i, N_i, Q_i, S_i, F_i \ldots). \quad (2)$$

There are a number of important questions about this production process pertaining to what is the impact of policies on cognitive achievement through health, nutrition, school quality and perhaps other channels and what are the interactions between these factors and other factors that enter into the production of cognitive achievement or how is the marginal product of program changes affected by other individual or family or school attributes. How much do resources devoted to schooling and other educational programs improve cognitive achievement? Are resources devoted to schools more effective in their impact on cognitive achievement if a student has better health and nutrition? Greater abilities? Comes from a better family background?

If analysts, whether they be sensible policymakers or high-powered econometricians, had good information on all of the inputs and outcomes of these production processes, considerable light could be shed on the answers to questions such as are raised in the previous paragraph by estimating the critical parameters in these basic production processes. But there are at least three basic data problems in undertaking such estimates: (i) some critical variables are not likely to be observed by analysts (e.g. individual abilities, parents's abilities, parents' time devoted to child education, some aspects of school quality) which not only means that it is very difficult to assess their impact, but also means that it may be difficult to assess the impact of observed variables; (ii) some critical variables are likely to be observed only very poorly by analysts, which, for right-side variables results in underestimates of the true effects of human resource investments if the measurement errors are random; and (iii) production processes are dynamic so longitudinal data on all the relevant variables may be required to understand well the production processes, but often data are available only at a point in time or for very limited time periods.

2. MICRO FRAMEWORK FOR EVALUATING POLICIES RELATED TO SCHOOLING AND ECONOMIC DEVELOPMENT

Often analyses of policies related to schooling and economic development are undertaken without consideration of the general rationale for policies. It is just presumed that policies that say, increase schooling must be good. But such analyses are of little help in convincing skeptics that scarce resources should be allocated for these purposes, given the many competing alternative uses. Moreover they may not provide much in the way of guidelines for choosing among policy alternatives. Therefore it is useful to begin by asking why policy interventions (related to schooling) might be desirable. For most economists the two possible justifications for governmental policy interventions in these and other areas are: (1) to increase efficiency/productivity and (2) to redistribute resources. As I discuss below in Section 2.4, these two justifications include some other common concerns about policies, such as questions of access and quality of services and sustainability of overall economic development and of particular programs.

The policy justifications based on efficiency and on distribution are both firmly rooted in micro dimensions of behaviors as outlined in Section 1. That is the case because *both of these standard economic motivations for policy are concerned ultimately with the welfare of individuals as judged by those individuals*. I emphasize this last statement by placing it in italics because economic efficiency is viewed by some as a concern about allocation of things and technical and financial concerns, but *not* a concern about people. But this reflects a fundamental misunderstanding. To the contrary, economic efficiency ultimately is a concern with the welfare of people as judged not by policy makers or international experts, but by the individual decision makers involved. There, in addition is a separate important concern about the distribution of decision making powers. But it is important to recognize that the efficiency motive for policy, far from being purely a mechanical or technical concern of "dismal scientists" devoid of concerns about people, is based fundamentally in people's perceptions of their own welfare.

2.1. Efficiency/Productivity

Resources are used efficiently in the economic sense of the term if they are used to obtain the most product possible given the quantities of the resources and the available production technologies at a point of time and over time and if the composition of that product increases the welfare of members of society as much as is possible given the resource and technological constraints and the distribution of resource ownership. It is important to note that efficiency is not just a concern about the static use of resources at a point in time, but also is a concern about the use of resources over time and thus productivity and productivity growth over time. An human resource investment (or expenditure) related to schooling is efficient if the marginal *social* benefit of the last unit of that human resource investment just equals its marginal *social* cost.[5] If the marginal social benefit of a particular human resource investment is greater (less) than the marginal social cost, society is not investing enough (is investing too much) and would benefit from increasing (decreasing) the level of human resource investment until the marginal social benefits and costs are equalized.

Although applying the above rule maximizes social gains, private maximizing behavior leads to human resource investments at the level at which the marginal *private* benefit of the human resource investment equals its marginal *private* cost under the assumption that, given the information available to them and the constraints that they face, individuals act in what they perceive to be their best interests, as is discussed in Section 1. Figure 1 provides an illustration for human resource investments related to one individual decision maker.

Now consider what happens if the private incentives for human resource investments related to schooling differ from the social incentives for such human resource investments, first with respect to the marginal benefits and then with respect to the marginal costs.

Let the dashed line in Fig. 2 now represent the marginal *social* benefits for human resource investments related to schooling that are drawn to be greater than the marginal *private* benefits (rather than the changed marginal private benefits as in the discussion in Section 1).[6] In this case the private incentives are to invest at level H*, which is less than the socially-optimal (efficient) level of human resource investment at level H**. Therefore there is an efficiency argument for policies to induce or to require private human resource investments at level H** instead of level H*.

Why might marginal social benefits exceed marginal private benefits for human resource investments related to schooling? Among the most frequent answers to this question are:

(1) Human resource investments in education are thought to have not only private benefits to the person being educated through increasing his\her productivity in various activities, but, by adding to society's stock of knowledge, social benefits beyond the private benefits.[7]
(2) Information on which private human resource investment decisions are made may misrepresent the private rates of return to these human resource investments because it is incomplete or incorrect. This is an important consideration in the case of human resource investments related to schooling because actual and potential consumers are often not well informed about many of the private benefits and costs, and thus may underinvest in comparison with what they would do with better information.[8]
(3) There may be social gains beyond the private gains to human resource investments related to schooling if such changes are viewed by society as inherently valuable in themselves (e.g. to facilitate catching up with other

rapidly growing economies) in addition to their effects on individual economic productivity, consumption and welfare.

(4) The combination of uncertainty, risk aversion and imperfect insurance markets may result in private incentives to underinvest in human, financial and physical assets that enhance economic development from a social point of view because from a social point of view the risks are pooled.[9]

Now let the dashed line in Fig. 3 represent the marginal social costs for human resource investments related to schooling (rather than alternative marginal private costs, as in Section 1) that are drawn to be less than the marginal private costs.[10] In this case the private incentives are to invest at level H*, which is less than the socially-optimal (efficient) level of human resource investment at level H***. Therefore there is an efficiency argument to consider the possibility for policies to induce or to require private human resource investments at level H*** instead of at level H*.

Why might marginal social costs be less than marginal private costs for human resource investments related to schooling? Among the most frequent answers to this question are:

(1) There may be capital market imperfections for human resource investments (in part because human capital is not recognized as collateral) such that the marginal private costs for such human resource investments exceed their true marginal social costs, probably more so for individuals from poorer families who cannot relatively easily self-finance such human resource investments. This may be a problem, for example, for risky human resource investments in such areas as some types of education that involve substantial outlays with some of the benefits reaped only with considerable delays under conditions in which capital and insurance markets do not function well or are nonexistent.

(2) The sectors that provide schooling and other human-resource-related services may produce inefficiently because institutional arrangements and regulations do not induce efficient production of an efficient basket of commodities. This may be mainly a problem where services are provided directly by governments that operate under incentive schemes that do not encourage efficiency, or are heavily regulated; but it can also be an outcome of monopoly pricing particularly in isolated communities.

2.2. Distribution

Distribution is a major policy motive distinct from efficiency. Distributional concerns, at least officially, often focus on the command over resources of the poorer members of society. Society well might want to assure, for example, that everyone has basic social services even at some cost in reduced efficiency. Sometimes schooling (as well as family planning, health and nutrition services) is viewed as a "merit good" that is socially desirable in itself.

Though distributional concerns often are characterized by focus on the distribution of income or other resources *among* households, there also may be important distributional considerations *within* households. Household decision makers are not likely to consider equally the preferences of all household members in allocating household resources. For example, if women have preferences for fewer or more widely-spaced children, safer childbirth or using more resources to invest in children than do their husbands, these preferences may not be weighed equally as those of their husbands in decisions made by their husbands. Moreover, even if some households as aggregates have sufficient resources to cover what society considers to be basic needs, certain types of individuals in households may not be allocated what society considers to be sufficient resources for their individual satisfaction of basic needs. For example, there are many suggestions that women (and even more so widows), infants and children (particularly females and high birth orders) and the elderly may have relatively limited resources allocated to them by household decision makers.

2.3. Policy Choices to Increase Efficiency and to Improve Distribution

Consider first efficiency. If all other markets in the economy are operating efficiently and there are differences between marginal private and social incentives in markets related to schooling so that private incentives are to invest at level H* instead of at level H** in Fig. 2 or level H*** in Fig. 3, policies that increase the human resource investment to the socially efficient levels increase efficiency. If all other markets in the economy are *not* operating efficiently, then policies that narrow the differences between private and social incentives in markets related to schooling do *not* necessarily increase efficiency and productivity. It is conceivable that the distortion between private and social incentives in the particular market that results in the discrepancy between the private and the social optimal human resource investments in Figs 2 or 3 is just offsetting some distortion elsewhere, so that the economy will become less efficient and less productive if this distortion by itself is lessened (these considerations are part of the "theory of the second best"). Because there always are distortions between private and social incentives elsewhere in the economy, this may seem to be bad news for establishing an efficiency/productivity basis for policy. But, in the absence of specific information to the contrary, such as the existence of two counterbalancing distortions, a reasonable operating presumption is that lessening any one distortion between social and private incentives is likely to increase efficiency and productivity. With perfect information, one could know for sure and only make policy changes that in fact improve efficiency. In the real world in which policy makers (and everyone else) have very imperfect information, instead probabilistic statements must be made about policy changes that are likely to improve efficiency.

That still does not indicate what policies would be best to induce human resource investments related to schooling at level H** in Fig. 2 or level H*** in Fig. 3. There is a large set of possibilities, including governmental fiats, governmental provision of social services such as schools at subsidized prices, price incentives in the markets for human resource related services, price incentives in other markets, and changing institutional arrangements in various markets. To choose among alternatives, there are two important considerations.

First, it is necessary to realize that policies have costs. These costs include the direct costs of implementing and monitoring policies and the distortionary costs introduced by policies that may encourage socially inefficient behavior (including rent-seeking[11] by both public and private entities). Often policy makers focus only on the direct costs and ignore the distortionary costs that may be much greater because only the direct costs have obvious and visible direct ramifications for governmental budgets. In fact the costs may be sufficiently high that it is not desirable to try to offset some market failures by policies.[12] But, if it is desirable to do so, there is a case generally for making policy changes that are directed as specifically as possible to the distortion of concern because that tends to lessen the distortion costs. A *policy hierarchy* can be defined in which alternative policies to attain the same improvement in efficiency in a particular market are ranked according to their social marginal costs, including direct and distortion costs. A policy is higher in the policy hierarchy if it has lower social marginal costs for the same direct effect. This hierarchy indicates the preferential ordering of policies to deal with particular divergences between private and social incentives. For example, it sometimes is argued that female schooling should be subsidized because more-schooled women have fewer children which relieves budgetary pressures on subsidized schooling and health services. But it is not clear that increasing female schooling through such subsidizes is high in the policy hierarchy. It would seem, for instance, that higher in the policy hierarchy might be the elimination of any public subsidies that are not warranted by the marginal social benefits exceeding the marginal private benefits.

Second, there are tremendous information problems regarding exactly what effects policies have, particularly in a rapidly changing world. This is an argument in favor of policies that are as transparent as possible, which generally means higher in the policy hierarchy with regard at least to distortion costs because more direct policies are likely to be more transparent.[13] Information problems also provide an argument for price policies (taxes or subsidies) because if there are shifts in the underlying demand and supply relations they are likely to be more visible in a more timely fashion to policymakers if they have impact on the governmental budget than if they only change the distortions faced by private entities as tends to happen with quantitative policies such as quotas and restrictions on production or use.[14] Finally information problems in the presence of heterogeneities across communities point to the possible desirability of decentralization and empowerment of users of social services in order to increase the efficacy of the provision of those services, though such considerations must be balanced off against possible economies of scale, higher quality of staff and possibly lower levels of corruption at more centralized levels, as well as intercommunity distributional concerns.

Thus, for efficiency/productivity reasons, particularly given that in the real world information is imperfect and changes are frequent, there is an argument generally for choosing policies as high as possible in the policy hierarchy defined by the extent of marginal direct and distortionary costs – and thereby using interventions that are as focused directly on the problem as possible. Note in particular that this means that if there is a good efficiency reason for public support for human resource investments related to schooling, that does *not* mean that the best way to provide that support is through governmental provision of the relevant services. Higher in the policy hierarchy than direct governmental provision of such services, for example, may be subsidies or taxes that create incentives for the efficient provision of these services whether the actual providers are public, private or some mixture. On the other hand policies that discriminate against one type of provider – for example, by making the availability of such subsidies dependent on whether the provider is public – are generally likely to be lower in the policy hierarchy than policies that do not have such conditions.

Now consider distribution. Generally speaking the subsidization of specific goods and services (and even less, the direct provision by governments of goods and services at subsidized prices) is *not* a very efficient way of lessening distributional problems. Because subsidies are designed to lower prices to consumers, they induce inefficient consumption behavior (i.e. consumption at a point to the right of point H* in Fig. 1). Instead, it generally is more efficient (and thus less costly in terms of alternative resource uses) to redistribute income to consumers, allowing them to allocate the income in ways that lead to efficient patterns of consumption.[15] Nevertheless, there are some cases in which subsidization of selected goods and services may be defensible to attain distributional objectives. For example, in cases where it is difficult (and therefore costly) to target the poor, subsidizing certain goods and services that are mainly consumed by the poor may be the most efficient policy alternative. A second example is if policy makers believe that there is a serious problem of intra-household distribution (e.g. women and children are disadvantaged relative to adult males). In such circumstances, subsidies directed to goods and services consumed mainly by those who are viewed as disadvantaged within households may serve to improve their welfare more than cash payments directed to the household.[16]

Rather than being concerned with the general command over resources of its poorer members, as noted above, society may deem it desirable that everyone enjoy basic human resource related (and other) services, including basic schooling.[17] That is, society may decide to behave paternalistically by designing policies to assure consumption of certain goods and services even if their consumption results in less welfare for the private decision makers affected than would consumption of other goods and services requiring the same aggregate resources – perhaps because of concern about intrahousehold distribution.[18] Such an objective might be obtained through many means. But presumably it is desirable to assure that everyone have these basic options at as little cost in terms of productivity (and therefore economic development) as possible. Therefore, rather than ignoring efficiency considerations, it is desirable to choose policies as high as possible in the policy hierarchy and still assure that the basic service objectives are met. Efficiency goals thus play an important role in interaction with the pursuit of distributional goals, not as independent considerations.

2.4. A Priori Considerations for Selected Policy Issues Related to Schooling

As Sections 3 and 4 illustrate, good empirical information about micro aspects of schooling and development are difficult to obtain. Therefore conditional predictions and policy analysis must be made under considerable uncertainty – for which reason a priori models (whether called that or not) must provide guidance. Therefore I discuss briefly in this section the relations between selected issues that are prominent in policy discussions relating to schooling and other social services on one hand and the frameworks of Sections 1 and 2 above on the other hand – namely, (1) pricing and cost recovery in programs related to human resource development, (2) access to programs and quality of programs related to human resource development, and (3) overall and program sustainability.

Pricing and Cost Recovery in Programs Related to Human Resource Development

Many human resource programs provide services free of charge or with substantial subsidies to the entire population. Financing these programs has become burdensome in many developing countries. One response has been to consider charging fees for these services. Although charging fees has the potential to recover a significant share of costs, concern is often expressed that it will reduce use levels and impose a burden on the poor. Other issues relate to how high fees should be (i.e. what percentage of costs should be recovered) and how fees should be set for different services (e.g. for different types and levels of schooling).

The policy framework presented above in this section suggests that from an efficiency perspective all human resource related services (including schooling), whether produced by the government or by the private sector, should be priced so that the marginal social benefit of the last unit consumed is equal to its marginal social cost. Unless situations such as those depicted in Figs 2 or 3 exist, the normal operation of competitive markets should ensure that this occurs. If not, methods such as those illustrated in Behrman and Knowles (1998, Section 3.1) should be used to find an appropriate level of subsidy (or tax). If a case can be made for subsidizing governmental services on efficiency grounds, the same subsidy should be provided to private providers. Failure to extend the same subsidies to the private sector as the public sector receives puts the private sector at a competitive disadvantage and risks losing the potential gains in efficiency that may be obtained by strengthening competition between public and private sector producers.

The same guidelines should be used to determine subsidies, if any, from an efficiency perspective for different services and goods, such as different types of schooling. In general, these guidelines imply that prices (and subsidies) for different options should be roughly proportional to their marginal cost, so, for example, the fee charged for tertiary schooling should in most cases be significantly higher than that for basic schooling. For a given option, the same absolute subsidy should be given to each provider (including private providers). Actual pricing policy often departs from these principles, however; more expensive options and services typically receive higher subsidies to make them affordable. In many programs, all services regardless of cost are provided free. Varying the subsidies from one option or source to another, other than dictated by applying the principles presented in the policy framework above, creates the wrong incentives for consumers and encourages inefficient choices.[19,20] It can also exacerbate distributional inequities, as is illustrated by the large subsidies per students in many societies for upper levels of education that are attained primarily by students from middle- and upper-income families.

It is certainly true that pricing schooling and other human-resource-related services according to efficiency guidelines may hurt the poor in comparison with free provision (just as competitive pricing of food, housing, and clothing may hurt the poor in comparison with free provision). They will be hurt whether they reduce their consumption of the service or not. At times it is suggested that they will be hurt more if they reduce their use of such services a lot due to price increases (e.g. Gertler, Locay & Sanderson, 1987). But in fact the implication of the framework in Section 1 is that they will be hurt more by increased prices due to "cost recovery" applied to services that induce no or little reduction in utilization following the price increase (i.e. demand is inelastic) than they are by price increases that result in large reductions of quantities demanded (i.e. for which there are close substitutes so demand is elastic). If the poor can readily substitute other goods and services (e.g. additional food purchases) for human resource services if the price of the latter increases, this may be of concern to the education specialist and teachers and school staff. But it may be of less concern to the poor themselves because they are able to substitute consumption of other goods and services with little loss of welfare.

That the poor may reduce their use of human resource services because of price increases as part of cost recovery efforts, however, does not imply that continuing to subsidize all such services necessarily helps the poor. If certain services have a high income elasticity of demand, for example, most of the subsidy will be captured by upper-income groups, as for tertiary schooling in most countries. In such cases, the poor would be better off if the funds were used to subsidize the goods and services of which they consume proportionately more.

Access to Programs and Quality of Programs Related to Population Change and Human Resource Development

The human resource communities, including those concerned with schooling, frequently identify "access" and "quality" as program objectives in addition to efficiency, distribution (or equity), and "sustainability" (discussed below). "Access" is usually defined as the absence of economic and physical barriers that keep a prospective client away from a service delivery point; "quality" is usually defined as the factors that determine whether the care is effective and acceptable to the client once she/he walks through the door. Optimal levels of access and quality are defined in this literature to be those that maximize utilization and impact. Thus, for example, schooling services are delivered free of charge to a client's neighborhood (taking access to the limits), and governmental human resource programs are criticized for their poor quality if they do not provide a full range of options or highly trained staff, citing studies that utilization increases with quality defined by such indicators.

I do not see, however, that these concepts add to the perspective of the framework above. "Access" and "quality" considerations already are included in the concept of efficiency. The framework presented in Section 2 is consistent with improved access (e.g. bringing services closer to clients) up to the point where the marginal social benefits of doing so are equal to the marginal social cost. Goods and services of varying qualities are effectively distinct goods and services, and the normal functioning of competitive-like markets (assuming consumers are well informed) provides an optimal mix of goods and services of different qualities (i.e. that mix that responds to consumer demand). Unless a situation such as that depicted in Figs 2 or 3 is present, the normal operation of competitive markets should result in optimal "access" and "quality". If this is not the case, the optimal levels of subsidies (or taxes) that should then be applied to all producers, public and private. If consumers cannot be relied upon to make "good" quality choices in markets for human resource services because they do not have as much information as is warranted by the same efficiency guidelines (i.e. to the point at which the marginal social benefit equals the marginal social cost of information), subsidization of more information is likely to be higher in the policy hierarchy than emphasis on access and quality per se.

If governments have an effective monopoly in the provision of certain services, which can easily occur if they are financed mainly out of governmental budgets, problems of poor access and quality often arise. Decisions about the location of governmental human resource and other facilities, for example, are often made on the basis of political considerations, rather than on the basis of efficiency and distribution. Governmental providers have little incentive to provide good quality services; and they tend to provide a more limited range of service quality than do private providers. The poor quality of governmental services often is a consequence of a lack of competition from private providers. Policies that subsidize governmental providers, and not private providers, are the source of the problem. Providing subsidies, if any are warranted on efficiency or distributional grounds, equally to all providers is the solution. Special quality improvement programs directed to governmental providers are not nearly as high in the policy hierarchy as solutions to problems of poor-quality governmental services.

Sustainability

"Sustainability" seems to have two different meanings in the policy and development literatures.

The first meaning of sustainability concerns the overall process of development and whether it is sustainable in the sense that it is supported by the evolving resource base of an economy. This might be referred to as overall sustainability of the development process to distinguish it from the second meaning discussed below. There is concern, for example, that the development process in some contexts can not be continued in the long run because of environmental degradation in many forms ranging from water pollution to solid waste disposal to soil runoffs. Many seem to think that this concept of sustainability adds another basic policy motivation beyond efficiency and equity.

But, again, I do not understand why. All of the concerns that I understand are covered by this meaning of sustainability are already incorporated in the concept of efficiency, including efficiency over time (or dynamic efficiency). If an economy is efficient but the rate of development is not maintainable I do not see what insights the concept of overall sustainability has to offer. Any changes suggested by this concept would, be definition, only make people worse off by reducing welfare. If an economy is not efficient, then – whether or not the rate of development is maintainable (sustainable) – welfare improvements can be made (at least in the probabilistic sense discussed above) by making it more efficient, by bringing marginal private incentives more into alignment with marginal socially desirable incentives. If there is a divergence between the private and the social marginal benefits or costs, for example, because there are knowledge externalities or pollution or resource degradation that are not incorporated into private calculus at the margin, the implication of the efficiency motive for policy is that there would be gains from changing the private behavior (perhaps by changing prices through taxes and subsidies) so that the optimal private decision (e.g. H* in Figs 2 and 3) coincides with the optimal social decision (e.g. H** or H*** in these figures). What does the concern with overall sustainability have to add beyond that?

The second meaning of sustainability that is encountered frequently in the human resource and some other development communities is much narrower. It concerns the capacity of a donor-funded project to continue producing the benefits it is expected to provide (for example, those that were projected in the project's cost-benefit analysis) beyond the period for which financing and other inputs are provided by a donor. This might be characterized as project sustainability to distinguish it from the overall sustainability discussed above. For example, if the project provides funds for a capital investment, will the beneficiary agency (e.g. usually a government or NGO) be able to mobilize the recurrent expenditure to operate the capital investment when project support terminates? Although the primary focus of this meaning of sustainability is financial, the concept has been broadened to consider the beneficiary's institutional and managerial capacity to continue generating projected benefits of an investment beyond the life of an externally-funded project period. It is widely accepted that projects that provide support for recurrent expenditures are more difficult to sustain than those that provide support only for capital investments.

Sustainability is a major problem for many development projects and programs, including those directly related to human resource investments. One reason for such concerns is that in the early stages of new governmental programs support is often relatively weak, and there is no presumption that the necessary recurrent expenditure would be forthcoming in the absence of donor funding. Given ongoing concerns for the sustainability of such programs, some donors have funded entire projects designed to promote sustainability. The problem is perceived to be particularly acute in the case of NGOs, which tend to receive a very high share of their recurrent budgets from donor funding and have only limited opportunities available for obtaining alternative sources of funding. Cost recovery is often advocated or strengthened as one way to replace donor funding. One problem is that as fees rise, the characteristics of the clients served tend to change, with the result that a smaller share of the remaining subsidies reach the poor. "Cross subsidization" is another common sustainability strategy frequently supported by sustainability projects. An example is for an NGO to establish a school in a relatively high-income location and to charge fees there that more than cover its costs, using the surplus to support its operations in low-income areas. Cross-subsidization activities, although they are sometimes generously funded by donors, rarely succeed (Janowitz & Gould, 1993).[21]

In the context of the policy framework presented above, however, project or program sustainability poses less serious problems. To begin with, governments should not in most cases necessarily be providing schooling related services directly.[22] Even from a distributional perspective, the policy options are not likely to be high in the policy hierarchy that have heavily subsidized governmental providers servicing the poor while a self-sufficient private sector services the rich and middle-income groups. If governmental financing is justified on either efficiency or distribution grounds, the preference should be for demand-side subsidies (e.g. vouchers) provided directly to target groups of consumers (e.g. the poor). In addition to fostering competition and therefore efficiency, the use of targeted demand-side subsidies also helps to resolve the problem of sustainability. In the case of NGOs, for example, those that are able to control their costs and compete effectively for consumers' business (including the business of those consumers who are equipped with vouchers) will survive; the rest will fail. Governmental programs probably should disappear if they show themselves less able to compete with commercial and NGO providers on a level playing field.

3. RELATIONS TO BE ESTIMATED TO ASCERTAIN THE DETERMINANTS OF AND THE IMPACT OF HUMAN RESOURCE INVESTMENTS AND THE ROLE OF POLICIES

To obtain better knowledge of the relevant empirical magnitudes of the micro determinants of schooling and of the impact of schooling and the policy implications, micro relations must be estimated. In this section I first discuss these and then I turn to estimation problems and possible resolutions.

3.1. Relations to be Estimated

The implied relations for investigating the *determinants* of family behaviors related to human resource investments and the *impact* of such investments are (i) to estimate directly the underlying structural relations that determine human resource investments or their impact (e.g. human resource and wage production functions analogous to relation 2) and (ii) to estimate dynamic decision rules or demand relations for human resource investments or for inputs used to produce human resource investments or for their impact (e.g. expressions that are analogous to relation 1).

Structural relations – production functions: Structural relations are the basic underlying relations in the models of family behaviors in Section 1. The most commonly-estimated structural relations are production functions. A linear or log-linear approximation to a general production function of the type discussed in relation (2) above can be represented with cognitive achievement (CA_i) produced by two categories of variables relating to the ith individual and his/her household (XI) and to the sth school (XS) and by an explicit stochastic disturbance term (Ui):[23]

$$CA_i = a_{XI}XI + a_{XS}XS + U_i, \qquad (2A)$$

where the "a"s are parameters to be estimated that indicate the respective effects on CA_i of changing the respective inputs. The stochastic term captures random effects that are not correlated with any of the other predetermined right-side variables. Generally each of the types of variables may be vectors that represent a whole set of variables. It is useful for the discussion below of estimation issues to distinguish among four different subgroups of variables each: the superscripts "o" and "u" refer to "observed in the data used" and "unobserved in the data used", the superscript "b" refers to variables that are behaviorally determined within the model used, and the superscript "p" refers to variables that are predetermined within the model used. With these distinctions, the variable list in the general production function relation is XI^{ob}, XI^{ub}, XI^{op}, XI^{up}, XS^{ob}, XS^{ub}, XS^{op}, XS^{up}, U_i. If these were substituted into (2A) each would have its own coefficient "a" with an appropriate superscript to indicate its impact on CA_i. The distinctions among these different variables are important because some of the most important and most pervasive estimation problems arise from unobserved variables or behaviorally determined variables, as discussed in Section 3.2 below

The parameters ("a"s) in the production function give the direct impact of the right-side variables – some of which may reflect directly policies. With good estimates of the appropriate production functions the direct determinants of many outcomes determined by household behaviors and the direct impacts of many human capital investments could be evaluated with considerable confidence to answer many of the questions raised above. Good production function estimates may be difficult to obtain, however, because of estimation problems discussed below.

Reduced-form dynamic decision rule or "demand" relations: A second set of relations that can be estimated to explore the determinants of and the impact of family and firm behaviors and of policies related to human resource investments are dynamic decision rules or "demand" relations. These relations give some behavioral outcome in the current period as dependent on all predetermined (from the point of view of the entity – family, or whatever – making the decisions) prices and resources and on the parameters in the underlying production functions and preferences. These are the relations that most commonly are estimated in the literature. These demand functions in principle are derived explicitly from the constrained maximization behavior of families that is discussed above in Section1. As such they incorporate all of the underlying structural (e.g. production function) parameters that are involved in that process. But all of the choice variables during the period of interest are substituted out, so the demand functions are not underlying structural relations such as production functions, but so-called reduced-form relations because the maximizing behavior that determines such variables has been combined and "reduced" to the relations that give the behavioral outcomes as a function of purely predetermined and expected prices and resources. In some empirical studies the underlying structural parameters can be identified from estimation of the demand relations. In most cases, however, demand functions are just posited to result from constrained maximization and the underlying structural parameters are not identified in the estimates, though the demand parameters still are some combinations of these parameters. In such cases demand functions permit the estimation of the total effects of predetermined variables on the behavioral variables of concern, but not estimation of the exact mechanisms through which preferences and technical production functions affect the behavioral outcomes.

On a general level demand functions can be written with a vector of behavioral outcomes (Z) dependent on a vector of prices broadly-defined (P) and a vector of resources (R) – with the relevant prices and resources depending for what entity is the demand function. If there are uncertainties regarding relevant future prices, policies and shocks, then the characteristics known at the time of the decision of interest regarding the distributions of those outcomes should be included instead of their realized values. A linear approximation to the demand function for a family facing prices PF and with resources RF_f and a vector of stochastic terms (V_f) is:

$$Z_f = b_{PF}PF + b_{RF}RF_f + V_f, \qquad (1A)$$

where the "b"s are the parameters to be estimated and indicate the impact of the variables for which they are coefficients on the demands for Z_f. The stochastic term in each relation includes all the effects of all the stochastic terms in all of the production activities in which the family is engaged (i.e. all of the elements of the vector U_i), plus perhaps other chance events. Both prices and resources may be observed or unobserved in the data, so it is useful to indicate that distinction here as above in the discussion of production function inputs (again, using superscripts "o" and "u"). There is one such demand relation (or one element in the vector Z_f) for every behavioral outcome of the family (and similarly for firms or other entities), including all human resource investments and all behavioral inputs that affect human resource investments through production relations such as (2A). Each of these demand relations conceptually includes the same identical right-side predetermined variables, reflecting that there may be important cross-effects (e.g. the nature of health services may affect schooling demand, and vice versa) so that, as noted, a nutrition or health policy may affect educational (and other) outcomes, and vice versa. That means that any predetermined variable that affects any one behavioral outcome may affect any or all other behavioral outcomes.

Conditional demand functions: Conditional demand functions, as contrasted with the unconditional ones in relation (1A), include among the right-side variables some variable(s) that are determined by the behavioral model for the decision-making unit – the family – under examination. If the included behavioral variable(s) on the right side is determined at least in part by past behavior, its effect may be estimated in such relations by including its start-of-the current period value instead of the past prices, resources and stochastic terms that determine this value.[24] For example, the current period cognitive achievement function might be posited to depend on the start-of-the-period stocks of human resources, which depend on all past behavioral decisions related to human resource investments. In such a case, the start-of-the-period stocks of different human resources are just other right-side variables that are predetermined with respect to current prices and (other) resources in the demand relation in (1A). That relation also includes *all* of the current prices. An unbiased estimate of the coefficients of the stocks of different human resource investments gives the impact of the start-of-the-period stocks of ith's human resources on the current-period cognitive achievement. That the start-of-the-period stocks of human resources were determined by past behavior poses some estimation problems discussed below, but the interpretation of unbiased coefficients of these variables as reflecting their impact on current cognitive achievements is clear.

Mincer-type earnings functions: The most common framework for estimating the private rate of return to time spent in school in terms of labor market outcomes is due to Mincer (1974). In the simplest case once one finishes school and starts to work, earnings depend only on years of schooling. Let Y_O be wages with no schooling and Y_S be wages with S years of schooling, but with the earnings stream starting S years later due to the time spent in school. In equilibrium, the present discounted values of the earnings from the two options are equated, which implies a semilog earning or wage function:

$$\ln Y_S = \ln Y_O + rS, \tag{3}$$

where r is the private rate of return to time spent in school. To estimate this relation (or more extended versions of it) a stochastic term (e) is added (usually without comment). One possible justification for such a stochastic term would be random measurement error in the dependent variable (see Section 3.2). Another would be that some individuals have better "market luck" than others, independent of their schooling. There are numerous estimates of such expressions that purport to measure the private rate of return to time spent in schooling. Most of these studies include some measure of "work experience" (X) that often is interpreted to reflect the wage impact of on-the-job training (most typically in quadratic form). How does relation (3) and various extensions thereof relate to the production function for wages for the general form of relation (1A)? Implicitly relation (3) assumes that the impacts of schooling and of experience are independent of all other inputs (e.g. ability, motivation, other inputs, technology). Thus strong assumptions are placed on the production technology, such that ability, motivation and family connections have impact on wages only through their effects on schooling and experience.[25,26]

An alternative interpretation of relation (3) is as a "hedonic" (so the market values depend on the individual's characteristics) wage function. In this case the coefficients are the market valuations of attributes such as schooling and experience, again under the (usually implicit) assumption that there are no effects of variables that are not controlled in the estimation but that may have an effect on both human capital human resource investments and on wages such as ability, motivation and family connections.

3.2. Estimation Issues in Ascertaining the Determinants of and the Impact of Human Resource Investments and the Impact of Policies

There are a number of possible problems in obtaining good estimates of the determinants of and the impact of human resource investments and the role of policies.. Therefore what are presented as estimates of relations such as those that are discussed in Section 1 may be biased. These estimation problems share a common characteristic: the disturbance term in the relation actually estimated is *not* simply an element in U_i or V_f that is distributed independently of all the right-side variables in the relation being estimated, but instead is correlated with right-side variables (e.g. because it is a compound disturbance term that includes unobserved variables as well as U_i or V_f or because of the way that U_i or V_f is defined for the sample used in the estimates).

Measurement error: Measurement error may contaminate any of the observed variables used for estimates of the relations (1A), (2A) and (3). Random measurement error occurs if what is observed is not the true variable, but the true variable plus a random error.[27] Random measurement error is what usually is emphasized, and is what is discussed here. Measurement error also may be systematically related to the true variable, with implications that depend on exactly what is the nature of the systematic relation. Random measurement error in a right-side variable causes bias in the coefficient estimates of interest towards zero. The bias is greater the larger is the variance in the measurement error relative to the variance in the true value. For example, random measurement error occurs if what is observed (S*) is not the true schooling variable (S), but the true variable plus a random error (w):[28]

$$S^* = S + w. \tag{4}$$

Whether random measurement error affects the estimates depends on whether the variable measured with error is the dependent or a right-side variable.

Random measurement error in the dependent variable merely adds to the stochastic disturbance term, but does not bias the coefficient estimate of the right-side variable. Therefore random measurement in schooling does not cause biases in the estimates of the determinants of schooling.

Random measurement error in a right-side variable causes bias in the coefficient estimates of interest. Consider, for example, the case in which schooling S is a right-side variable that is posited to affect wages (Y). The true relation is:

$$Y = bW + v, \tag{1B}$$

and the relation estimated is:

$$Y = bS^* + v + bw. \tag{1C}$$

Intuitively, because the observed schooling is a noisy measure of the true schooling, the true dependence of wages on schooling is masked, and the result is an underestimate of the schooling effect. Formally, the estimate of the schooling impact is:[29]

$$\hat{b} = b\sigma_S^2/(\sigma_S^2 + \sigma_v^2). \tag{1D}$$

Because $\sigma_S^2/(\sigma_S^2 + \sigma_v^2)$ is less than one, the estimate $\hat{b}$ is biased towards zero. The bias is greater the larger is the variance in the measurement error relative to the variance in the true value.

Random measurement error can be reduced with better measurements of the desired concept. Random measurement error can be controlled with instrumental variable estimates if the error in the instrument used is independent of the error in the observed variables of interest, as may occur with different measures of the same concept. For example, schooling reports from adult siblings or from adult children could be used for this purpose as in some recent studies (Ashenfelter & Krueger, 1994; Behrman, Rosenzweig & Taubman, 1994).

Omitted variables: In both production function estimates and demand function estimates there may be variables that should be included among the right-side variables but that are not observed and therefore not included. For the production function estimates, for example, there may be unobserved inputs such as inherent ability, motivation, and management capabilities. In terms of (2a) with the subcategories of variables, the basic estimation problem is that the observed right-side variables (XI^{ob}, XI^{op}, XF^{ob}, XF^{op}) may be correlated with the unobserved variables (XI^{ub}, XI^{up}, XF^{ub}, XF^{up}) that are included in the compound disturbance term with U_I.[30] Therefore the estimates of the impact of the observed variables include not only their true effects but also part of the effects of any correlated unobserved variables. For the demand relations (1A) the compound disturbance term includes, in addition to V_f, the other unobserved variables (PF^u, RF^u). If any of the observed variables on the right-side of relation (1A) is correlated with any of the unobserved variables, its coefficient estimate is biased because, in addition to its own effects, it is representing in part the effect(s) of the correlated unobserved variable(s). Further, in conditional demand functions in which some variable such as the start-of-the-period health stock discussed above is included among the right-side variables, such predetermined behavioral variables may be correlated with any of the right-side unobserved variables because such variables entered into the determination of these predetermined behavioral variables in earlier periods. Finally, in Mincer earnings functions the disturbance term may contain variables such as ability, motivation and family connection that are not orthogonal to schooling, despite the maintained assumption to the contrary.

The sign and magnitude of omitted variable bias depends on the effect of the omitted variable(s) and on its correlation with included variables. Consider, for example, the case of a simple wage production relation in which schooling (S_i) and ability (A_i) together with a stochastic term (v_i) determine wages for individual i (Y_i):

$$Y_i = aS_i + bA_i + v_i. \tag{1E}$$

If ability is not observed, the compound disturbance term is $bA_i + v$ and the estimate for the coefficient of schooling is:

$$\hat{a} = a + br_{S,A}, \tag{1F}$$

where $r_{S,A}$ is the coefficient in the regression of ability on schooling. The sign and magnitude of the omitted variable bias is given by the last term. It depends on the sign and magnitude of the true impact of the omitted variable (b) and on the regression coefficient of the omitted variable on the observed right-side variable ($r_{S,A}$). If ability has direct positive effects on wages and on schooling, in the present case the bias is upwards.

Five means of dealing with omitted variable bias are (i) to measure variables that often are unmeasured (e.g. "ability" could be measured using Raven's tests, as in Knight & Sabot, 1990), (ii) to use fixed effects to control for unobserved variables (which requires multiple observations at the level of aggregation at which the fixed effects are used), (iii) to replace right-side variables with their instrumented values by using identifying instruments that do not appear in the relation being estimated and are not correlated with the disturbance term in the relation being estimated, (iv) to compare the behaviors or the "before and after" changes in behaviors of beneficiaries of policies with those of individuals who would be eligible but are in the control group (perhaps with matching on observed characteristics as in Heckman, Ichimura and Todd 1997), and (v) to use experimental data. All of these approaches have their limitations. Some variables are very hard to measure at reasonable costs. Fixed effects do not permit the estimation of the linear effects of observed variables at the same level of aggregation as the fixed effects (though the effects of interactions among observed variables, such as family background and program characteristics, can be estimated) and exacerbate the impact of measurement errors. It often is difficult to find identifying instruments, though lagged price and other shocks are candidates for panel data.[31] The control group comparisons depend on good assignments of households to the actual and potential beneficiary groups versus those who are not eligible; if there are incorrect assignments, misleading comparisons may be made (though matching may reduce this problem considerably). Experiments often are costly, hard to maintain (i.e. keeping control and treatment groups separate), and is some cases not politically possible or ethical.[32]

Simultaneity: Simultaneity bias occurs when a variable that is determined in the current period within the model appears as a right-side variable in some other relation. Among the relations discussed above, production functions are the ones for which simultaneity might be a problem because the right-side variables include some behavioral inputs (e.g. health and nutritional status in relation 2A). Included on the right side of that relation is a stochastic term (V_f) that, as noted above in Section 3.1, includes the stochastic terms from all of the production function relations in the model – including those for health and nutritional status. This results in a correlation between health and nutritional status in the current period and the stochastic term in the cognitive achievement production function that causes biases in the estimated impact of health and nutritional status on cognitive achievement production. The sign and the magnitude of the bias depends upon the exact structure of the model. If there are no unobserved behavioral inputs in the production function, prices and any other variables that enter into the reduced-from demand relations in (1A) but not directly in the production function can served as identifying instruments for controlling for simultaneity.

Selectivity: Selectivity bias may result if observations are available only for a selected subset of the sample. Figure 4 provides a simple illustration of selectivity bias. If there were observations on wages and schooling for everyone in the sample, the true wage relation in relation (3A) could be estimated. The solid line in this figure represents this true relation between wages and schooling, with a slope of g. Of course the observations for most individuals do not fall exactly on this line. For a given level of schooling (e.g. S_1) some observations are above the line (e.g. point A), some are below (e.g. point B) and some may be on the line (e.g. point C). But the true line best summarizes the cloud of points of wage-schooling combinations for all the individuals in the whole sample.

However wages are not observed for everyone in the sample. In fact wages are only observed for individuals whose labor market wages exceed the opportunity cost of spending time in activities such as in household and family enterprise activities instead of in wage labor. For simplicity, assume that this opportunity cost is the horizontal dotted line at R on the vertical axis in Fig. 4. Therefore the wage estimate is based only on the wage-schooling subsample above this horizontal line. The problem is that this subsample is not randomly selected. For schooling levels below the level at which the true relation crosses R (i.e. S_2 in the figure), only individuals who have sufficiently positive disturbances to get them to R or above are selected into the subsample (e.g. for S_1, the individual at point A, but not the ones at points C and B). For schooling levels above the level at which the true relation crosses R, only individuals who have sufficiently negative disturbances to get them below R are selected out of the subsample (e.g. for S_3, the individual at point F is selected out of the subsample, but those at points E and D are selected into the subsample). Thus the subsample selection procedure, with its systematic relation between the disturbance term in the true relation and schooling, creates a correlation between the disturbance term and schooling for the subsample for which estimates of the wage relation can be made. As a result, if the wage relation is estimated using only this subsample, something like the dashed slope line is obtained, with a biased estimate of the true relation between wages and schooling.

The standard means of controlling for selectivity is to incorporate the behaviors that cause selectivity explicitly into the model, though in some cases identification may be difficult. There are various ways of incorporating the sample decision rule into the analysis. Perhaps the most widely-used method is Heckman's two-step method. In this method in the first step the selectivity rule is estimated and in the second step the relation of interest is estimated including among the right-side variables a statistic calculated from the first-stage

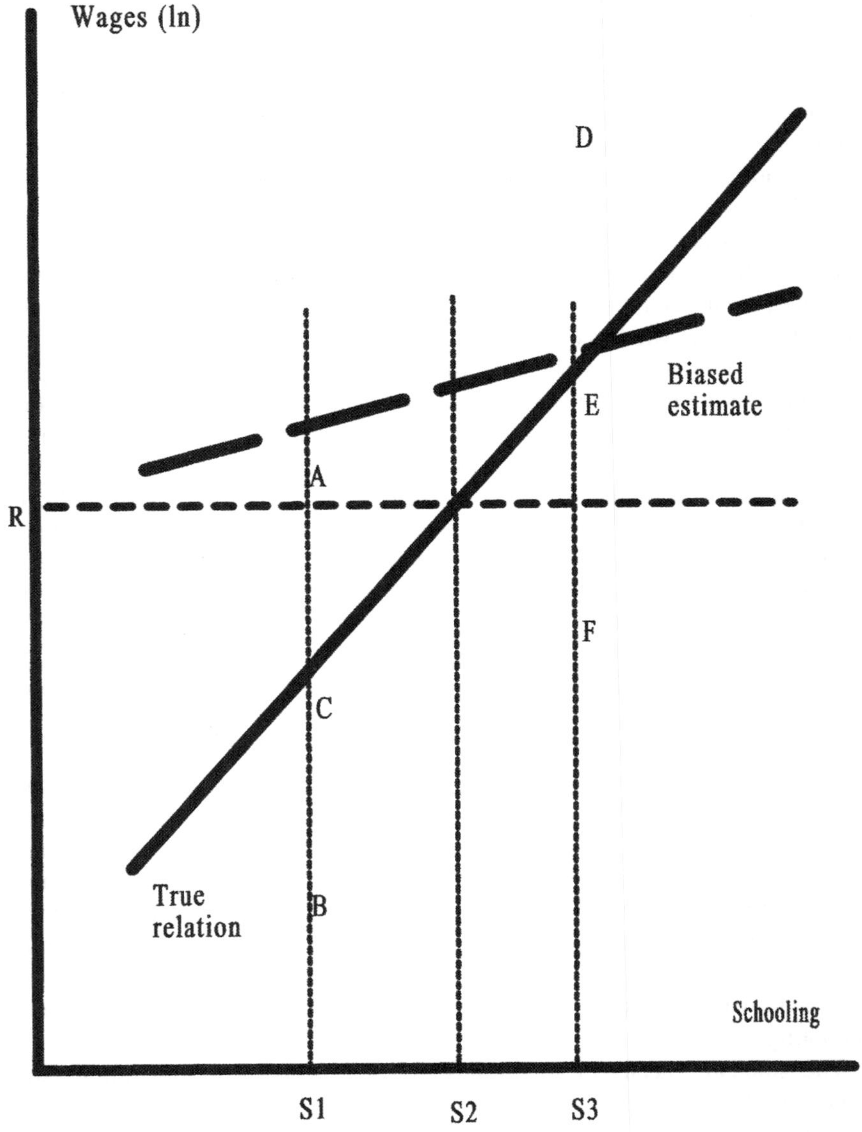

Fig. 4. Selectivity Bias in Estimated Relation of Wages to Schooling.

residuals that corrects for the effect of sample selectivity on the original disturbance term (so that it no longer is correlated with the observed right-side variables).[33]

4. SELECTED MICRO EMPIRICAL STUDIES FOR THE ASIAN CONTEXT

I now summarize a few recent selected micro studies of aspects of the determination of and the impact of schooling in the Asian context. These studies have been selected because they build on the considerations in the previous three sections to go beyond the simple multivariate descriptive associations that predominate in the literature.

4.1. Determinants of and Associations with Schooling

Credit markets and schooling in rural India: One central implication if the framework sketched out in Section 1 is that the nature of markets are critical for understanding the determinants of schooling. To the extent that schooling is an investment rather than current consumption, as is emphasized in that presentation and in much of the development literature, one critical market is the capital market. There are frequent claims that imperfect capital markets cause there to be inefficiently little investment in schooling, but very little in the way of direct empirical evidence on this topic.

Jacoby and Skoufias (1997) explore the link between incomplete financial markets and human capital accumulation in the form of schooling in rural India. They note that there has been considerable emphasis on both financial markets and human capital as major factors in development, but not on their interaction. They also note that there have been a number of recent studies to test the implications of incomplete financial markets in both developing and developed economies, but that most of these studies shed little light on the mechanisms by which consumption smoothing is attained. They investigate how child school attendance responds to seasonal income fluctuations in agrarian Indian households using panel data. They study responses to aggregate and household idiosyncratic and anticipated and unanticipated income shocks. They posit a dynamic model of school attendance with different degrees of financial market completeness and note that with incomplete markets consumption and schooling investment decisions are not separable. Their estimation strategy is to relax successively restrictions on the relationships between school attendance and income shocks implied by successively more incomplete financial markets, all with control for unobserved household heterogeneity of the sort emphasized in Section 3.

Their results indicate that seasonal variations in school attendance are a form of self-insurance that significantly reduce the schooling of children in households that are vulnerable to risk, which is likely to be a costly form of insurance, particularly for poorer households. The results have a number of potentially important policy implications, such that expanding schools without understanding the nature of financial risks and market constraints may have more limited effects on education than expected, effective compulsory schooling laws or restrictions against child labor may have substantial negative effects on household welfare, and improved short-term credit and insurance markets may have important long-term benefits in the form of greater human capital investments.

Mothers' schooling and home education of children in rural India: Most of the micro evidence that has been used to support the importance of schooling in augmenting incomes in low-income countries comes mainly from data describing the returns to schooling for men (e.g. Psacharopoulos, 1994). Given the relatively low rates of participation by women in formal-sector labor markets in such countries, information on the potential contribution of women's schooling to income is less available and where found problematical to interpret due to labor market selectivity. Advocates of development and poverty reduction policies that emphasize investments in female schooling, however, suggest that significant returns to women's schooling are to be found in the household sector, where the schooling of women has important effects on the human capital of future generations (Haveman & Wolfe, 1995; Hill & King, 1993; Schultz, 1993a,b; Summers, 1992; UNDP, 1996; World Bank, 1990, 1991). One argument of development strategists, in particular, is that better educated mothers are superior teachers in the home, so that investments in women's human capital complement those in schools.

In principle, estimates of the technology of human capital production provide direct evidence on the extent to which mothers with higher levels of schooling are more productive home teachers. That is, for given investments in children, more educated mothers produce children with higher levels of human capital. Because of the difficulty of obtaining credible information on all human capital inputs, however, almost all of the estimates in the literature do not attempt to identify the production technology but describe the relationship between maternal schooling and child schooling net of the influence of other, more readily measurable, family characteristics, such as income and paternal schooling.[34] The estimated relationship between maternal and child schooling thus reflects not only home technology but also the constraints and preferences of household members. The interpretation of these behavioral relationships therefore depends on the structure of household decision-making.

An important alternative interpretation of the maternal-child schooling association, based on conceptions of households in which individuals optimize and bargain (Lundberg & Pollak, 1996), is that mothers with higher levels of schooling have superior options outside the household that confer to them greater command of resources within the household, which they choose to allocate to children at higher levels than would men (Folbre, 1984, 1986; Haddad, Hoddinott & Alderman, 1996; Thomas, 1990). While this view is not incompatible with the hypothesis that schooling actually augments home skills for women, it presupposes that women's schooling has returns outside the household. More importantly, it implies that the expansion of options for women in the labor market along with enhanced investments in women's schooling is necessary to achieve greater investments in children. However, growth in female employment opportunities, which may be difficult to affect via specific program interventions in a short period of time, is not a necessary condition for achieving greater schooling investments if schooling enhances women's productivity in the home production of human capital and there are returns to schooling men.

Of course, the observed positive associations between the schooling of mothers and their children admit to a number of other interpretations within the framework of Section 1. More schooled women may contribute more income to the household, in environments in which their schooling has earning returns, which may lead to increased investments in child schooling even if all household incomes are pooled and schooling has no in-home productivity effects. In this latter case, just as in the bargaining interpretation, a necessary condition for female schooling to induce increased child schooling is that schooling for women reap a return in the labor market. Finally, men with greater preferences for schooling may marry women with higher levels of schooling and invest more heavily in their children's schooling. The schooling of wives and children may be more highly correlated with men's preferences for schooling than men's own schooling in settings where own schooling is principally determined by parents.

Behrman, Foster, Rosenzweig and Vashishtha (1999) develop a model of household decision-making in order to assess empirically the contribution of maternal schooling to investments in children's schooling while taking into account the roles of preferences for schooling in the home and in the marriage market; the effects of schooling on home productivity, household bargaining power, and the time costs of household activities; and differential returns to schooling for men and women in the labor market. The framework is used to demonstrate, among other results, that the returns to female schooling in the marriage market mirror those in the labor market and in the home. These linkages are exploited using data describing the demand for educated wives and household investments in schooling in rural India before and during the "green revolution," a time when the returns to men's but not women's schooling rose substantially in the farming sector, investments in the schooling of both boys and girls increased at comparable rates, but in which the apparently limited role of women in agricultural decision-making or in rural formal-sector employment activities remained unchanged. They show that in this context, in which the schooling of women appears not to contribute directly to household earnings or to enhance their opportunities outside the household, it is possible to establish a number of tests, using information from both the marriage market and from the household, of the hypothesis that schooling enhances women's ability as home teachers that are robust to the influences of schooling on household bargaining and of unobserved preference heterogeneity.

They develop a model of household schooling investment incorporating individual decision-making consistent with household bargaining models, differential preferences for child schooling between men and women, and marital choice. In particular, they allow maternal schooling to be endogenously chosen, to affect household income, and to affect a woman's bargaining position, in addition to potentially augmenting child schooling in production. They specify the model in two stages, first in a one-period model in which maternal schooling is taken as exogenous and children's schooling does not directly affect household income. They then extend the model to two stages to allow marital choice and an income producing sector in which son's schooling has a return. The two-stage version incorporates key features of the Indian setting that they study to develop tests that identify the mechanisms by which increases in the schooling of women affect the schooling of children.

The estimates indicate that despite the absence of any evident increase in employment activities by women in sectors in which schooling is rewarded and the lack of participation by women in farm decisions associated with the new technologies, the demand for schooled, in particular literate, wives increased more rapidly in the high agricultural growth areas, where returns to evidently male-dominated farm management skills rose. Consistent with the interpretation of this as derived demand for female schooling as an input in the production of child schooling, estimates that exploit the extended structure of Indian households to reduce the influence of male preferences for schooling, variation in market returns to schooling, and wealth effects indicate significantly higher levels of study hours among children with literate mothers. Finally, estimates of the determinants of dowry values indicate that, consistent with female literacy having value to men rather than providing an improved post-marriage bargaining position for women, literate women command a premium in the marriage market. Schooling achievement by women beyond levels that enable literacy, however, are not associated with higher levels of child study nor with enhanced value in the marriage market. The results from the Indian green revolution experience, which suggest that literate mothers are better teachers for children in the home, thus imply not only that investments in female schooling payoff where there are returns to schooling anywhere in the market sector, no matter how segmented by sex, but help explain why after the onset of the green revolution in India there was increased investment in both boys and girls schooling at approximately the same rates, despite very low returns in the labor market to investments in girls schooling.

Gender gaps in cognitive achievement and household versus school policies in rural Pakistan: Girls lag markedly behind boys in schooling in many developing countries even though this gender gap has been declining in recent years (King & Hill, 1993; Behrman, 1993; World Bank, 1993). For all low-income countries as defined in World Bank (1993), for example, the number of girls per 100 boys enrolled in primary school was only 78 in 1990, though this represented an increase from 52 in 1965. Girls tend to receive less schooling than boys particularly in rural areas, low-income countries, and in South Asia.

The effect of schooling on productivity is central in the so-called "new economic growth models", and many empirical estimates of the rates of return to schooling are high. Since the rates of return to schooling are estimated generally to be no lower for women than for men, and the social benefits to changes in household behavior induced by educated women are also believed to be high, there is a presumption that lower investments in the education of girls than of boys slow economic growth in addition to increasing inequity.

Pakistan provides an extreme example of this gender gap. At independence, Pakistan's female primary school enrollment rate was 16% of the male rate (Government of Pakistan, 1991). By 1965, at 32%, it was among the ten lowest in the world. While the ratio of female to male enrollments in primary school increased markedly from 1965 to 1991, at 52% it remained well below the average of 77% for all low-income countries and below the levels of other major Asian countries. Indeed it was among the four lowest in the world, and Pakistan's ratio of female to male enrollments in secondary school of 41% was one of the seven lowest in the world (World Bank, 1994). Female enrollments in rural Pakistan have been even lower absolutely and relative to the male rate.

Differences in enrollment rates, however, may be a crude proxy for the gender gap in education. Time spent in school is but one of many inputs into the education production process – others include pre-school ability, school quality and out-of-school investments in human capital – and may be, therefore, a poor predictor of the output of that process. Two people with the same years of schooling may differ markedly in their level of cognitive skills. Likewise two countries with the same gender gap in enrollments may differ markedly in the gender gap in cognitive skills. It is the output of schooling that, in human capital models, is presumed to affect subsequent productivity – a presumption that is supported by the few available studies of the relation between cognitive achievement and wages in developing countries (e.g. Alderman, Behrman, Ross & Sabot, 1996b, for rural Pakistan).

Alderman, Behrman, Ross and Sabot (1996a) use indicators of an output of the education production process, cognitive skills, to characterize and to investigate the determinants of the educational gender gap. Their data indicate a large and significant gender gap in cognitive skills in rural Pakistan, which presumably results in a large gender gap in productivity and in the command over resources. For example, for the 20–24 age cohort in 1989 they find that the gender gap in cognitive achievement is 74%. That is, the mean score for women on a test of cognitive skills is only 26% of the male score – 15.2 for men and 3.8 for women out of a maximum of 72.

To what is this large gender gap in the product of schooling due? It could result from parental biases in preferences that favor boys over girls. Alternatively, as Rosenzweig and Schultz (1982) and others have argued, greater investment in boys than in girls may reflect a gender gap in the expected returns to those investments. This gap in total returns may be traced to: wage discrimination in labor markets; strong attachments to traditional gender roles that effectively limit female access to high productivity sectors, activities or occupations; and gender roles within the household that result in different shadow prices for out-of-school investments in human capital and for time in school.

Or it could be that expectations regarding the total benefits of schooling are gender neutral but that parents, who are not entirely altruistic in their investment decisions, expect to receive a higher proportion of the benefits reaped by men. Parents may have such expectations because men are more likely to support parents in their old age. There nevertheless may be high returns to schooling daughters if, for a given quality marriage, schooling reduces dowries sufficiently (e.g. Rao, 1993.)

The common thread in all of these explanations is that differences in the demand for schooling account for the gender gap in human capital accumulation. Indeed, Haddad, Carnoy, Rinaldi, and Regel (1990, p. 10) claim that the available evidence "suggests that family economic conditions are more important than school-related variables . . . in explaining this gap. . .". Schultz (1993a) also stresses the importance of household demand in the explanation of the gap in schooling enrollments generally in developing countries, while Burney and Irfan (1994) stress its importance in the case of Pakistan.

The gender gap, however, also could originate on the supply side of the market for education. There may be gender differences in school availability or in school quality. Where most schools are single sex, as in rural Pakistan, it is easy to observe the extreme form of discriminatory rationing in which schools are available only for boys. Parents might send their girls to school if only a school were available.

Both the social costs of the gender gap and the prospects for public policy having an ameliorative impact are likely to vary markedly with the extent to which the immediate source of the gender gap is school availability and thus more subject to direct governmental policies or on the demand side and, therefore, deeply imbedded in preferences and incentive structures for large numbers of individuals.

Alderman, Behrman, Ross and Sabot use a rich data set, which was specially designed for this study, to go beyond previous analyses of the gender gap in schooling. It contains, inter alia, measures of school availability, schooling output, pre-school ability and other inputs into the education production process. They use these data first to document the educational gender gap in rural Pakistan. They then estimate a set of behavioral relations underlying this gender gap and use these estimates to decompose the gender gap in cognitive achievement. Their findings suggest that a substantial portion of these gaps are due to gender differences in local school availability. Solely by eliminating the gender gap in local primary school availability the gender gap in total cognitive achievement for the 10–25 year-old cohort could have been reduced by a almost a third, and that in literacy by over 40%. Though disadvantaged in terms of school availability, however, girls do not appear to be disadvantaged by the quality of their schooling.

They interpret the importance of the gender gap in school availability to be good news. To reduce substantially the gender gap in cognitive achievement, policy need not address biases deeply embedded in incentive structures or household preferences. Rather, the focus can be on biases in public policies governing local school availability, which presumably are more amenable to change. The gender gap in rural school availability has declined in recent decades, but for 10–14 year olds in 1989 a large gap – over 45% according to their data – remained. Therefore, the potential for further reductions in the gender gap in cognitive achievement through policies that eliminate the local primary school availability gap is still great. Moreover, increasing local school availability for girls need not require large capital investments. For example, existing boys' schools could be opened to girls, either with co-education or with multiple shifts. Though eliminating the gender gap in local school availability will help, large gender gaps in cognitive skills will remain unless gender gaps in household demand to have children start school also are reduced. Their estimates suggest that this demand is, perhaps surprisingly, quite responsive to incentives – including travel times and out-of-pocket costs. Where prices currently favor boys, such as those related to out-of-pocket costs, closing the gender gap in cognitive achievement by "getting prices right" would seem warranted from an efficiency point of view, although this prescription depends on getting a better understanding of why out-of-pocket costs differ by gender. Because of the greater responsiveness of girls to travel times, reducing travel times for both boys and girls (e.g. by making more primary schools truly local) also would lessen the gender gap in cognitive achievement, but this would entail higher public investment in schooling.

Impact of nutrition on schooling enrollments in the Philippines and Pakistan: Improved child health and nutrition are welfare-enhancing in themselves. Better child health and nutrition, in addition, are widely thought to improve various dimensions of child school performance, and therefore subsequent post-school productivity. There are many studies in the literature that report significant associations between child health/nutrition and child schooling performance (see Pollitt 1990 and Behrman 1996a for references).

But the associations between child health/nutrition and child school performance based on socioeconomic behavioral data in the literature may not accurately portray the *causal* impact of child health and nutrition on school performance. Within the framework of Section 1, child health/nutrition and schooling performance *both* reflect household decisions regarding investments in children's human capital. Almost all of the literature on the impact of child health/nutrition on child schooling performance based on behavioral data ignores this implication of household models of human resource investment behavior.[35] As a result, the estimated impact of child health/nutrition on child schooling in these studies may be biased and therefore misleading as a guide for understanding behavior or for policy formation, perhaps substantially. Moreover, depending on the nature of the intrahousehold allocation processes, the bias may be either upwards or downwards (Behrman & Lavy, 1998). That is, the standard estimates that do not control for the behavioral choices underlying child health/nutrition can not be confidently assumed to give a known (i.e. known to be upper or known to be lower) bound on the true impact of child health/nutrition on child school success.

Glewwe, Jacoby and King (1996) use longitudinal data within an explicit dynamic framework to explore the impact of child nutrition on child schooling decisions and academic achievement, including the age of initial school enrollment, grade repetition and cognitive achievement for longitudinal data from Cebu in the Philippines. They find that malnourished children enter school later and perform more poorly than their better nourished counterparts, but that parental behavior compensates partially by delaying initial school enrollment until malnourished children are older. They control for unobserved family heterogeneity by using within-sibling estimates, which in itself reduces the estimated (inverse) impact of child height-for-age on age of initial primary school enrollment by about a third. These estimates, thus, suggest that height-for-age may be proxying in part for unobserved family characteristics in the "naive" (GLS) estimates or that measurement error is causing a large downward bias in the within-family estimates. Glewwe, Jacoby and King in separate estimates also control for the endogeneity of child height-for-age in the within-family estimates by using the height-for-age of the older sibling at birth as an instrument in one alternative instrumental variable (IV) estimate and height-for-age of the older sibling at birth, at 12 months, and at 24 months in a second IV estimate. The point estimates for child height-for-age are greater in both of these IV estimates than in the within-family estimates without control for the endogeneity of child-specific height, and that for the second IV estimate is about 50% greater than in the naive estimates. On the basis of the latter estimate they conclude that "estimates of the impact of nutrition on learning that fail to correct for child-specific heterogeneity . . . are severely downward biased" (abstract).

However the specific conclusion about the estimate being "severely downward biased" without control for child-specific heterogeneities is conditional on strong assumptions: (i) that heights-for-age at birth, at 12 months and at 24 months do *not* embody unobserved (by analysts) endowments to which parents directly respond in subsequent schooling decisions for the same child (in addition to any indirect responses through child nutrition/health status); (ii) that heights-for-age at 12 months and at 24 months do *not* embody the effects of nutrient/health investments in the child by parents that are correlated with subsequent nutrient/health investments in the same child by the age of schooling decisions;[36] and (iii) that height-for-age for the same child a few years later at the time of the schooling enrollment decision *do* embody either the unobserved endowments to which parents directly respond in schooling decisions (in addition to any indirect responses to the child's nutrition/health status) or behavioral choices regarding nutrient/health investments in that child that are correlated with the unobserved endowments to which parents directly respond in the schooling decisions.[37] The conjunction of assumptions that there is no estimation problem with using child height-for-age for the first two years of the child life but that there is a problem with using child height-for-age when the same child is a few years older is puzzling and problematic, particularly in light of the standard interpretation of child height-for-age as an indicator of longer-run persistent nutrition/health status that is affected importantly by infant and early childhood nutrition and health investments and the perception that it is longer-run child nutrition/health status that is particularly important in schooling decisions.

Alderman, Behrman, Lavy and Menon (2001) also employ longitudinal data to investigate the impact of child health/nutrition on school enrollments in rural Pakistan. These longitudinal data do not cover as long a time period as do the Cebu data, nor do they include as many indicators of schooling success. But they have one important advantage for exploring the impact of child health/ nutrition on school enrollment. With these longitudinal data they are able to use price shocks when children are of pre-school age to control for the behavior determining the child health/nutrition stock measure. This would be harder for the Cebu data because that sample is all from one metropolitan area with little in the way of cross-sectional price variation. Such price shocks are uncorrelated with price shocks at the subsequent ages at which decisions are made whether or not to enroll children in school. Therefore this procedure permits estimation of the impact of child health/nutrition on child school enrollment without contamination from unobserved behaviors such as time allocation decisions for other household members that are concurrent with the enrollment decision. They explore how much the estimates differ with this preferred procedure in comparison with a range of alternatives, including a "naïve" model in which (like most of the literature) child health/nutrition is treated as exogenous as well as other alternatives with current price shocks and current and lagged price levels used for the first-stage estimates to attempt to control for the behavior determining child health/nutrition.[38]

Their results indicate that the estimation strategy and related assumptions make a considerable difference in the estimated impact of child health/nutrition status on enrollment. Their preferred point estimates, guided by explicit modeling of human resource dynamic decision rules, indicate that child health/ nutrition is three times as important for enrollment than suggested by "naïve estimates" that assume child health/nutrition is predetermined rather than determined by household choices in the presence of unobserved factors such as preferences and health endowments. Their results also illustrate that how the behavioral determinants of child health/nutrition are treated statistically makes a considerable difference in the estimated effects. If current or lagged price *levels* are used to identify the impact of child health/nutrition on schooling enrollment under the ad hoc assumption that long-run expected prices do not directly affect the schooling enrollment decision, the regressions do *not* yield significantly positive estimated effects of child health/nutrition on enrollment.

The results of this study, therefore, reinforce strongly the importance of using estimation methods that are consistent with the economic theory of households to explore the impact of some choice variables on others using socioeconomic behavioral data. In this case the preferred estimates indicate a much more powerful effect of child health/nutrition on school enrollment than do some of the a priori less satisfactory alternatives in the previous literature on the impact of child health/nutrition on child school performance. Therefore private behaviors and policies that affect child health/nutrition have much greater effects on school enrollments and on eventual productivities than suggested by the methods used in the previous literature.

Schooling and income associations in Viet Nam: The studies just summarized attempt to uncover underlying behavioral relations that play roles in determining schooling within particular contexts of imperfect markets within the general framework of Section 1. Behrman and Knowles (1999), in contrast, explicitly estimate the *association* between schooling and household income, without attempting to identify causal relations. Their motivation is related to the distributional motive for policy that is discussed in Section 2. The stronger is the association between household income and child schooling, the less is intergenerational social mobility and the less equal is opportunity. Their study estimates household income-child school success associations in Viet Nam. The estimates indicate that the income associations for child school success are fairly considerable – e.g. the income elasticity of completed grades is five times the median in 42 earlier studies. In part they obtain much higher estimates than in previous studies because they use a permanent income measure of household resources rather than just current income (or expenditures) as in most previous studies, which seems plausible in light of the considerations of Section 1 if there is some fungibility over time in resources. This association, moreover, is strongest *not* for completed grades on which most of the previous literature has focused, but for grades completed successfully per year of school. There are some gender differences, with the most important one being a smaller income association with grades completed per year of schooling for males than for females, which implies that schooling of females is treated as more of a "luxury" (less of a "necessity) than is schooling of males.

Exploration of the sensitivity of the estimates indicates that the estimated associations with income are significantly larger with predicted income to represent longer-run income as noted above, with control for censoring, and with representation of additional channels through which there may be effects beyond just the number of completed grades. Therefore, much of the previous literature may understate the true associations of household income with child school success and overstate true intergenerational social mobility and equality of opportunity.

Finally, some dimensions of how policies relate to household incomes are investigated. School fees are progressive, but school fees are only about a third of what households pay directly to schools, and a much smaller proportion of total household school-related expenditures. All household expenditures paid directly to schools increase with household income much less proportionately than do school fees alone, so the overall structure of such payments is much less progressive than is the structure of school fees alone. Because school enrollments are positively related with household income, moreover, the extent of progressivity in school fee structures is less for the entire population than it is for the selected subset that has children enrolled in school. Further, the two school quality measures that are most strongly positively associated with child school success are allocated so that they are significantly more available for higher-income households, which means that part of the higher household school expenditures of higher-income households is because they are obtaining higher-quality schooling and not because of progressive charges for the same quality schooling across income classes.

4.2. Impact of Schooling

There are literally hundreds of micro studies, many for Asia, that purport to investigate the impact of schooling on economic (e.g. wages, agricultural productivity) productivity and nonmarket productivity in developing countries (see the surveys in Behrman, 1990a, b, 1996b; King & Hill, 1993; Psacharopoulos, 1994; Schultz, 1988; Strauss & Thomas, 1995). An effective way to summarize many of the studies on wages has been through the calculation of the real rates of return to the costs incurred in schooling. This has been effective because rates of returns permit comparisons among a wide range of human resource investments, both within the schooling sector and elsewhere in the economy. Typically these rates of return have been calculated by comparing the direct economic outcomes for individuals with different amounts or types of schooling (or for different types of individuals) and calculating the rate of return to the private costs (primarily the time costs but perhaps also tuition, books and materials and other private costs) and to the social costs (the private costs plus public subsidies) to obtain, respectively, the so-called "private" and "social" rates of return to schooling.[39] These estimates are widely interpreted to imply that in developing countries: (1) the rates of return to schooling are high,[40] (2) they do not decline very rapidly with the level of development, (3) the impact of schooling, particularly for females, on nonmarket outcomes is considerable and generally greater than that of males, (4) the social rates of return decline with schooling levels (though the private rates of return do not necessarily do so because of relatively high per student subsidies to higher schooling levels), are higher for general as opposed to technical vocational schooling, and at least as high on average for female as for male schooling, (5) variability in schooling is associated with the variability in income distribution and more schooling is associated with less probability of being below the poverty line, and (6) there is not likely to be an equity-productivity trade-off in expanding schooling in the most productive way because the social returns are highest for basic (primary, then secondary) schooling for which further expansion is likely primarily to enroll more children from very poor families and the total returns are higher for females than for males. Under the assumption that wages are strongly associated with productivities, these conclusions generally are interpreted to carry over to the impact of schooling on productivity.[41] Conclusions of this sort are used to guide policy recommendations by many, including the World Bank (1991), for increasing policy support for schooling, particularly for basic (primary and lower secondary) schooling and particularly for females.

But the information basis for these policy recommendations is weak in a number of respects. There are a number of well-known possible problems with the methodology sketched out above for estimating the impact of schooling in most of these studies. Though most studies in this genre do not attempt to control for these problems, some of those that do report that such controls make considerable differences in the estimated impact of schooling. Most of the existing studies do not control well for the behavioral decisions that determine who goes to what type of school for how long with what degree of success. Simple analytical frameworks for school human resource investments as in Section 1 above, as well as casual observations, suggest that individuals with higher human resource investments in schooling are likely to be individuals with more ability and more motivation who come from family and community backgrounds that provide more reinforcement for such human resource investments and who have lower marginal private costs for such human resource investments and lower discount rates for the returns from those human resource investments and who are likely to have access to higher quality schools. Therefore most studies that (usually implicitly) assume that schooling is distributed randomly among sample members probably suffer from omitted variable biases. The association of schooling with outcomes such as wage rates, agricultural productivity, fertility, and child health does not necessarily represent causality because in most estimates years of schooling is representing not only time in school, but also factors that are correlated with years of school such as abilities, discount rates, family backgrounds, and schooling qualities. To obtain insight into the impact of years of school on such outcomes, one needs to control for these other factors, as do to a certain extent some – but not many – of the existing studies.

Several "revisionist" studies for developing countries, parallel to a similar literature for developed economies,[42] have explored the impact in the Asian context of some of these estimation problems on estimated schooling returns with data or specification modifications of the standard earnings function framework and in some cases have extended the analysis. I now review a few examples.

Schooling quality impact in rural Pakistan: While interest in estimating the impact of indicators of school quality on earnings has increased, most research on this topic has had important limitations: ignoring behavioral decisions regarding schooling; omitting individual and family background characteristics; using excessively aggregated school quality measures; and relying on crude indicators of teacher quality. These limitations may explain why the micro evidence of important school quality effects on earnings has been scant. Moreover, the question of the relative rates of return to improving school quality versus increasing school quantity hardly has been addressed.

The data demands for estimating such rates of return are considerable. Behrman, Ross, Sabot and Tromp (1997) developed a conceptual framework for undertaking such estimates, collected the necessary data for rural Pakistan, and made estimates with methods that overcome the major limitations noted.

In rural Pakistan, increasing the quantity of schooling a child receives raises his or her cognitive skills. Improving school quality has the same effect. Increasing the quantity of schooling – by providing a primary education to children who otherwise would not go to school, or by providing a middle school education to children who otherwise would leave school upon the completion of primary school – entails substantial costs. Similarly, social direct costs rise with the quality of primary schools. However, in this case, there is little or no change in the opportunity cost of student time – a large component of the total cost of schooling.

Higher cognitive skills are rewarded with higher wages in rural Pakistan, presumably because more skilled workers are more productive. Because they are more skilled, graduates of even low-quality primary schools earn more than uneducated workers. In a like manner, graduates of high-quality primary schools and graduates of middle schools who attended low-quality primary schools earn more than students who complete only low-quality primary schools.

Increasing the quantity and improving the quality of schooling are alternative means of increasing the productivity and earnings of the labor force. Their estimates of social rates of return indicate that improving the quality of primary schooling may be preferable to increasing access to middle schooling: the rate of return to improving the quality of primary schooling is substantially greater than the rate of return to increasing access to middle school, although it is somewhat lower than the rate of return to expanding enrollment in low-quality primary schools.

In this context, it appears that productivity and equity concerns both point towards expanding primary schools, even if they are of lower quality. And, because few boys now lack access to a primary school, girls will benefit particularly.

Returns to male and female schooling in Indonesian labor markets: Women are thought to be disadvantaged in developing countries. One of the major respects in which they are conjectured to be disadvantaged is that labor market rewards to their schooling are less than those for males. Behrman and Deolalikar (1995) investigate whether there are gender differentials in Indonesian wage and earnings relations. The results indicate that, although the returns to work experience are greater for males than for females, the marginal rates of return to schooling beyond the primary level are significantly greater for females. This paper goes beyond most previous studies in this area by (i) estimating wage rate, in addition to earnings, functions, (ii) controlling for schooling repetition and dropout rates, (iii) testing for fixed versus random effects, and (iv) controlling for unobserved household and community background characteristics.

The estimated age-gender interactions tend to favor males, and may in part reflect gender differences in labor market experience, because the associated male-female wage gap increases over the initial decades of work experience. But this gap is significant also for very young workers, which suggests that it is not due to differential experience alone. It is difficult to tell from the data whether such differences are due to different distributions of skills rewarded in the labor market by gender, different work experience and on-the-job training by gender, or labor market discrimination against women.

The empirical results concerning associations of schooling with wages may be surprising to many because they indicate that the marginal increases of wage rates and earnings with post-primary schooling are *greater* in percentage terms for females than for males in Indonesia once there is control for the estimation problems noted above. Thus, these estimates do *not* suggest that females face strong discrimination in the form of lower rates of return to post-primary schooling investments, as is sometimes claimed. This result *may* indicate that private rates of return to schooling investments in females are higher than are those to males. Private rates of return to schooling may be higher for females than for males because of higher interest costs of resources for schooling investments in females, higher risk premia for investing in schooling of females, disequilibrium in labor markets with greater lags in schooling investment adjustments for females, or less discrimination against women in occupations demanding more schooling. If any of these explanations are important, lessening the gender gap in schooling would be advantageous in terms of efficiency in the wage labor market, in addition to the gains in terms of equity. The means of best doing so would depend on identifying which of these factors underlie the gender differences in the estimated rates of return to schooling. However, the discussion in this paper also indicates other reasons, related to differential nonpecuniary returns and selectivity by gender, why these and other estimates may be misleading regarding gender differences in the estimated private pecuniary rates of returns to schooling.

Adoption and adaptation of technological innovations in rural India and schooling externalities: A striking aspect of almost all of the empirical literature on the impact of schooling, even including studies such as the two just reviewed that address a number of problems that usually are ignored in the literature and that may cause biases in the usual estimates of the impact of schooling such as unobserved household heterogeneity, is that they provide no or virtually no information with which to judge the efficiency rational for policy interventions related to schooling that is discussed in Section 2 above. That is the case because most of these studies are focused on obtaining better estimates of the private returns to schooling. But knowing whether the private returns to schooling are high or low is not informative regarding efficiency reasons for policy interventions, though it may be useful for considering whether schooling policies may be effective in obtaining distributional goals. The so-called "social rates of return" that are surveyed in Psacharopoulos (1994) and elsewhere do not in fact provide information regarding the efficiency motive for policies supporting schooling because they only adjust the private rate of returns estimates to include social costs through public subsidies, but do not include any externalities.[43]

A few recent studies based on data from the "Green Revolution" adoption of new agricultural seeds analyze the causal effect of education on the adoption of new technologies, as well as on the costs incurred due to imperfect information about these techniques and schooling externalities (Besley & Case, 1994; Foster & Rosenzweig, 1995, 1996; Munshi, 1997; Rosenzweig, 1995).

Evidence from the green revolution is significant because it exemplifies the case in which there was a role for information acquisition and learning. Learning about appropriate allocations of inputs was a challenge to farmers formerly engaged in traditional practices. The continual introduction of new seeds permanently raised the returns from skills in information decoding. The new technologies were more complex than previous technologies because the new seeds were considerably more sensitive to appropriate use of fertilizer and water than were traditional varieties. So the margin for error was greater.

Analysis of a national sample of farmers in 250 Indian villages confirms that initial HYV adoption in 1969–1971 was consistent with a learning model of optimal input use. Learning was based on own and on neighbors' experiences with the new seeds and learning was enhanced by schooling. The results from studies based on the data provide critical insights because longitudinal data are available that follow the same farmers over the critical initial adoption period and because the use of new high-yielding seeds is easily identified, which permits the measurement of adoption rates over time and comparisons of before and after-adoption profits for the same farmers.

Farmers planted only 3% of their area in the HYV seeds in the first year and on the average incurred initial loses due to misuse of inputs from a lack of experience. The costs of slow and inappropriate adoption due to imperfect knowledge were considerable. Over the first five years, the average loss due to slow adoption and misuse of the new technologies in comparison with their subsequent experienced use averaged 3.7 times the pre-green revolution annual income. which suggests that imperfect information has large costs, and thus there are important gains from speeding information flows or accelerating learning.

Farmers learned how to use the new technology both from their own experience and from the experience of neighboring farmers, with the latter causing a "free rider" problem in which farmers held back on their own use until they could learn more from their neighbors. Farmers with primary schooling had a very small initial advantage over farmers with no schooling that translated into about 5% greater profits in the first year. But farmers with primary school had more rapidly increasing profits with experience than did those with no schooling – an 18% greater effect on profits in the second year per acre-year of prior experience. Moreover there are spilloer effects of schooling, with an unschooled farmer reaching full specialization in the new technologies in four years rather than five if he has schooled rather than unschooled neighbors.

These studies also indicate that the schooling return is on average augmented by exogenous technical change. Moreover, the increase in the return to schooling is greater the higher is the rate of growth in the area; having a primary education increased the impact of technical change on profits by 70%. An approximate dynamic schooling decision rule is estimated that makes use of the profit-function estimates of district-level technical change. In conformity with the profit function estimates, the demand for schooling is found to increase with the level of technical change. It is also found to increase with the availability of schools which implies that policies that promote technical change are complementary to investments in schools.

Thus, these studies provide evidence on two important facts about the returns to schooling. First, the rates of return are high only if there are technological and market opportunities so learning, which is enhanced by schooling, has a payoff. In terms of Fig. 1, the expected marginal benefits curve is high because of an expected stream of ongoing learning opportunities. Learning is not likely to be very great, on the other hand, in a stagnant economy. Second, there are social returns beyond the private returns, as in Fig. 2, so the private incentives are to underinvest in schooling from a social point of view.

5. CONCLUSIONS

Schooling is emphasized as a critical factor in obtaining efficiency/productivity and distributional goals in Asia and in other developing areas. There are many empirical studies that purport to demonstrate that schooling has positive effects. Yet these studies are not nearly as persuasive as they usually are claimed to be. Most of them are not based on explicit models of underlying behaviors, and thus are difficult to interpret and are likely to be characterized by biased estimates. Very few of them indicate awareness of what is required to establish and efficiency basis for policies supporting schooling.

Recently there have been a small number of studies that investigate the determinants of schooling and the impact of schooling within an Asian context with explicit modeling of the underlying behaviors and use of econometric techniques and often special data (particularly longitudinal data) to deal with the estimation problems. These studies indicate substantially different magnitudes for estimated determinants of schooling and impacts of schooling than are obtained if the usual techniques were used with the same data. They indicate how in a number of respects market imperfections both affect the determinants of schooling and the impact of schooling, and thus efficiency justifications for policy interventions as well as better understanding of distributional justifications for policy interventions.

With respect to the determinants of schooling, for example, they suggest that seasonal variations and related capital market imperfections, mother's home schooling, child nutrition, supply-side factors, and expected returns to schooling in terms of learning about new technologies and markets all are likely to play more important roles than would be revealed in studies that use the ususal methods, and that associations with household income are much strongly than indicated in studies that use the usual methods. Such results suggest that improving markets – particularly financial and insurance markets, but also product and labor markets – may have more important long-run effects on schooling and that regulations such as on child labor and compulsory schooling may have greater household welfare costs than suggested by standard studies.

With respect to the impact of school, they suggest that in the usual studies the failure to control for family background may overstate the returns to schooling and understate the related returns to female versus male schooling at higher levels, that the failure to control for school quality may suggest too large gains from extending schooling to more years as opposed to improving schooling quality, and that the failure to include possible externalities such as learning more from more-schooled neighbors about the appropriateness of new technologies may miss important evidence regarding efficiency arguments for public subsidies for schooling.

Only by ongoing careful studies such as these, with careful attention to the implications of the underlying behaviors for modeling and estimation and policy interpretation, is our understanding of the determinants and the impact of schooling and the policy implications thereof in Asia likely to continue to improve.

NOTES

1. Behrman is the William R. Kenan, Jr. Professor of Economics and Director of the Population Studies Center at the University of Pennsylvania. Correspondence regarding this paper should be sent to Jere R. Behrman, Economics, McNeil 160, 3718 Locust Walk, University of Pennsylvania, Philadelphia, PA 19104-6297, USA; telephone 215 898 7704, fax 215 898 2124, e-mail jbehrman@econ.sas.upenn.edu.

2. For simplicity I am also assuming risk neutrality. If there is risk aversion, the analysis needs to be recast in terms of the present discounted of marginal utilities, but at the level of generality of this presentation the basic points that are made below hold.

3. I come back to this assumption in Section 3.

4. This can be seen by considering first a level of investment H′ to the left of H*, at which investment level the marginal benefit R'_B exceeds the marginal cost R'_C; therefore welfare gains can be made by increasing the investment to H*. The net gain for a small increase in investment from H′ to H* is equal to the total additional benefit minus the total additional cost, which is the triangle ABC in the figure. Consider next a level of investment H″ to the right of H*, at which investment level the marginal benefit R''_B is less than the marginal cost R''_C; therefore welfare gains can be made by decreasing the

investment to H* with the net gain for a small decrease in investment from H″ to H* equal to the triangle ADE in the figure.

5. Three points should be noted. First, economic efficiency is not the same as engineering efficiency because of the incorporation of marginal benefits and marginal costs rather than focusing exclusively on technological efficiency. Second, these marginal conditions for efficiency may not hold if there are, for example, large discontinuities in production processes. In such cases choices may have to be made among a number of different alternatives, using an explicit welfare function to compare among the alternatives. Third, the considerations noted above in Section 1 about expected present discounted values and about uncertainty hold for this discussion as well.

6. The marginal social benefits also could be lower than the marginal private benefits so that the marginal social benefits curve are below the marginal private benefits curve, and policies to attain efficiency would have to reduce the private incentives to the social levels.

7. When human resource investments in one individual's education affect the productivity and welfare of others other than in ways that are reflected in market prices, this human resource investment is said to generate "externalities" – i.e. effects that are external to what is transmitted through markets. If at least some individuals do not fully consider the external effects of their own decisions on the welfare of others, a purely private market solution leads to under-human resource investment from society's standpoint in activities that generate positive externalities (e.g. knowledge, immunizations) and an over-human resource investment in the case of negative externalities (e.g. use of antibiotics). Such "externalities" are one case of a broader category of "market failures" (other examples are discussed below).

8. The "public good" nature of information (i.e. that the marginal cost of providing information to another consumer is virtually zero) leads to under-production of information from a social point of view by private markets because private providers cannot cover their costs if they price information at the social marginal cost as required for efficiency.

9. These are major reasons for a divergence to arise between private and social marginal benefits for human resource investments related to schooling though there may be other factors as well (e.g. the social discount rate may be lower than the private discount rate, wage and price rigidities may preclude wages and prices from reflecting social marginal benefits and costs, income taxes may cause private marginal returns to human capital human resource investments to be lower than social marginal returns).

10. The marginal social costs also could be higher than the marginal private costs, in which case the marginal social cost curve would be above the marginal private cost curve, and policies to attain efficiency would have to reduce the private incentives to the social levels.

11. Economic "rents" are returns to fixed supplies. Because the supplies are fixed, the rents gained by the owners of them do not affect the quantities supplied. If supplies are fixed, for example, by regulations, rents are created by the supply restrictions, so there are incentives for various private and public entities to try to obtain these rents ("rent-seeking"). Rent-seeking uses resources and transfers income among various groups, but does not add to production.

12. If the policies involve public expenditures as do most policies, it is important to consider the cost of raising the necessary tax revenue to finance the policy. In the U.S.,

for example, it has been estimated that the distortionary cost (often called the "deadweight loss") of raising a dollar of tax revenue ranges from $0.17 to $1.00, depending on the type of tax used (e.g. Feldstein, 1995).

13. This is also an argument for considering an experimental approach to evaluating policy alternatives when possibly – e.g. rather than introducing a reform country-wide, introduce variants of reforms for schools (and other social services) in randomly selected sites with careful monitoring of the results for both the experimental groups and the control groups.

14. Nevertheless there are likely to be some cases, such as providing information regarding the quality of goods and services related to human resource investments related to schooling, for which quantitative regulations may be higher in the policy hierarchy than price policies because of the nature of the information requirements.

15. However, even redistributing income may lead to inefficiency because it can affect the work effort of those on both the tax-paying and tax-receiving sides.

16. However, household decision-makers may reallocate resources so that the intended beneficiaries of such policies receive much smaller benefits than intended (e.g. children receiving subsidized food at school may be fed less at home).

17. In the context of such an objective it is important for governments to decide whether success is to be measured in outcomes (i.e. cognitive achievement and other outcomes rewarded in labor markets), utilization (school enrollment or attendance), or access (ability to consume basic school services at all income levels).

18. For example, society might decide that more resources should go to children and less to adult goods such as jewelry and alcohol than the parents would choose.

19. In some cases there may be a strong rationale in terms of the framework in Section 2 for subsidizing certain options more heavily. The condom, for example, provides protection against STDs and AIDS. Its use generates significant externalities to other consumers in the form of reduced risk of contracting these illnesses.

20. Even if paternalistic policy makers decide that the basic policy objective should not be increasing welfare of the populace but some goal such as reducing population growth and long-term contraceptive methods are perceived to be more cost-effective in reducing fertility, the underlying private household behaviors that are discussed in Section 1.1 and the behaviors of suppliers need to be taken into account in devising the structure of subsidies. A subsidy will be ineffective if demand for or supply of the method is highly inelastic (i.e. a price reduction does not encourage much additional demand or much additional supply, Hammer & Berman, 1995).

21. There is little reason to expect they would succeed because in competitive (and even monopolistically competitive) markets there is no reason to expect producers to enjoy long-run profits. If activities in high-income markets are profitable (i.e. generate revenue above actual costs), other investors are attracted into the market and the increased supply drives prices down to the point where price is equal to average cost (i.e. zero profit). This has in fact been the experience of many donor-funded cross-subsidizing ventures.

22. If governments do provide these services directly (and most do), they should be made to compete with the private sector on an equal basis, i.e. the private sector should receive the same level of subsidies received by governmental providers.

23. Linear approximations are used here because they are the simplest forms but they still permit characterization of various estimation issues. Log-linear forms in which all of the variables are replaced by the logarithms of their values (which implies

interactions among all the right-side variables) are identical in representation once the variables are redefined. In empirical studies linear and log-linear specifications are very common, but other functional forms also are used at times. For other functional forms the essence of the estimation issues is the same. If the functional form that is used is not a good approximation to the true functional form, there is misspecification error that is akin to omitted variable bias discussed below (with the unobserved variable being the variable that would have to be added to transform the assumed specification to the true functional form).

24. It is not possible, however, to estimate the effect of one behavioral variable determined in the current period on another with demand relations. This point is not always understood, so it is developed briefly in this note. For example, current cognitive achievement may be posited to be determined simultaneously with current child nutrient intakes. Consider the following two demand relations for current cognitive achievement (Z_{CA}) and for current child nutrient intakes (Z_N) as dependent on two prices (PF_1, PF_2), predetermined resources (RF_f), and stochastic terms (V_C, V_N):

$$Z_{CA} = b_{11}PF_1 + b_{12}PF_2 + b_{13}RF_f + V_C \quad (1A\text{-}1)$$

and

$$Z_N = b_{21}PF_1 + b_{22}PF_2 + ba_{23}RF_f + V_N. \quad (1A\text{-}2)$$

These two relations imply:

$$Z_{CA} = b_{31}Z_N + b_{32}PF_2 + b_{33}RF_f + V'_C. \quad (1A\text{-}3)$$

This might appear to be a conditional demand function with which the impact of current child nutrient intakes on current child cognitive achievement could be explored. But the coefficient of current child nutrient intakes is merely the ratio of the effect of the price that has been eliminated in relation (1A-1) relative to the effect of that price in relation (1A-2) (i.e. $b_{31} = b_{11}/b_{21}$). Thus the coefficient of current child nutrient intake in relation (1A-3) does *not* reveal anything interesting about the effect of current child nutrient intakes on current child cognitive achievement, but only of the relative price effects (for PF_1) in the two demand relations.

25. A number of, but not all, recent studies report evidence using twins or other sibling data or direct measures of ability that suggest that usually unobserved endowments have direct effects on wages (e.g. Behrman & Rosenzweig, 1998, review most of these studies).

26. To be able to learn about the *productivity* rather than the earnings impact of schooling, moreover, there must be knowledge of the relation between private earnings and marginal productivities.

27. Random measurement error is what usually is emphasized, and is what is discussed here. Measurement error also may be systematically related to the true variable, with implications that depend on exactly what is the nature of the systematic relation.

28. Random measurement error is what usually is emphasized, and is what I discuss here. Measurement error also may be systematically related to the true variable, with implications that depend on exactly what is the nature of the systematic relation.

29. σ_T^2 is standard statistical notation for the variance in T, which is simply a measure of how much the individual observations of T differ from the average (mean) value of T.

30. The compound disturbance term includes all the unobserved variables unless their effects are controlled in some way.

31. Good identifying instruments have the following properties: (a) they do not appear in the relation being estimated, (b) there is at least one such observed instrument for every observed behavioral right-side variable in the relation being estimated, (c) they are independent of the compound disturbance term in the relation being estimated (which includes all of the unobserved variables), and (d) they are sufficiently correlated with the observed right-side behavioral variables. If instruments are used that have these properties, the correlations between the observed right-side behavioral variables and the compound disturbance term are eliminated by using for the former their estimated values based on the instruments. Primary candidates for identifying instruments in a production function are the right-side variables in the demand relations for the right-side behavioral variables that do not appear in the relation being estimated – basically the predetermined prices and resources broadly defined (with the latter subject to condition a) in the demand relations (2A) discussed above. For demand function estimates, any variables that are included in the demand functions do not satisfy condition (a). For conditional demand functions with variables predetermined from an earlier period the natural instruments for these variables are the subset of their past determinants that are independent of the current right-side predetermined variables, such as past price shocks or other shocks. In practice finding such instruments usually is not easy. In particular for production functions, if there are unobserved behavioral variables (XI^{ub}, XF^{ub}) in the compound disturbance term, these depend on the same predetermined variables in the demand relations as do the observed variables, so that set of predetermined variables does not satisfy condition (c). Therefore most instrumental variable estimates of production functions are unbiased conditional on the (often implicit) assumption that there are no unobserved behavioral inputs. But this generally is likely to be a strong assumption. This particular problem does not exist for demand relations. But for demand relations as for production functions the past data necessary to instrument variables that are determined prior to the period of estimation are not available in many data sets. For such reasons, instrumenting for variables determined within the model and measurement error and fixed effects for controlling for unobserved fixed variables are combined in a number of studies.

32. It is difficult in many contexts to conduct such experiments – though people interested in schooling might be randomly selected into experimental and control groups, those who are not interested cannot be so selected so the population for which an experimental sample is useful is a selected sample. It also is difficult to assure that the experimental "treatment" in fact is received by all of the experimental group and not (nor some close substitute) by the "control" group. Moreover, "experimental effects" may cause distortions, and it is difficult to devise experiments relating to schooling in which there is a placebo that can be administered that appears to participants and to administrators to be identical to the experimental treatment of interest. Further, cases questions of fairness or of ethics may preclude a number of experiments. Finally, conducting experiments usually is relatively costly and often difficult to arrange because of pressures on those administering the experimental treatment to include others than just those in the experimental group. A pragmatic compromise that sometimes is adopted is to promise members of the control group the experimental treatment with a lag; but within the model presented in Section 1 this affects current behavior by changing future expected prices and policies.

33. Implementation of this method requires that the statistic used from the first stage (the "inverse Mills ratio") be identified by variables that enter into the decision rule but not directly into the second stage or by functional form. For example, land holding may affect whether an individual participates in the labor market, but not the wage received for those who do participate.

34. Recent examples for low-income countries in Asia include Alderman, Behrman, Ross and Sabot (1996b) for Pakistan; Deolalikar (1993) for Indonesia; Lillard and Willis (1993) for Malaysia; and Parish and Willis (1993) for Taiwan.

35. See, for example, Jamison (1986), Moock and Leslie (1986) (two well-known studies on the Asian experience) and the other studies based on behavioral data that are surveyed in Pollitt (1990) and Behrman (1996a).

36. Only with these first two assumptions for height-for-age at 12 and 24 months and then only in some models, do the IV estimates differ significantly from the naïve estimates.

37. If (i) and (ii) do not hold, child heights-for-age at birth, 12 months and 24 months are *not* appropriate instruments for the subsequent child nutrition/health status because they are correlated with the unobserved child endowments or the subsequent parental nutrition/health investment decisions that cause the estimation problem. If (iii) does not hold, there is no estimation problem.

38. Unfortunately they do not have data on anthropometrics for these children during their first two years of life so they cannot make direct comparisons of our estimates using price shocks with what would be obtained using the procedure in Glewwe, Jacoby and King (1996).

39. Sometimes the Mincerian (1974) semilog relation between wages and schooling with control for post-schooling experience (or age) is used to calculate the private rate of return to time spent in schooling instead of in the labor market under the Mincerian assumptions (e.g. there is equilibrium so that individuals are indifferent among various schooling levels and characteristics such as ability and family background enter into the wage determination function so they are not correlated with schooling). The Mincerian formulation assumes that there are opportunity costs to schooling at all ages, an assumption that Psacharopoulos (1994) and some others criticize.

40. Such estimates imply, in fact, that human resource investment in schooling is such a high return human resource investment that they are not completely credible on these grounds alone. Human resource investments with a real annual rate of return of 16–24% (the social rate of return to primary school in the four developing regions given in Psacharopoulos 1994) and with reinvestment of the proceeds of such human resource investment imply that society can double the real invested assets in 2.9 to 4.3 years, and the social real rate of 11–18% on secondary schooling implies the possibility of doubling real assets in 3.8 to 6.3 years. These estimates, moreover, understate the true social rates of return and overstate the true time that social assets could be doubled by marginal schooling human resource investments if there are positive externalities to schooling as often is claimed. If developing countries have available such human resource investment opportunities on a fairly broad scale (i.e. in most of its children), it would seem that much higher economic growth would be observed than ever has been experienced for any sustained period of time.

41. The third conclusion and a small subset of the studies underlying the other conclusions use direct measures of productivity, not wages, as the dependent variables.

42. See, for example, Behrman and Rosenzweig (1998), Card (1995), and the references therein.

43. This means that these "social" rates of returns are always less than the associated private rates of return, which implies from an efficiency perspective that public subsidies should be reduced to zero.

REFERENCES

Alderman, H., Behrman, J. R., Lavy, V., & Menon, R. (2001). Child Nutrition, Child Health, and School Enrollment: A Longitudinal Analysis. *Journal of Human Resources, 36*(1), 185–205.

Alderman, H., Behrman, J. R., Ross, D., & Sabot, R. (1996a). Decomposing the Gender Gap in Cognitive Skills in a Poor Rural Economy. *Journal of Human Resources, 31*(1), 229–254.

Alderman, H., Behrman, J. R., Ross, D., & Sabot, R. (1996b). The Returns to Endogenous Human Capital in Pakistan's Rural Wage Labor Market, *Oxford Bulletin of Economics and Statistics, 58*(1), 29–55.

Ashenfelter, O., & Krueger, A. (1994). Estimates of the Economic Return to Schooling from a New Sample of Twins. *American Economic Review, 84*(5), 1157–1174.

Becker, G. S. (1967). Human Capital and the Personal Distribution of Income: An Analytical Approach, Ann Arbor: University of Michigan, Woytinsky Lecture. In: G. S. Becker (Ed.), *Human Capital* (2nd edition 1975, pp. 94–117). New York: NBER.

Behrman, J. R. (1990a). *Human Resource Led Development?* New Delhi, India: ARTEP/ILO.

Behrman, J. R. (1990b). *The Action of Human Resources and Poverty on One Another: What We Have Yet to Learn.* Washington, D.C.: Population and Human Resources Department, World Bank.

Behrman, J. R. (1993). Investing in Female Education for Development. *Journal of Educational Planning and Administration, 7*(4), 393–412.

Behrman, J. R. (1996a). Impact of Health and Nutrition on Education. *World Bank Research Observer, 11*(1), 23–37.

Behrman, J. R. (1996b). Measuring the Cost-Effectiveness of Schooling Policies in Developing Countries: Revisiting Issues of Methodology. *Economics of Education Review, 15*(4), 345–364.

Behrman, J. R., & Deolalikar, A. B. (1991). School Repetition, Dropouts and the Returns to Schooling: The Case of Indonesia. *Oxford Bulletin of Economics and Statistics, 53*(4), 467–480.

Behrman, J. R., & Deolalikar, A. B. (1993). Unobserved Household and Community Heterogeneity and the Labor Market Impact of Schooling: A Case Study for Indonesia. *Economic Development and Cultural Change, 41*(3), 461–488.

Behrman, J. R., & Deolalikar, A. B. (1995). Are There Differential Returns to Schooling by Gender? The Case of Indonesian Labor Markets. *Oxford Bulletin of Economics and Statistics, 57*(1), 97–118.

Behrman, J. R., Foster, A., Rosenzweig, M. R., & Vashishtha, P. (1999). Women's Schooling, Home Teaching, and Economic Growth. *Journal of Political Economy.*

Behrman, J. R., & Knowles, J. C. (1998). Population and Reproductive Health: An Economic Framework for Policy Evaluation. *Population and Development Review* (December).

Behrman, J. R., & Knowles, J. C. (1999). Household Income and Child Schooling in Vietnam? *World Bank Economic Review.*

Behrman, J. R., & Lavy, V. (1998). Child Health and Schooling Achievement: Association, Causality and Household Allocations. Philadelphia, PA: University of Pennsylvania, mimeo.

Behrman, J. R., & Rosenzweig, M. R. (1998). 'Ability' Biases in Schooling Returns and Twins: A Test and New Estimates. *Economics of Education Review*, (forthcoming).

Behrman, J. R., Rosenzweig, M. R., & Taubman, P. (1994). Endowments and the Allocation of Schooling in the Family and in the Marriage Market: The Twins Experiment. *Journal of Political Economy, 102*(6), 1131–1174.

Behrman, J. R., Ross, D., Sabot, R., & Tropp, M. (1997). Improving the Quality Versus Increasing the Quantity of Schooling. Philadelphia, PA: University of Pennsylvania, mimeo.

Besley, T., & Case, A. (1994). Diffusion as a Learning Process: Evidence from HYV Cotton. Princeton, NJ: Princeton University, mimeo.

Burney, N. A., & Irfan, M. (1994). Determinants of Child School-Enrolment: Evidence from LDCs Using Choice-Theoretic Approach. Islamabad: Pakistan Institute of Development Economics, mimeo.

Card, D. E. (1995). Earnings, Schooling, and Ability Revisited. *Research in Labor Economics, 14*, 23–48.

Deolalikar, A. B. (1993). Gender Differences in the Returns to Schooling and School Enrollment Rates in Indonesia. *Journal of Human Resources, 28*(4), 899–932.

Folbre, N. (1984). Comment on 'Market Opportunities, Genetic Endowments, and Intrafamily Resource Distribution. *American Economic Review, 74* (June), 518–520.

Folbre, N. (1986). Cleaning House: New Perspectives on Households and Economic Development. *Journal of Development Economics, 22*(1), 5–40.

Foster, A., & Rosenzweig, M. R. (1995). Learning by Doing and Learning from Others: Human Capital and Technical Change in Agriculture. *Journal of Political Economy, 103*(6), 1176–1209.

Foster, A. D., & Rosenzweig, M. R. (1996). Technical Change and Human-Capital Returns and Investments: Evidence from the Green Revolution. *American Economic Review, 86*(4), 931–953.

Glewwe, P., Jacoby, H., & King, E. (1996). An Economic Model of Nutrition and Learning: Evidence from Longitudinal Data. Washington, D.C.: World Bank, mimeo.

Government of Pakistan (1991). Review of Secondary and Intermediate Education.

Haddad, W. D., Carnoy, M., Rinaldi, R., & Regel, O. (1990). *Education and Development: Evidence for New Priorities*. Washington, D.C.: World Bank Discussion Paper 95.

Haddad, L., Hoddinott, J., & Alderman, H. (Eds) (1996). *Intrahousehold Resource Allocation: Methods, Models, and Policy*. Baltimore, MD: The Johns Hopkins University Press for the International Food Policy Research Institute.

Haveman, R., & Wolfe, B. (1995). The Determinants of Children's Attainments: A Review of Methods and Findings. *Journal of Economic Literature, 33*(4), 1829–1878.

Heckman, J., Ichimura, H., & Todd, P. (1997). Matching as an Econometric Evaluation Estimator: Evidence from Evaluating a Job Training Program. *Review of Economic Studies, 64*(4), 605–654.

Hill, A. M., King, E. M. (1993). Women's Education in Developing Countries: An Overview. In: E. M. King & M. A. Hill (Eds), *Women's Education in Developing Countries: Barriers, Benefits, and Policies* (pp. 1–50). Baltimore and London: The Johns Hopkins University Press, Published for the World Bank.

Jacoby, H., & Skoufias, E. (1997). Risk, Financial Markets and Human Capital in a Developing Country. *Review of Economic Studies, 64*(3), 311–335.

Jamison, D. T. (1986). Child malnutrition and school performance in China. *Journal of Development Economics*, *20*(2), 299–231.

King, E. M., & Hill, M. A. (Eds) (1993). *Women's Education in Developing Countries: Barriers, Benefits, and Policies*. Baltimore and London: The Johns Hopkins University Press, Published for the World Bank.

Knight, J. B., & Sabot, R. H. (1990). *Educational Productivity and Inequality: The East African Natural Experiment*. New York: Oxford University Press.

Lillard, L., & Willis, R. (1994). Intergenerational Education Mobility: Effects of Family and State in Malaysia. *Journal of Human Resources*, *29*(4), 1126–1166.

Lundberg, S., & Pollak, R. A. (1996). Bargaining and Distribution in Marriage. *The Journal of Economic Perspectives*, *10*(4), 139–158.

Mincer, J. B. (1974). *Schooling, Experience, and Earnings*. New York: NBER.

Moock, P. R., & Leslie, J. (1986). Childhood malnutrition and schooling in the Terai region of Nepal. *Journal of Development Economics*, *20*(1), 33–52.

Munshi, K. D. (1997). Farmers as Econometricians: Social Learning and Technology Diffusion in the Indian Green Revolution. Boston: Boston University, mimeo.

Parish, W. L., & Willis, R. J. (1993). Daughters, Education, and Family Budgets: Taiwan Experiences. *Journal of Human Resources*, *28*(4), 863–898.

Pollitt, E. (1990). *Malnutrition and Infection in the Classroom*. Paris: Unesco.

Psacharopoulos, G. (1994). Returns to Investment in Education: A Global Update. *World Development*, *22*(9), 1325–1344.

Rao, V. (1993). The Rising Price of Husbands: A Hedonic Analysis of Dowry Increases in Rural India. *Journal of Political Economy*, *4*(101), 666–677.

Rosenzweig, M. R. (1995). Why Are There Returns in Schooling? *American Economic Review*, *85*(2), 153–158.

Rosenzweig, M. R., & Schultz, T. P. (1982). Market Opportunities, Genetic Endowments, and Intrafamily Resource Distribution: Child Survival in Rural India, *American Economic Review*, *72*(4), 803–815.

Rosenzweig, M. R., & Wolpin, K. J. (1982). Governmental Interventions and Household Behavior in a Developing Country: Anticipating the Unanticipated Consequences of Social Programs. *Journal of Development Economics*, *10*(2), 209–226.

Schultz, T. P. (1988). Education Investments and Returns. In: H. Chenery & T. N. Srinivasan (Eds), *Handbook of Development Economics* (pp. 543–630). Amsterdam: North-Holland Publishing Company.

Schultz, T. P. (1993a). Returns to Women's Education. In: E. M. King & M. A. Hill (Eds), *Women's Education in Developing Countries: Barriers, Benefits, and Policies* (pp. 51–99). Baltimore and London: The Johns Hopkins University Press, Published for the World Bank.

Schultz, T. P. (1993b). Investments in the Schooling and Health of Women and Men: Quantities and Returns. *Journal of Human Resources*, *28*(4), 694–734.

Schultz, T. W. (1975). The Value of the Ability to Deal with Disequilibria. *Journal of Economic Literature*, *13*(3), 827–846.

Strauss, J., & Thomas, D. (1995). Human Resources: Empirical Modeling of Household and Family Decisions. In: J. R. Behrman & T. N. Srinivasan (Eds), *Handbook of Development Economics* (Vol. 3A, pp. 1883–2024). Amsterdam: North-Holland Publishing Company.

Summers, L. H. (1992). Investing in All the People. *Pakistan Development Review*, *31*(4), 367–406.

Thomas, D. (1990). Intrahousehold Resource Allocation: An Inferential Approach. *Journal of Human Resources*, *25*(4), 635–664.

UNDP (1996). *Human Development Report 1996*. New York: UNDP.
Welch, F. (1970). Education in Production, *Journal of Political Economy, 78*(1), 35–59.
World Bank (1990). *World Development Report: Poverty*. Oxford: Oxford University Press.
World Bank (1991). *World Development Report: The Challenge of Development*. Oxford: Oxford University Press.

8. GROWTH, CRISIS AND POVERTY INCIDENCE IN SOUTHEAST ASIA[1]

Peter G. Warr

1. INTRODUCTION

East Asia's currency crisis of 1997 led to a deep economic contraction. The contraction ended a boom which had been enjoyed by most of East Asia since the late 1980s. During this period of prolonged boom, impressive progress was made in poverty reduction in each of the countries which experienced rapid growth. But in the late 1990s these same countries were now experiencing deep recessions. Economic hardship was being felt at all levels of the income distribution in these countries but the implications for the poorest people has rightly been a concern in popular discussion and in the planning of the international aid community. The present paper attempts to contribute to this discussion. It focuses on four large countries of Southeast Asia, all of which were badly affected by the economic crisis: Thailand, the first to succumb; Indonesia, which has proven to be the most severely affected; Malaysia, where the crisis produced the most radical macroeconomic policy responses; and the Philippines, where the pre-crisis boom had been least significant but where the crisis itself was nevertheless a serious event.

Information about the level of poverty incidence is available for each of the above four countries in the form of household income and expenditure surveys, conducted by the national statistical agencies of these counties and using large samples of households. The greatest problem in using these surveys for policy planning is that the surveys are conducted only periodically and, more significantly, that the results typically become available only after delays of

Restructuring Asian Economics for the New Millennium, Volume 9A, pages 165–176.
2001 by Elsevier Science B.V.
ISBN: 0-7623-0709-9

some years. Substantial delays in analyzing the results of such large surveys are inevitable, but policy decisions often cannot wait. The response to major events like the crisis of 1997 and its aftermath must be planned and implemented long before the results of household surveys are analyzed and released.

The case of Thailand is illustrative. For systematic study of poverty incidence and inequality, the periodic Socio-Economic Surveys (SES) conducted by the Thai government's National Statistical Office (NSO) provide virtually the sole source of reliable information that can be compared over time (Krongkaew, 1993). Household surveys were conducted periodically prior to 1975, when the present form of the survey commenced. Another survey was conducted in 1981 and since 1986 these surveys have been conducted every two years. Processing the data has normally taken at least two years. For example, in late 1998, data were released for the SES survey results for the pre-crisis year of 1996, but survey-based information about the impact of the crisis on poverty incidence must await release of the 1998 SES results.

Similar accounts apply to each of the other major Southeast Asian countries. Policy responses to the crisis must be in place long before reliable, survey-based data on the impact of the crisis become available. In the absence of these data, various alternative methods of projecting the impact of the crisis have been attempted. These include mini-surveys, which suffer from the problem of unknown representativeness, general equilibrium modeling approaches and approaches based on statistical analyses of existing data. The present paper falls in the last of these categories. It attempts to draw inferences about the impact of the crisis using information which is readily available, that relating to current aggregate output. The core of the paper is an analysis of the relationship between poverty incidence and aggregate output as revealed by past data.

Part 2 of the paper summarizes available data on poverty incidence in Southeast Asia. Part 3 sets out the analytical framework for the study. Part 4 describes the statistical method and the results and Part 5 concludes.

2. POVERTY INCIDENCE IN SOUTHEAST ASIA

Available data on poverty incidence in the four Southeast Asian countries listed above are summarized in Tables 1a to 1d. The data are presented as aggregate poverty incidence and its rural and urban components. The aggregate poverty data are summarized in Figs 1a to 1d. For the time being, the reader should disregard the data points indicated by the large squares in these figures and focus on those indicated by the smaller diamond shaped symbols.

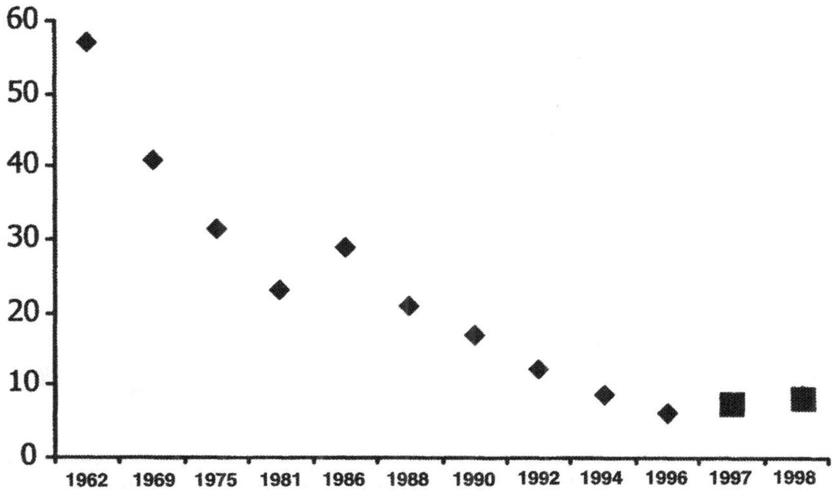

Fig. 1a. Thailand: Actual Poverty Incidence and Projections of Crisis Impact (per cent of total population).

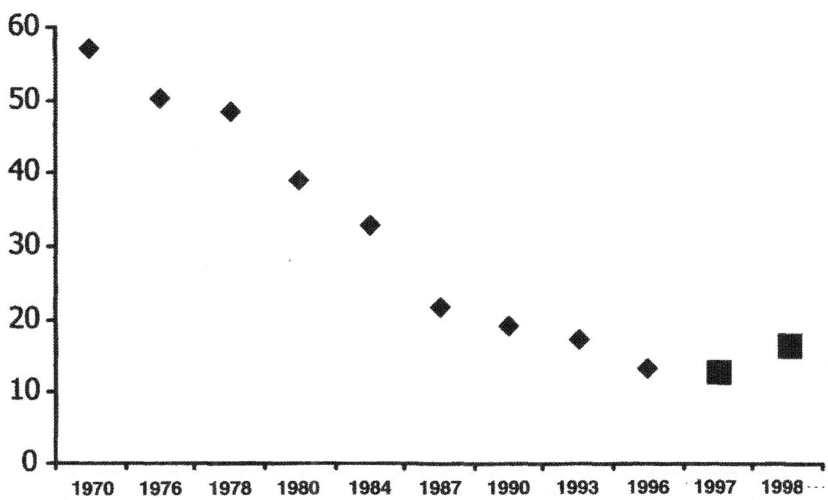

Fig. 1b. Indonesia: Actual Poverty Incidence and Projections of Crisis Impact (per cent of total population).

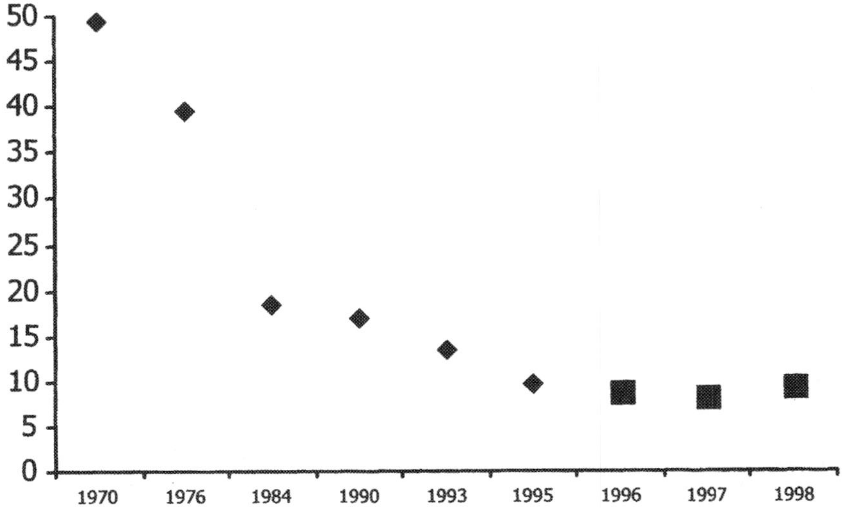

Fig. 1c. Malaysia: Actual Poverty Incidence and Projections of Crisis Impact (per cent of total population).

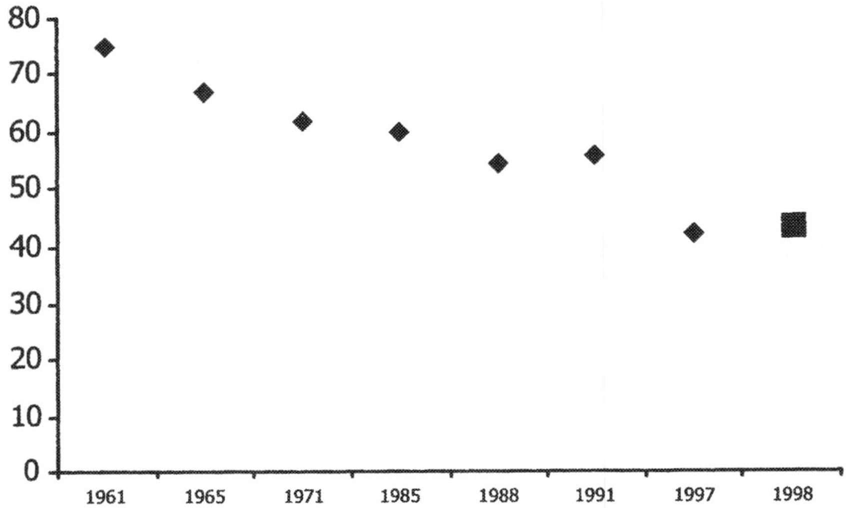

Fig. 1d. The Philippines: Actual Poverty Incidence and Projections of Crisis Impact (per cent of total population).

Poverty incidence and its changes over time obviously depends on many factors, of which economic variables are only part of the story, and among the economic variables many issues aside from simply the overall rate of growth will be relevant. Changes in commodity prices will play a role, along with tax policies, labor market policies, technological change, and so forth. The sectoral composition of growth may also be important. Nevertheless, cursory examination of the data suggests that the overall rate of growth may be central to the story. Large reductions in poverty have been achieved in each of the four countries but the rate of reduction was lowest in the Philippines, where the average rate of growth was also lowest. At the level of individual countries, a relationship between the rate of growth and the rate of poverty reduction also seems possible. For example, in Thailand poverty incidence fell throughout the period indicated except for the recession period of the early 1980s, when measured poverty incidence actually increased. We shall explore this statistical relationship in more detail below.

3. ANALYTICAL FRAMEWORK

Aggregate, Rural and Urban Poverty Incidence

We shall review first the relationship between aggregate, rural and urban poverty incidence and then turn to the manner in which each of these measures is affected by economic growth. Changes in aggregate poverty incidence may be decomposed into rural and urban components, as follows. We shall write N, N^R and N^U for the total, rural and urban populations, respectively, where $N = N^R + N^U$. We write $n^R = N^R/N$ and $n^U = N^U/N$ for the rural and urban shares of the total population, respectively, where $n^R + n^U = 1$. The total number of people in poverty is given by $N_P = N_P^R + N_P^U$, where N_P^R and N_P^U denote the numbers in poverty in rural and urban areas, respectively. Aggregate poverty incidence is given by

$$P = N_p/N = (N_P^R + N_P^U)/N = n^R P^R + n^U P^U, \qquad (1)$$

where $P^R = N_P^R/N^R$ denotes the proportion of the rural population that is in poverty and $P^U = N_P^U/N^U$ the corresponding incidence of poverty in urban areas.

Now, differentiating (1) totally and dividing by P, we obtain a key relationship,

$$dP/P = d\ln P = s^R d\ln P^R + s^U d\ln P^U + (P^R - P^U)(N^R/N_P)\, d\ln n^R, \qquad (2)$$

where $s^R = N_P^R/N_P = n^R P^R/P$ denotes the share of the rural poor in the total number of poor people and s^U denotes the corresponding share for urban areas. Obviously, $s^R + s^U = 1$.

From (2), the proportional change in aggregate poverty incidence may be decomposed into three parts, corresponding in turn to each of the right hand terms of (2): (i) the change in rural poverty incidence, weighted by the share of rural areas in total poverty, (ii) the change in urban poverty incidence weighted by he share of urban areas in total poverty, and (iii) the movement of population from rural to urban areas weighted by the difference in poverty incidence between these two areas.

The last of these terms is described by Anand and Kanbur (1985) and by Ravallion and Datt (1996) as the "Kuznets effect". As the population moves from rural to urban areas, a change in aggregate poverty incidence will occur even at constant levels of rural and urban poverty incidence, provided that the levels of poverty incidence in these two sectors are different. In growing economies, we expect to find that the rural population share is falling ($d \ln n^R < 0$) and that the incidence of poverty in rural areas typically exceeds that in urban areas ($(P^R - P^U) > 0$). Finally, since $(N^R/N_P = n^R/P) > 0$, the expected sign of $(P^R - P^U)(N^R/N_P)\, d \ln n^R$ is negative. How important the Kuznets effect is as a determinant of overall poverty reduction is, of course, an empirical matter.

Poverty and Aggregate Growth

We now turn to the manner in which poverty incidence is affected by economic growth and, for simplicity, we hypothesize that the proportion of households in poverty, P, depends on the aggregate level of real income, Y, and the size of the population, N. Thus

$$P = \varphi(Y, N). \tag{3}$$

Totally differentiating this equation,

$$dP = \varphi_Y dY + \varphi_N dN. \tag{4}$$

We now use lower case Roman letters to represent the proportional changes of variables represented in levels by upper case Roman letters. Thus $p = dP/P = d \ln P$, $y = dY/Y$ and $n = dN/N$ denote the proportional change in total poverty incidence, the growth rate of aggregate real income and the growth rate of population, respectively. The proportional change in total poverty incidence is now given by

$$p = (\varphi_Y Y/P)y + (\varphi_N N/P)n. \tag{5}$$

In the special case where per capita income is the determinant of aggregate poverty incidence, (5) reduces to an expression in $y-n$. We shall not be imposing this assumption, but shall instead estimate relationships of the kind

$$p = a^1 + b^1 y + c^1 n, \tag{6}$$

and test whether the coefficient b^1 is significantly greater than zero. We shall also test whether $b^1 = -c^1$, that is, whether the growth of per capita income is the determinant of the change in poverty incidence, or whether population growth affects the reduction in poverty incidence in some other way.

We wish to study the way economic growth affects each of the components of the change in aggregate poverty incidence, as given by (2), which in our notation is now

$$p = s^R p^R + s^U p^U + (P^R - P^U)(N^R/N_P)\, d \ln n^R. \tag{7}$$

Ravallion and Datt apply an ingenious method for estimating decomposed equations systems of this kind. We have a four equation system, consisting of (6) and:

$$s^R p^R = a^2 + b^2 y + c^2 n, \tag{8}$$

$$s^U p^U = a^3 + b^3 y + c^3 n \tag{9}$$

and

$$(P^R - P^U)(N^R/N_P) \ln n^R = a^4 + b^4 y + c^4 n. \tag{10}$$

From the identity given by (2), these equations are linearly dependent. Equation (6) is identically the sum of equations (8), (9) and (10). Of these four equations, only three need to be estimated. The parameters of the fourth can be computed from (7). We shall estimate equations (6), (8) and (9) and infer the parameters of (10) from the identities $a^4 = a^1 - a^2 - a^3$, $b^4 = b^1 - b^2 - b^3$ and $c^4 = c^1 - c^2 - c^3$.

4. STATISTICAL METHOD AND RESULTS

The limited availability of data which may support statistical analysis has been an impediment to the systematic study of poverty incidence. Some recent studies have attempted to explore the relationships involved by analyzing cross sectional data sets across countries, or across regions or households for individual countries, while others have attempted to assemble long term time series data sets on poverty incidence for individual countries. This paper begins with the time series approach. The problem is that the number of data points for any one of the four countries is insufficient to support formal statistical analysis. We therefore pool the data for all four countries. Since the meaning of the poverty lines is different in each of the countries and also since the structure of the economies is different, we should not assume that the same relationship between poverty incidence and aggregate growth will necessarily exist in all these countries. We therefore experiment with intercept and slope dummy variables to determine the relationship most consistent with the data.

Data are assembled for the dependent variables $d \ln P$, $s^R d \ln P^R$ and $s^U d \ln P^U$. Each interval between the data points indicated in Table 1 is used to construct the values of these dependent variables, with the calculated value divided by the number of years corresponding to that time interval, giving an annual rate of change for the variable concerned. These annualized rates of change then become the variables used in the regression analysis.

Equations (6), (8) and (9) were estimated and the results are summarized in Tables 1 and 2. The regressions were run with and without the population variable (Tables 1 and 2, respectively). The restriction that population enters through per capita growth was not rejected. Nevertheless, the equations in which the population variable was deleted performed slightly better and these equations (Table 2) were used for the projections phase of the study. For this purpose, the equations with country intercept dummy variables (results shown in the left panel of Table 2) rather than slope dummies (results shown in the right panel) were selected, although their statistical performance is similar. Surprisingly, neither country slope nor intercept dummies were statistically significant. The set of equations with intercept dummies was selected because we are interested in using the estimated equations to project poverty incidence in the neighborhood corresponding to zero and negative growth rates, which are at the lower end and even outside the range of growth rates observed in the data used for estimation, and therefore close to the intercepts.

Estimates of GDP growth rates for 1997 to 1999 were then used to project the estimated equation for these years, using the estimated coefficients, including the country-specific intercept dummy variables shown in the left panel of Table 2. These estimates are summarized in Tables 2a to 2d. The poverty projections for 1997 are based on actual GDP data for that year, derived from official sources. Those for 1998 are based on preliminary estimates of 1998 GDP growth rates, also from official sources. The projections of aggregate poverty incidence corresponding to these years are shown by the large squares in Figs 1a to 1d. The growth projections for 1999 shown in Tables 2a to 2d are hypothetical. For each country except Indonesia the range of hypothetical estimates which seems most probable is from zero to –5%. Because Indonesia's crisis is so much deeper, a wider range seems appropriate and values of zero to-10 are considered.

Table 1.

(a) Thailand: Poverty Incidence, 1962 to 1996 (per cent)

	Aggregate poverty (P)	Rural poverty (P^R)	Urban poverty (P^U)
1962	57	61	38
1969	40.73	44	26
1975	31.41	36.2	12.5
1981	22.98	27.3	7.5
1986	29.02	35.8	5.9
1988	21.05	25.5	6.1
1990	16.97	20.5	5.3
1992	12.38	15.5	2.4
1994	8.8	11	1.9
1996	6.15	7.73	1.4

Source: National statistical data from government sources.
Note: Aggregate poverty is the percentage of the total population whose incomes fall below a poverty line held constant over time in real terms; rural poverty is the percentage of the rural population whose incomes fall below a poverty line held constant over time in real terms, and so forth.

(b) Indonesia: Poverty Incidence, 1970 to 1996 (per cent)

	Aggregate poverty (P)	Rural poverty (P^R)	Urban poverty (P^U)
1970	57.2	58.5	50.7
1976	50.18	54.5	31.5
1978	48.5	54	25.7
1980	39.2	44.6	19.7
1984	33	39.4	12.8
1987	21.6	26.8	7.3
1990	19.3	23.3	10.6
1993	17.3	21.3	9
1996	13.49	18.98	6.47

Sources and notes: As in Table 1a.

(c) Malaysia: Poverty Incidence, 1970 to 1995 (per cent)

	Aggregate poverty (P)	Rural poverty (P^R)	Urban poverty (P^U)
1970	49.3	58.6	25.5
1976	39.6	47.8	17.9
1984	18.4	24.7	8.2
1990	17.1	21.08	7.5
1993	13.5	18.6	5.3
1995	9.6	16.1	4.1

Sources and notes: As in Table 1a.

Table 1. Continued.

(d) The Philippines: Poverty Incidence, 1961 to 1997 (per cent)

	Aggregate poverty (P)	Rural poverty (P^R)	Urban poverty (P^U)
1961	75.02	80.19	65
1965	67.08	71.15	57.43
1971	61.63	66.08	51.32
1985	59.65	63.3	51.18
1988	54.16	61	43.01
1991	55.77	64.5	47.08
1994	49.06	56.74	42.28
1997	42.1	43.15	32.6

Sources and notes: As in Table 1a.

The striking feature of the results is that the projected increases in absolute poverty are small relative to the reductions in poverty incidence that occurred over the past two decades. According to these results, the crisis is a set back for poverty reduction in Southeast Asia, but it is far from the social catastrophe that may observers have predicted.

5. CONCLUSIONS

Analysts of the economic crisis in Southeast Asia have rightly focused upon its implications for poverty incidence. Our estimates in this paper indicate that poverty incidence will indeed rise but that the numbers of people falling below the poverty line as a result of the crisis may be less that is often suggested. For example, for Thailand our estimates suggest that by the end of 1998 the incidence of poverty will have increased by around 1.5% of the total population compared with its value in 1996. This corresponds to about one million people whose real incomes fell below the official poverty line. This increase in poverty incidence is serious and warrants an urgent policy response, as it is indeed receiving. But it is considerably smaller than many previous estimates.

Our estimate of the increase in poverty incidence for Indonesia over the same period is larger, at around 3.3% of the population. This estimate corresponds to around 6.5 million Indonesian people whose real incomes fell below the poverty line during this period. An increase in poverty incidence of this magnitude is extremely serious and carries with it grave social dangers, but it is much less that the 80 million Indonesian people falling into poverty that was claimed by some earlier reports.

Table 2.

(a) Thailand: Actual and Projected Poverty Incidence, 1995 to 1999

	1996	1997	1998	1999(1)	1999(2)	1999(3)
Growth rate of real GDP	5.5	–0.4	*–7.5*	*0*	*–2.5*	*–5*
Aggregate poverty	6.15	*6.50*	*7.58*	*7.97*	*8.26*	*8.56*
Rural poverty	7.73	*7.97*	*8.85*	*9.09*	*9.34*	*9.59*
Urban poverty	1.4	*1.40*	*1.40*	*1.40*	*1.40*	*1.40*
Rural population share	75.0	*77.6*	*83.0*	*85.4*	*86.5*	*87.4*

Note: Actual data shown in regular type; projections in italics.
Units: Growth measures, per cent annual growth rate of real GDP.
Poverty measures, headcount measure of poverty incidence as per cent of population.
Rural population share: rural population as share of total population.

(b) Indonesia: Actual and Projected Poverty Incidence, 1995 to 1999

	1996	1997	1998	1999(1)	1999(2)	1999(3)
Growth rate of real GDP	8	4.7	–15	*0*	*–5*	*–10*
Aggregate poverty	13.49	*13.13*	*16.79*	*17.56*	*18.87*	*20.17*
Rural poverty	19	*18.80*	*22.08*	*22.81*	*23.85*	*24.89*
Urban poverty	6.5	*6.49*	*6.51*	*6.50*	*6.51*	*6.52*
Rural population share	55.9	*53.9*	*66.0*	*67.8*	*71.2*	*74.3*

Notes and units: As in Table 2a.

(c) Malaysia: Actual and Projected Poverty Incidence, 1995 to 1999

	1995	1996	1997	1998	1999(1)	1999(2)	1999(3)
Growth rate of real GDP	9.40	8.60	8.00	–5.90	*0.00*	*–2.50*	*–5.00*
Aggregate poverty	9.60	*8.82*	*8.19*	*9.36*	*9.85*	*10.22*	*10.58*
Rural poverty	16.10	*15.54*	*15.08*	*16.56*	*17.28*	*17.66*	*18.04*
Urban poverty	4.10	*4.09*	*4.08*	*4.08*	*4.08*	*4.08*	*4.08*
Rural population share	45.8	*41.3*	*37.4*	*42.4*	*43.8*	*45.2*	*46.6*

Notes and units: As in Table 2a.

(d) The Philippines: Actual and Projected Poverty Incidence, 1995 to 1999

	1995	1996	1997	1998	1999(1)	1999(2)	1999(3)
Growth rate of real GDP	4.80	5.50	5.10	1.00	*0.00*	–2.50	–5.00
Aggregate poverty	44.71	43.25	42.10	*43.66*	*45.96*	*47.65*	*49.35*
Rural poverty	46.09	44.41	43.15	*43.96*	*45.10*	*45.86*	*46.62*
Urban poverty	33.37	32.96	32.60	*32.60*	*32.62*	*32.66*	*32.70*
Rural population share	89.2	89.9	90.0	*97.3*	*106.9*	*113.6*	*119.6*

Notes and units: As in Table 2a.

NOTE

1. Paper presented to the American Committee on East Asian Economic Studies – Chulalongkorn University Conference Bangkok, 16–18 December 1998. The able research assistance of Xinpeng Xu and Haiying Zhang is gratefully acknowledged.

REFERENCES

Anand, S., & Kanbur, S. M. R. (1985). Poverty Under the Kuznets Process. *Economic Journal, 95* (March), 42–50.

Fields, G. S. (1980). *Poverty, Inequality and Development.* Cambridge, U.K.: Cambridge University Press.

Krongkaew, M. (1993). Income Distribution and Poverty. In: P. G. Warr (Ed.), *The Thai Economy in Transition.* Cambridge: Cambridge University Press.

Lewis, W. A. (1954). Economic Development with Unlimited Supplies of Labour. *Manchester School, 2,* 139–191.

Moll, T. (1996). Mickey Mouse Numbers and Inequality Research in Developing Countries. *Journal of Development Studies, 28* (July), 689–704.

Ravallion, M., & Datt, G. (1996). How Important to India's Poor Is the Sectoral Composition of Economic Growth? *World Bank Economic Review, 10,* 1–25.

Sen, A. (1976). Poverty: An Ordinal Approach to Measurement. *Econometrica, 44*(2), 219–231.

Takayama, N. Poverty, Income Inequality and Their Measure: Professor Sen's Axiomatic Approach Reconsidered. *Econometrica, 47*(3), 747–759.

Warr, P. G. (1993). The Thai economy. In: P. G. Warr (Ed.), *The Thai Economy in Transition* (pp. 1–80). Cambridge: Cambridge University Press.

Warr, P. G., & Nidhiprabha, B. (1996). *Thailand's Macroeconomic Miracle: Stable Adjustment and Sustained Growth.* Washington, D.C. and Kuala Lumpur: World Bank and Oxford University Press.

World Bank (1993). *East Asia's Economic Miracle.* New York: Oxford University Press.

World Bank (1996). *Poverty Reduction and the World Bank: Progress and Challenges in the 1990s.* Washington, D.C.: World Bank.

9. MACROECONOMIC FUNDAMENTALS OF POSTWAR ECONOMIC GROWTH IN JAPAN – A GREAT SUCCESS AND RECENT FRUSTRATION LESSONS IN ASIAN ECONOMIES[1]

Hiromitsu Ishi

1. INTRODUCTION – THE SIGNIFICANCE OF JAPAN'S MODE

The most remarkable feature of Japan's postwar economic performance was the rapid surge of economic growth from a state of devastation and defeat in World War II, although Japan is now facing the worst depression during the postwar era. Particular attention should be paid to the I950s and 1960s. This period is often referred to as an "economic miracle" and continues to attract worldwide attention. In the process of rapid growth, Japan succeeded in catching up with the Western powers by the late 1960s, reaching the forefront of major industrialized countries in the 1980s. Since such a remarkable record of economic growth is unparalleled in history, many relevant lessons may be gleaned by other countries from "Japan's model". However, Japan's model inevitably had to incur recent frustration in prolonged recession. Thus, we

Restructuring Asian Economics for the New Millennium, Volume 9A, pages 177–195.
2001 by Elsevier Science B.V.
ISBN: 0-7623-0709-9

should note both the good and bad sides of Japan's model. With this in mind, the following potential application of the Japanese experience is stressed.

The Japanese experience during the earlier postwar period must be of great heuristic value and interest to East Asian countries, because their economies have successfully taken off from an underdeveloped stage to achieve annual growth rates of over 10%. Similar to Japan's earlier postwar years, many have dubbed this recent phenomenon in the East Asian region "an Asian miracle". This has been true at least until recent years: "an Asian crisis" is now prevalent in this region. Interestingly enough, Japan had achieved economic growth front the mid-1950s to 1973 at a rate comparable to that of the East Asian economics up to the outbreak of recent economic difficulties. What is more, during the period under consideration, the Japanese economy encountered the same problems that some East Asian economics experience nowadays, such us overheating, a shortage of foreign reserves, etc., and the government had to tackle the difficult task of engineering a "soft landing" (for a general discussion, see Ito, 1996).

Thus, the growth process of Japan in the immediate two decades after the end of World War II seems to contain special policy implications and is an interesting case study as Japan's model. It is often pointed out that the government has played a vital role in promoting rapid growth by intervening in the market-based activity of postwar Japan. Japan's institutions might have been unique as part of the growth-oriented policy, and they were certainly instrumental in bringing about an exceptional pattern of economic growth. However, looking over the recent trends of rapid growth in other East Asian economics, the earlier Japanese experience may no longer be considered exceptional. Nonetheless, in view of investigating the future trend of East Asia, Japan's model would be worth reconsidering as a fresh case study. As such, it a necessary to clarify the function of government intervention in the growth process that Japan has performed well in the past.

2. ECONOMIC RECONSTRUCTION AND MACROECONOMIC FUNDAMENTALS AFTER WORLD WAR II

2.1. Occupation and Economic Restoration

Japan was utterly defeated in World War II, and the economy was completely destroyed beyond any hope of recovery. As a result of American bombing around the end of the war, its industries were at a standstill, and its infrastructure was virtually demolished. In August 1945, when World War II came to an end, no one could ever imagine that Japan mould become the second largest industrial power in the world by the late 1960s.

However, Japan desperately arose from its hopelessly depressed stale of the immediate postwar period and eventually reached the top group of industrial nations in terms of per capita gross national product (GNP) within less than five decades. The term "the rise of the phoenix from the ashes" is most appropriate to express such a process of resurgence from the ashes of war and swift economic recovery with the high-speed growth that Japan has successfully achieved (see Patrick, 1970; Tsuru, 1993). As a result of defeat in the war, Japan was occupied militarily by the victorious Allies. Under the Occupation, General MacArthur, Supreme Commander of the Allied Powers (SCAP),[2] was responsible for decision-making of any consequence as well as implementation of occupation policy.

Rising up from the ashes, Japan regained economic power within a relatively short period, Fig. 1 depicts the trends of real GDP at 1934–1936 prices for 1930–1960. Obviously, a large gap can be observed between the prewar and postwar periods as the GDP fell sharply from 20.634 million in 1944 to 11,594 million in 1946, a decrease of 44%. It was not until 1954 that the 1939 level of real GNP, the highest of the prewar period, was restored.

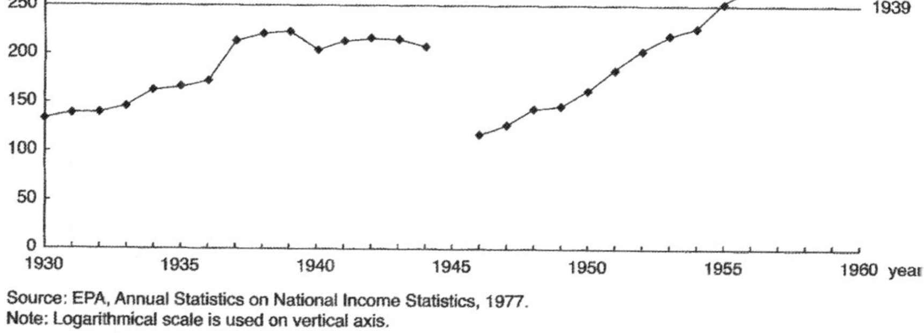

Source: EPA, Annual Statistics on National Income Statistics, 1977.
Note: Logarithmical scale is used on vertical axis.

Fig. 1. Trend of Real GDP at 1934–1936 Prices Compared with the Prewar and Postwar Period.

As a result of strenuous efforts attempted through successive reforms by SCAP and the Japanese government, economic recovery progressed steadily. In fact, in growth terms, economic recovery was swift; i.e. a real GNP growth rate of nearly 10% had been achieved between 1947 and 1955. In 1949, the growth rate stagnated at only 2.2%, reflecting the deflationary effects of the SCAP anti-inflation policy, but the depressed state of the economy was soon stifled. One chief reason behind this was the outbreak of the Korean War in June 1950. In fact, an unexpected boom to Japan's economy was provided by the special procurement needs of the Korean War, which lasted for roughly three years. Thus, real growth rates turned upward from 1950. Recovery was progressing at a substantial speed, but the economy began from such a low initial level in 1945–1946 that even rapid rates of growth were not sufficient to reconstruct the Japanese economy promptly. Generally speaking, Japan was unable to regain its prewar level (1934–1936)[3] of per capita income and production until 1953 or 1954. Thus, Japan lost roughly 18 years of economic growth in the reckless adventure of World War II, as it would take that long to return to prewar levels of economic performance.

2.2. Unprecedented Growth Patterns: 1955–1973

Contrary to the gloomy prospects immediately after the end of World War II, the Japanese economy started its genuine economic growth from 1955, based upon a number of major socio-economic reforms, such as the Zaibatsu dissolution, antimonopoly legislation, land reform, and a renewed labor union movement, during the occupation period. In fact, it continued its rapid surge of economic growth at the highest rate among many advanced countries up to the early 1970s, when the first oil shock occurred. This high growth pattern over along period of almost 20 years has drawn the attention of people around the world. They were very keen to learn the process and causes of this "economic miracle" story.

We can ascertain the pattern of Japan's growth performance compared with those of G7 countries in Table 1. Obviously, the Japanese economy continued its upsurge of growth until the early 1970s at a rate far greater than that of any other major country. Of most importance was the accelerated tempo of real growth rates in both the second half of the 1950s and the 1960s. Indeed, the average growth rates of real GNP in Japan for 1956–1973 achieved 10.2%. followed by 5.5% in France, whereas the United States and the United Kingdom grew only at rates of 3.8 and 3.2%, respectively. Almost half of the 18 years recorded two-digit rates of real growth in Japan, and Japan's real GNP

Table 1. Growth Patterns of Real GNP Among G7 Countries: 1956–1973.

Year	1956–1960	1961–1965	1966–1970	193.871–1973	1956–1973
Japan	9.8	10.1	12.1	8.6	10.2
U.S.	2.2	4.9	3.2	5.0	3.8
U.K.	2.6	3.3	2.4	4.4	3.2
W. Germany	6.4	5.0	4.8	3.9	5.0
France	4.8	5.9	5.8	5.6	5.5
Italy	5.5	5.3	5.9	3.8	5.1
Canada	3.3	6.0	4.8	6.1	5.1

in 1972 was 5.1 times larger than it was in 1955. This growth pattern was really unprecedented.

What are the main factors that explain the remarkable achievement of this rapid economic growth? It would be necessary to analyze them from both sides of economic supply and demand. Let us begin with the discussion of the supply-side factors of the economy. Several factors are worth enumerating, but it is difficult to reach an all-inclusive explanation of Japan's postwar economic growth. In what follows, the focus is on three interrelated factors: labor, capital, and technology (see Patrick, 1970; Patrick & Rosovsky, 1976; T. Nakamura, 1981, ch. 3).

First, one key factor to rapid growth was no doubt an ample supply of good-quality labor forces (for general discussion, see Galenson & Odaka, 1976). In the earlier postwar period, Japan was a country endowed with a relatively abundant labor supply, reflecting the great number of repatriations from overseas territories and the rapid growth of population due to a "baby boom". As a result, an enormous stock of labor had built up in agricultural and other low-productivity sectors (e.g. small enterprises, retailing, and other services). It was easy to shift such an abundant labor force to large-scale manufacturing industries like the heavy-chemical industries, which played a major role in the ensuing rapid growth. This favorable condition continued into the 1960s. In addition to an ample labor supply, Japan's labor force was well trained, industrious, and hard working.

Second, the high rate of capital formation and savings was certainly another key factor to promote rapid growth. Without a high rate of new investment in plants and equipment, innovation could not have been prevalent and the growth of labor productivity not feasible. As a result, it would have been impossible to absorb excess labor forces into higher productivity sectors. Throughout the 1950s and the 1960s, Japanese enterprises in the private sector were very eager to invest to expand productive capacity and output. Although investment usually had to pay off in a relatively short period, business and managerial optimism about the growth of the Japanese economy were widespread with a strong perception of profit opportunities in the future. Most Japanese businessmen looked ahead with a bullish sentiment that contrasted with the prudent attitudes of their foreign counterparts. Investment must be financed by savings, either domestic or abroad. Fortunately, no reliance was placed on the borrowed savings of foreigners in Japan. In other words, the extraordinarily high rate of investment was well matched by that of domestic savings.

Third, technology was equally important for achieving rapid economic growth. As a result of incorporating new and innovative technology, Japan's high investment rate and labor supply became more productive. It is widely pointed out that the postwar economic growth in Japan greatly depended on the importation of foreign technology; i.e. "borrowed technology". In the 1950s and the 1960s, a wide gap remained between Japan's level of technology and the highest level of Western technology. It was greatly beneficial to fill such a gap, because Japan obtained the best new ideas and inventions at a relatively low cost and used its own skills to embody them swiftly in attractive new products. This may be called the strategy of a No. 2. Under such a strategy, the process of catching up with the advanced technology greatly benefited Japan in its achievement of postwar economic growth. Many of the major export goods, like automobiles, cameras, TV sets, etc., were all produced through a No. 2 strategy.

Parallel with the expansion of productive capability from the supply side of the economy, the demand-side effects played an equally important role in postwar economic performance by the early 1970s. Japanese economic growth since World War II had been exceedingly rapid. interrupted only by a series of short-term recessions, as will be argued later. Strong aggregate demand constantly prevented the actual level of GNP from dropping by a considerable extent. No doubt one of the most marked phenomena about rapid growth was the strength of aggregate demand. The following statement is worth noting: "Western economists who study the postwar Japanese economy (especially, perhaps, those from the United States or Britain) cannot fail to be struck by what they may describe as the great secular vigor or buoyancy of aggregate demand in Japan, seemingly very unlike the usual situation of their own economies" (see Ackley & Ishi, 1976, pp. 162–163).

With no special stimulus, aggregate demand in Japan seemed to be adequate enough to maintain reasonably full realization of potential output. Given the strength of aggregate demand, the government deliberately took restrictive actions to prevent the economy from overheating, resulting in policy-induced recessions, as will be explained later. This situation was quite different from that in mature economies like the United States and the United Kingdom over the same period. The aggregate demands in both these countries were met with full realization of potential output only during an occasional period of boom, which usually could not be sustained. Recessions were frequent and in the absence of continuing expansionary policies, the GNP remained stagnant. Thus, policy stimuli were always needed to sustain an adequate level of aggregate demand to promote economic growth.

Of course, aggregate demand was sometimes either excessive or deficient in Japan as well as in the West. Unwanted gaps between actual and potential output in Japan had probably been smaller and less persistent than in other major countries, reflecting the consistent buoyancy of aggregate demand. As a consequence, little or no active stimulus by the government was required during the period under consideration.

2.3. Growth-Oriented Policy and the Role of Government

Several factors previously mentioned were important in Japan's unique growth performance. However, greater importance should be placed on the policy background behind rapid growth and the factors that contributed to this surge of growth.

Because the economy shifted into full recovery in the early 1950s, the Japanese government had placed domestic economic growth at the top of its macroeconomic policy priorities. In Japan the direction of nation's macroeconomic policies is not always stated on a formal basis, but economic planning usually tended to convey a verbalization of policy goals to the nation along with their quantification. While the plan was merely indicative and had no compulsory nature. it was very effective in providing information and obtaining a national consensus (see Komiya, 1975; Komine, 1993). In fact, economic planning for the duration of multiyear periods (e.g. 5 or 7) involved fully articulated goals that were of great help in providing a good guideline for the government's position for the purpose of private economic performance.

The first multiyear plan drawn up by the government appeared in 1955, titled the Five-Year Plan for Economic Independence *(Keizai Jiritsu Gokanen Keikaku).* The goals of Japan's postwar economic policy were vividly pronounced in that plan by the use of three key words: growth, investments, and exports. The main idea of the total interlinkage of these policy goals was summarized as follows. A rapid growth of output and employment in the modern industrial sectors of the economy was regarded as absolutely necessary to absorb the large numbers of underemployed workers in traditional low-productivity sectors like agriculture and small enterprise. Obviously, a rapid expansion of modern capital-intensive sectors required massive investment and in turn a high rate of national savings. Furthermore, the increased production of modern sectors essentially needed rapidly expanding markets, a substantial part of which had to be consumed overseas. Thus, a rapid growth in exports was an equally important goal, partly because the expanding output of modern industries could not be fully absorbed into the domestic markets and partly because huge amounts of imports of raw materials required the means to pay by exports.

It is very important to stress that the growth-investment-export triad policy goals were well understood among various individuals and sectors in Japan. Indeed, such growth-oriented policies were advocated and popularized to the general public in Japan by politicians, businessmen. and the press.[4] The goals of government policies were widely accepted, partly because they were designed to use the incentives of private self-interest and partly because they were identified and observed by the public and viewed as important to the national interest. No doubt business was the front runner in promoting these policy goals among several groups. Major industries successfully improved their productivity, increasing the quality and diversity of their manufactured products while reducing costs. As a consequence, their competitiveness was greatly strengthened in both the domestic and overseas markets. In particular, big business embraced an active management style oriented toward growth performance and closely tied to government policies. Given the strength of aggregate demand, technological innovations combined with buoyant investment activity, and good labor supply, companies greatly improved their productivity and competitiveness under an aggressive growth-oriented policy initiated through close cooperation between government and big business.

Under the basic objective of growth-oriented policy, the main role was persistently played by private business, not government. The government merely supported the vigorous activity of the private sector behind the screen of growth performance. Indeed, the government pursued a passive, balanced budget policy in which the MOF (Ministry of Finance) consistently managed the government fiscal activity from a standpoint of budget orthodoxy before fiscal 1965. The balanced budget had been the dominant characteristic of fiscal polity in postwar Japan during the period of 1950–1965 at the national government level. The basis for this lies in the traditional view of "sound" finance, that is, all current expenditures must be financed by current revenues in the government sector. In addition, monetary policy tended to keep interest rates relatively lower to stimulate economic activities in the private sector.

To summarize, fiscal and monetary policies had never played an aggressive role in promoting rapid economic growth, and they remained relatively neutral or passive in the growth process of the 1950s and the 1960s. We must pay more attention to the active use of other policy instruments, such as industrial policy or administrative guidance, based upon government intervention into market activities.

3. BETWEEN THE GOVERNMENT AND THE MARKET

3.1. Government Intervention in the Markets

Apart from the macroeconomic effects of government on the Japanese economy, it is equally important to explore the role of the government's growth-oriented policy from a microeconomic standpoint. The government presented big business from competing with each other in the market to a great extent. Generally speaking, the Japanese seem to have a unique view concerning the acceptable degree of competition in the market that is deemed to be unusual by foreigners. For example, in Japan more emphasis tends to be placed on the costs of a competitive system rather than the benefits of free competition.

Historically, most Japanese have not accepted the working of the markets invisible hand but have approved of the "visible hand" of government guidance and business cooperation (i.e. cartels) to attain satisfactory economic results. Thus, the Japanese government and leaders of big business were frequently concerned with "excess competition". which they believed should be regulated to some extent by the government. By contrast, in Western countries, particularly in the United States, it is widely believed that the role of government should ideally be restricted to a minimum level, with government acting only as referee, making the rules of the game for the preservation of vigorous competition in the market economy. In this respect, there was a marked difference between the United States and Japan.

In accordance with SCAP efforts for deconcentration, Japanese industry was certainly deconcentrated in the mid 1950s, more so than before the war. However, with the passage of time a gradual trend toward concentration of production began to emerge once again in many industries. By the late 1950s the decentralization trend reversed, and the top three or four firms in major industries such as automobile, glass, steel, and home appliance industries occupied an increasing share of total output in their relevant industries. For example, in 1963 the share of output concentration among the top three firms was 100% in both the film and glass industries, 97.1% in the beer industry,

62.2% in the automobile industry, 48.1% in the steel industry, etc. (see Matsashiro, 1970). Beginning in the mid-1960s, the tendency toward concentration accelerated due to the increasing number of mergers among large firms in major industries. For example, a giant merger emerged from the two top steel companies. Yawata and Fuji, creating Nippon Steel in 1969, and also between the two big city banks, Daiichi and Kangin, in 1971, creating the new top-ranked bank, Daiichi-Kangin.

Apparently, the past trend toward restriction of competition resulted from the process of growth-oriented policy, as noted above. What was important was the fact that the government had taken the initiatives to combine large firms and restrict competition in the markets. It is widely pointed out that the antimonopoly act has gradually been altered under the control of government, turning away from the original nature when it was established by SCAP. The predominant view is that removal of excess competition by government intervention contributed significantly to the success of the postwar growth process. However, on the basis of empirical evidence, no decisive judgment could be made to evaluate postwar policy performance in its ability to control competitive markets. On the reverse side of the coin, we can say that the restriction of competition to some extent destroyed the climate in which new, small, and independent businesses were able to start, grow, prosper, and become essential to a system of a capitalistic economy.[5]

3.2. The Role of MITI Policy

During the 1950s and the 1960s the Japanese government SCI industrialization and export promotion as the primary target of the economy. As is well known, the MITI (Ministry of International Trade and Industry) played a vital role in achieving both industrial and trade targets under the slogan of catching up with the Western powers (see Okimoto, 1992). The MITI, a highly activist bureaucracy, took initiatives in devising a new industrial policy and intended to reshape the fundamental structure of Japan's industry domestically and internationally in the earlier postwar period. During the period in which growth performance was extremely strong, the MITI did perform a great deal of guidance, assistance. and intervention in the private sector. A number of policies were adopted by the MITI: export promotion, import protection, controls on inward foreign investment and purchases of foreign technology, preferential tax incentives, and financial loans to selective industries. In addition, the MITI often employed administrative leadership to adjust excess investment and production in key industries. The role of MITI in allocation of resources among Japanese industries is generally identified will "industrial policy"; whose connotation is unique to Japan (for a more general discussion, see Suzumura, 1996). Industrial policy is regarded as a systematic selection of industries that the government should encourage or discourage (implicitly) by positive action or deliberate inaction in view of national interests. Since the main objective of the nation's policy was economic growth during the period under consideration, the selection criterion to assist particular industries was the relative productive efficiency. Japan's growing industries (defined above) were naturally concentrated in the basic industries in the early postwar period: coal, fertilizer, steel, electric power, and shipbuilding. Later, other industries, such as petrochemicals, petroleum refining, and automobiles, were added to the foregoing categories, but among the new growth industries not singled out by the MITI were radio, television, computers, motorbikes, and optical goods. The selected industries grew rapidly, assisted by MITI's industrial policy, leading economic expansion to spread across the industrial spectrum.[6]

No doubt industrial policy was closely tied to the intricate regulations on foreign trade, which in turn gave significant power to the MITI. Balance-of-payment deficits posed a serious problem until the end of the 1960s, and a top priority was placed on control of foreign trade and investment (for a more expanded discussion, see Krause and Sekiguchi, 1976, pp. 411–417). Because the MITI had to approve any foreign trade transaction on a case-by-case basis, it became closely involved in business decisions from the beginning, resulting in a very powerful ministry.

Import protection and related measures were the most important instruments among the industrial policy techniques available to the MITI. In fact, from the beginning of the occupation period, the government introduced a set of stringent controls on foreign trade and exchange transactions. Industrial materials and other imports were placed strictly under a quota system, the purchases of investment goods from abroad required licenses; overseas investment, both inward and outward, was restricted; and contracts for foreign technology were screened by government officials. This system remained almost untouched during the 1950s, without any serious modification, and many of its key elements were kept until the late 1960s. Essentially, the import rationing assigned by the MITI affected Japan's industrial structure for these years. The highest priority was assigned to imports of raw materials and advanced industrial equipment, particularly in the heavy and chemical industries, whereas consumer manufacturing was virtually prohibited. Special attention was paid to all technological imports, and a case-by-case inspection was given by relevant industries until the mid-1960s. Because the MITI acted in consultation with business interest groups in selecting the kinds of technology imports that would be desirable, an effective interrelation emerged from this process, and the MITI was able to adjust the terms of technology contracts on condition of approval.

In Japan, import quota allocation was frequently rationalized to state that domestic industries ought to be completely protected from import competition in view of national interests. Thus, infant industries were assiduously protected well into adulthood under the MITI responsibility, and this protective policy was applied to all growing industries that were less productive and cost-effective than foreign competitors. After they had gained total competitiveness, the MITI decided to open the door to foreign competitors in Japanese markets. This is the well-known story of "notorious MITI" that was frequently criticized by foreigners. By the end of the 1950s, when the economy had reached a higher growth path, import liberalization was emphasized as an important policy target in the international community. Thus, the Japanese government had to make a start in this direction.

Likewise, MITI's industrial policy targeted the promotion of exports, because the MITI was endowed with the power to select which industries were suitable for exporting or left to perish. Most of the growing industries were expected to become exporters on a large scale in order to gain foreign reserves. For this purpose, the MITI provided export credit and quarantines, and special tax measures for export promotion as part of its industrial policy (see Ishi, 1994). In addition, the MITI strictly restricted foreign direct investment into the Japanese domestic markets for a long time. From the standpoint of promoting industrial growth, such restrictions should have been fewer, but more importance was placed on the poor competitive position of Japans enterprises against the potential monopoly of gram foreign firms. In particular, there was a great deal of political need to protect small bushes from strong foreign competitors in domestic markets.

3.3. Administrative Guidance and the "Japan, Incorporated" Thesis

Related to the role of MITI described above, we must pay particular attention to the significance of administrative guidance in the postwar growth process. The term "administrative guidance" refers not to a policy but rather a method that the Japanese government has widely used to support or strengthen a number of macroeconomic and microeconomic policies. Although there is no accurate definition, administrative guidance is generally considered to encompass influence peddling, persuasion, and suggestion to induce finessed individuals to follow the government's wishes in particular ways. Of course these guidelines are given by government officials to the private sector, backed by substantial power, implicit, or explicit, arising from government prerogative to provide subsidies and tax concessions make favorable loans, give permission to import, acquire foreign exchange, and approve cartel arrangements. Accordingly, private business and individual, must, more or less, accept such guidance, but it is not accurate to regard administrative guidance as compulsory directives issued exclusively by the government – It is important to stress that government leadership was traditionally accepted by private firms and individuals in guiding them toward specific policy directions. Furthermore, it was widely recognized that government official had more knowledge, information, and experience than that available to the private sector. It was thus a thoughtful strategy for business leaders to adopt the sharing of values, belief, and political preferences with government officials. It can also be defended as a method that avoids red tape, complicated legal procedures, and an atmosphere of conflict between government and business. All contributed to the success of administrative guidance.

Going back to the 1950s and the 1900s, administrative guidance was extensively used for a wide variety of aims in Japan, but it has induced increasing criticism, particularly from outsiders. For instance, it is criticized partly for the reason that there is no specific legal authority and no safeguard of due process and partly on the grounds that vital decisions tend to be made more or less confidentially. Such criticism focuses mainly on small circles of bureaucrats and business leaders (often ex-government officials) under the unsound political aspects of a system. Given the fact that the Japanese economy had achieved virtually uninterrupted expansion for 20 years after the end of the occupation in 1952, it is not surprising that many observers have reached the conclusion that Japan's political and administrative system had been uniquely designed to accomplish the goal of rapid economic growth and that governments had special capabilities for economic decision-making and management. In relation to these views, one question is easily raised: who governs government policy? What immediately come to mind is the close and harmonious interrelation between government and business in Japan. Hence, it is worthwhile to examine the adaptability of the so-called "Japan, Incorporated", a thesis widely discussed around 1970.[7]

The term Japan, Incorporated implies that growth-oriented policies have been jointly decided and achieved by a tripartite elite of politicians, appointed bureaucrats, and big business, indicating that the Japanese government and big business successively had close links and that the government has vigorously intervened in the private sector (see Abeglen, 1970, pp. 8–13). As seen in Fig. 2, such analyses focus on the linkages of three groups, the LDP (Liberal Democratic Party), the central government bureaucracy, and big business, which are linked with each other in a one-way route by their own powers.

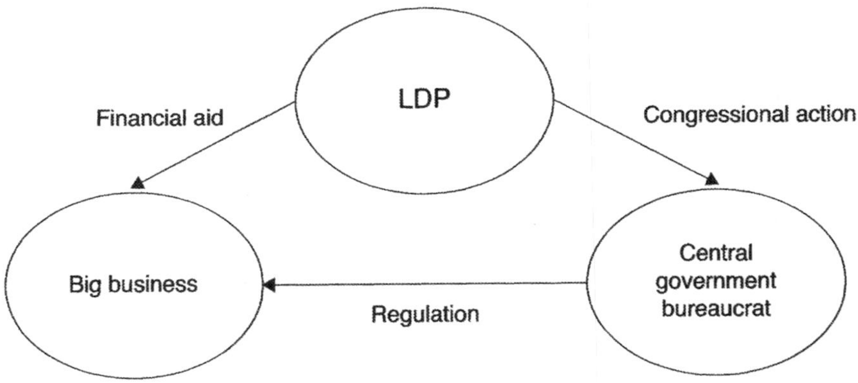

Fig. 2. The Linkage Among LDP, Bureaucrats, and Business.

When these considerations were taken together with observed rapid growth rates of the GDP, it was widely pointed out that the Japanese could used the energies and skills of private business within a centrally planned business and closely regulated private system. This combination of governing government policy played a vital role in promoting an ambitious target rapid economic growth targets very effectively. Obviously, both the political environment and government policies influenced economic events profoundly and in some sense crucially.

4. LESSONS FROM JAPAN'S MODEL

What kind of lessons can be derived from the Japanese experience, particularly for the growing Asian countries? Japan is the first country in Asia to catch up with the leading Western powers, and it seems that there are a number of positive lessons that the Japanese economy has to offer. Indeed, it is widely acknowledged that the postwar growth process of Japan in the 1950s and the 1960s can be regarded as a success story in which the government played an important role in promoting unprecedented economic growth. Several features should be stressed regarding the interaction between the good performance of the Japanese economy and the well organized function of government.

First, it was fortunate that Japan was occupied by the Americans after World War II and was forced to thoroughly complete a number of structural reforms that freed socio-economic institutions from the established system under pressure from the occupational authority of SCAP. Four major economic reforms in the package of occupation policies mentioned above greatly contributed to the establishment of a fundamental base for Japan's subsequent growth. How could these striking reforms be drastically achieved without any difficulty in terms of social disorder during the occupation period? Many Asian people may raise such a question. Interestingly enough, the occupation was considered the greatest social experiment attempted by Americans rather than a punishment for defeat in war. The Japanese were compelled to follow occupation policy and accept the result. Furthermore, because Japan was devastated and in a state of chaos, the need for tremendous sacrifices through reform, was easily accepted by all. As evidenced from land reform and Zaibatsu dissolution, the forced adjustment caused by structural reforms took the form of transferring economic power from a small minority to a large majority of Japanese who responded favorably. In the process, the previous owners of wealth and wielders of power could not sustain their vested interests, because they were under attack economically, socially, and politically. This was a very important factor, unique to Japan, and instrumental in the successful completion of this experiment. No doubt these sweeping structural reforms in the economy were of great advantage, enabling the Japanese government to take optimal growth strategies from a stabler initial condition. This is in sharp contrast to other Asian countries that are suffering from fiscal and external imbalances, uneven income distribution, incessant inflationary pressure, or external debt burdens.

Second, gradualism rather than shock therapy was predominant in the decision-making process of policy formation. The anti-inflationary measures taken by SCAP were the only exceptional case of shock therapy; other policy measures were made through a relatively gradual adjustment throughout the postwar period. For instance, tariffs were lowered step by step while import quotas were widened gradually. Foreign exchange transactions were restricted first on current and later on capital account transactions and afterward were removed one by one. While import protection was gradually lifted, the government targeted certain industries for growth in view of an import substitution strategy and provided them with many favorable subsidies or loans during the earlier postwar period. Thereafter, the government gradually phased out these favorable conditions over the years and attempted to strengthen competitiveness in the world market. Japan's successful attempt with the formation of an import substitution industry may contrast with the unsuccessful experience of many Asian countries. In addition, the government successfully encouraged export growth by taking special tax measures or making favorable loans to large-scale enterprises. Thus, emphasis should be placed on the government-led industrial policy and the gradual approach to liberalize and deregulate the markets in Japan. The speed of policy changes and economic reforms is obviously an important issue, in particular when we consider the shock-therapy disorders in the context of the continuing radical economic transformation in the former Soviet Union and Eastern Europe. The Japanese experience seems to suggest that if these countries had taken a more gradual approach to adjustment to economic reforms, the sharp economic collapse might have been avoided.

Third, macroeconomic policies performed well in encouraging rapid economic growth in collaboration with private sector activities. Fiscal policy kept the budget of the national government balanced until 1965, never generating an inflationary impetus caused by fiscal deficits. This neutral budgetary stance was important in establishing a sound policy framework that has not been seen in any other Asian country. On the other hand, monetary policy played a more positive role in accommodating macroeconomic fundamentals in the Japanese economy. In the earlier postwar period, the monetary authority was mainly adhering to a low-interest policy to encourage private investment, which was at the same time reinforced by favorable government loans. To combat an overheated situation, tight monetary policy was triggered in a timely fashion to calm down the inflationary pressures and stop the reduction in foreign reserves. This is the success story of government-led mini-recessions in the process of postwar economic growth in order that better growth performance could be sustained over along stretch of time. Overall, it is widely acknowledged that fiscal and monetary policies were well matched with the very brisk economic activities in the private sector.

Last but most important, we should note the debate over government intervention versus free market policies when we assess the past performance of Japan's growth-oriented policy taken by the government in the 1950s and the 1960s. Apart from the adoption of macroeconomic policies, the Japanese government successfully intervened in the markets not with an ordinary policy instrument but through a method of administrative guidance. Conversely, it is generally pointed out that a market-based growth strategy might not have best served a number of developing and transitional economies whose markets were not well developed in be past. Incipient markets needed to develop in a healthy manner, and the government should intervene to guide market development, correcting market failures. This is the Japanese example. In a manner similar to the Japanese model, in many East Asian countries, such as Korea, government intervention has figured prominently in the economy and succeeded in achieving better performance than those of other countries with market-based strategies. What is evident front these experiences is that more importance should be placed on government intervention at least in an earlier stage of a nascent market economy. In the process we can see a productive, not a disruptive, government intervention. The gradualist approach to government intervention employed in Japan seems to have more advantages than IMF-supported programs in which the IMF pushes for rapid adjustment such as price liberalization and privatization of government-owned enterprises on all fronts of the market economy in all countries. These adjustments pursued under the IMF programs by the European transition countries inevitably led to political changes, which took place in a great deal of socio-economic turmoil. If Japan had not grown swiftly in the 1950s and 1960s, it might have been bombarded with IMF missions that recommended adapting a more market-based strategy, which differs from an interventionist approach. Consequently, Japan's rapid growth produced in us a sort of arrogance that led us to believe that the normal rules of market economies did not apply.

Needless to say, another attention should be paid to the less-favorite side of the coin behind the success of economic growth in postwar Japan. The recent experience of the Japanese economy in the early 1990s after the collapse of the "bubble economy" offers certain negative lessons generated by pitfalls from both an inefficient public sector and regulated markets that have been steadily fostered throughout the postwar period. It is widely acknowledged that policy measures and institutions continued long after their original goal was complete. An entrenched policy stance is likely to have a life of its own once it has embarked, reflecting both a strong movement of vested interest groups and the persistent resistance of bureaucracies to reform. Typical examples are seen in the case of postal savings, government financial institutions, agricultural subsidies, public work projects, etc.

Government involvement in the economy under the triangle of ruling party, big business and central bureaucracy has recently been criticized for creating opportunities for inefficiency and rent-seeking in the policy-making process. Such involvement. which previously appeared efficient, has become extremely unproductive today and does not lead to the positive results seen in previous days of rapid growth. Once the Japanese economy achieved success in good economic performance based on a government interventionist approach, a great number of pitfalls involved in this approach come to the front all at once now. To conclude my own argument, 1 would say it perhaps now would be the time to reconsider long-term "shock therapy versus gradualism" in view of fiscal strategies for growth.

NOTES

1. This article is a short summary of my paper "The role of government in the postwar growth process of Japan," Journal of Asian Economics 10 (1999), 65–90.

ACAES – Chulalongkon University International Conference on a Macroeconomic Core of an Open Economy for Progressive Industrialization and Development in Asia in the New Millennium, 16–18, December 1998, Bangkok, Thailand.

2. SCAP usually referred both to MacArthur himself and the administrative bureaucracy of the occupation in general.

3. The average of 1934–1936 is often selected as a reference point of the prewar level because this was the last prosperous and peaceful period before the outbreak of the Sino-Japanese War in 1937.

4. Popular slogans such as "prosperity through exports and stability through savings" were often used to help the general public understand the government's position. We can also note an interesting argument of Huber by using the concept of "strategic economy" (see Huber, 1994).

5. Perhaps the truth exists in the middle of the two extremes. Even if the Japanese government continued to intervene in the markets. we must note that Sony, Honda, Matsushita, and the like appeared one after another in the 1950s from a group of small enterprises.

6. In practice, the Japan Development Bank also had a central role in the assistance of such prospective growth industries through provision of favorable loans.

7. Murakami states that "Western observers tend to argue that the interaction between culture and the political economy is consummated at the national level by means of the Japanese government's conscious and consistent policy. For example, the popular 'Japan, Ins.' model represents the idea that culture and the political economy are integrated at the national level rather than a lower level" (see Murakami, 1987, pp. 36–37).

REFERENCES

Abeglen, J. C. (Ed.) (1970). *Business Strategy for Japan*. Tokyo: Sophia University.

Ackley, G., & Ishi, H. (1976). Fiscal, monetary and other related policies. In: H. T. Patrick & H. Rosovsky (Eds), *Asia's New Giant: How the Japanese Economy Works*. Washington, D.C.: The Brookings Institution.

Ishi, H. (1994). Tax incentives for export promotion in Japan, 1953–1964. In: R. A. Musgrave, C. H. Chang & J. Riew (Eds), *Taxation and Economic Development Among Pacific Asian Countries*. Boulder: Westview Press.

Ito, T. (1996). Japan and the Asian economies: a miracle in transition. BPEA 2.

Komine, T. (1993). The role of economic planning in Japan. In: J. Teranishi & Y. Kosai (Eds), *The Japanese Experience of Economic Reform*.

Komiya, R. (1975). Economic planning in Japan. Challenge, May/June.

Jrause, L., & Sekiguchi, S. (1976). Japan and the world economy. In: H. T. Patrick & H. Rovosky (Eds), *Asia's New Giant: How the Japanese Economy in Japan*. Boulder: Westview Press.

Matsushiro, K. (1970). Industrial concentration and profit roles (in Japanese). Sanken Ronshu (Kangaku) March.

Ministry of Finance (1949, August). Monthly Report of Fiscal Monetary Statistics. Zaisei-kinku Tokei, Geppo.

Murakami, T. (1987). The Japanese model of political economy. In: K. Yamamura & Y. Yasuba (Eds), *The Political Economy of Japan* (Vol. 1). Stanford: Stanford University Press.

Nakamura, T. (1981). The Postwar Japanese Economy (J. Kaminski, Trans). Tokyo: University of Tokyo Press.

Okimoto, D. I. (1989). Between MITI and the Market. Stanford: Stanford University Press.

Patrick, H. (1970). The phoenix risen from the ashes: postwar Japan. In: J. B. Crowley (Ed.), *Modern East Asia: Essays in Interpretation*. New York: Harcourt Brace and World.

Patrick, H., & Rosovsky, H. (1975). Asia's New Giant: How the Japanese Economy in Japan. Washington D.C.: The Brookings Institution.

Suzumura, K. (1996). Industrial policy in developing market economics. In: E. Malinvaud & A. K. Sen (Eds), *Development Strategy and the Market Economy.* Oxford: Charendon.

Statistical Department, Bank of Japan (1991). *Economic Statistical Annual*, 1990.

Galenson, W., & Odaka, K. (1997). The Japanese labor Market. In: H. T. Patrick & H. Rovosky (Eds), *Asia's New Giant: How the Japanese Economy in Japan*. Boulder: Westview Press.

Tsuru, S. (1993). *Japanese Capitalism: Creative Defeat and Beyond.* Cambridge: Cambridge University Press.

10. KOREAN FINANCIAL CRISIS DURING 1997–1998: CAUSES AND CHALLENGES

Jang-Hee Yoo and Chul Woo Moon

1. ECONOMIC GROWTH OF KOREA (1960s-1990s)

Until the early 1960s, the Korean economic situation was that of a typical developing country: high rates of unemployment coupled with poverty. Then a radical shift occurred in the 1960s as Korea adopted an export-oriented economic strategy, and ever since the change Korea has logged impressive growth records. The new political leadership that emerged in the early 1900s committed itself to economic growth and embraced an expansionist development strategy. This was especially true under President Park Chung Hee, but the leaders after him also adhered at the basic precepts and encouraged the influx of foreign capital, primarily through external borrowing, which in turn financed the export industries and went into the construction of the necessary social infrastructure.

From the early 1960s the Korean government believed that developing an outward oriented economy, with growth as the top priority, would not only promote growth but also lay the foundation for enhancing equity and fair income distribution. It was also believed that export-led growth and industrialization, financed by foreign debt, would eventually generate a debt-servicing capability in the economy and stimulate domestic savings. To achieve export growth, the government initially took steps to finance the necessary funds required for the increase of exports. While promoting the use of foreign

Restructuring Asian Economics for the New Millennium, Volume 9A, pages 197–214.
2001 by Elsevier Science B.V.
ISBN: 0-7623-0709-9

capital, an export orientation included many specific policies, such as the devaluation of the Korean won, reduction of tariffs on inputs used for manufacturing exports, and other forms of export subsidies.

However, in the 1970s, the advanced countries. threatened by the rapid economic rise of the developing nations, raised import barriers against the labor-intensive products originated nations from the Third World. Confronted by such challenges, the Korean government altered courses and implemented policies to promote the heavy and chemical industries. During this period, the government intervened in resource allocation through taxation, finance, and restriction of imports so that the nations limited means could better bolster the nascent heavy and chemical industries.

During the worldwide recession of the early 1980s, the Korean government stablized factor prices such as wages, interest rates, and land prices. As a result, the rampant inflation rates of the late 1970s were tamed by 1982. The success of the government in stablizing prices had a positive effect on exports, and Korea's trade balances recorded surpluses from 1986 to 1989.[1] However, since 1989 the balance of trade has turned back into an annual deficit due to a number of new conditions, which include the increase of imported goods brought on by pressure from Korea's trading partners to open the domestic market, the appreciation of the won, and a steep rise in the cost of labor (see Fig. 1).

The export-oriented policy of the last three decades was carried out mainly throuelt intensive government intervention in the economy. Since 1962, seven Five-Year Economic Development Plans have been implemented. With a variety of economic measures, the Plans contributed to the unprecedented growth of the Korean economy. Consequently, the per capital gross national product (GNP) reached an impressive U.S.$$8.486 to 1994 compared with just U.S.$$67 in 1953 (see Fig. 2).

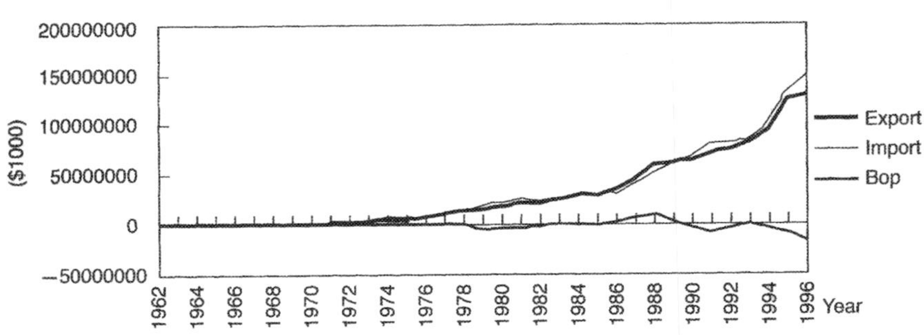

Fig. 1. Trade Balance (1962–1996). Source: Bank of Korea.

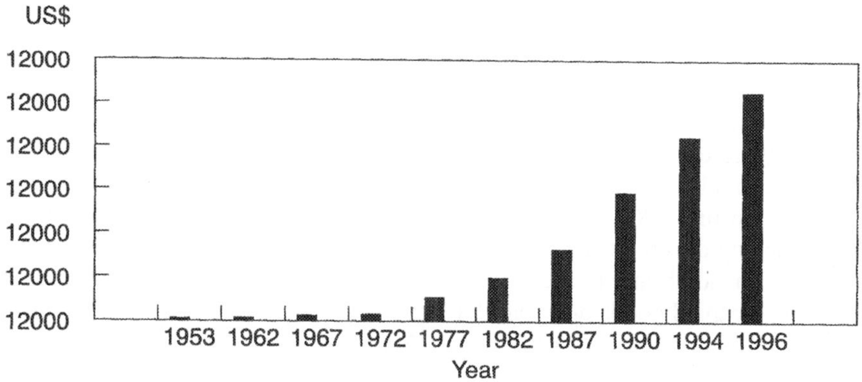

Fig. 2. Per Capita GNP (1953–1996). Source: Bank of Korea, 1994.

The rapid development of the Korean economy reflects the importance of the government's role in economic development, especially for a developing country. The government guided private and public investment based on long term performance, and in the early stage of economic development the Korean government even channeled limited capital resources into industries vital for continuous growth while restricting capital flows into less essential areas. With this strategy, Korea was able to maintain a high rate of growth and leap into the reeks of industrialized nations in a relatively short period of time.

However, the Korean government soon realized that its intervention, the main driving farce for economic development in the 1960s and 1970s, was not an effective mechanism for growth in the long run. Realizing this, the government began to transform its leadership in the economic system into one driven more by market forces. The major transformation had to be made at the financial sector, which had been strictly regulated as interest rates were artificially held below market levels, while the formation of new financial institutions was also restricted. Government intervention in the financial sector distorted the allocation of capital resources, thus bringing about the concentration of economic power.

Government intervention in the financial sector was one of the noticeable hallmarks of the Korean economy during the period the country implemented its economic development plan. Meanwhile, the government, still adhering to the export-oriented policy of this time, introduced several financial measures designed to foster export industries that were financed mostly by foreign loans.

In 1960, the government enacted the Foreign Capital Inducement Promotion Act to secure FDI (Foreign Direct Investment). Facing the unsatisfactory FDI level, the government began to give more emphasis to loans from foreign countries, and it established a new foundation for attracting foreign funds, especially foreign loans. To cope with the public and private demands for foreign capital, the Law for Payment Guarantee of Foreign Borrowing was conduced in July 1962. It facilitated both private and public foreign borrowing by enhancing the credit standing of Korean borrowers in international financial markets. In addition, the government introduced a special law to increase imports of capital goods on a long-term payment basis. These two laws and the Foreign Capital Inducement Promotion Act of 1960 were merged into a new law called the Foreign Capital Inducement Law of 1966. This law was intended not simply to integrate similar regulations and eliminate overlap but also make foreign capital inducements more effective by preventing "unnecessary" borrowing and improving governmental guarantee procedures for foreign loans. In addition, the measure concomitantly provided more incentives for FDI.[2]

Given that economic growth was to be achieved through an export-led strategy, it followed that the export sector and supporting social infrastructure had to be built through an appropriate level of investment. Accordingly, Korea borrowed heavily in the 1960s, 1970s, and 1980s and had become the fourth most heavily indebted nation in the world (Table 1).

2. CHAEBOLS

The chaebol started to emerge with the advent of the heavy and chemical industry policies of the 1970s, which were embedded in the Third Five-Year Development Plans (1971–1975).

In the early 1970, the Korean government thought it imperative to develop the heavy and chemical industries. There were several internal and external factors that contributed to this change in perspective. One internal factor was that Korea's comparative advantage in light industries was declining sharply as labor costs were rising faster than in other developing countries. In addition. the heavy dependence on imports in such fields as machinery, equipment. and raw materials weakened the competitiveness of the export industries. As for external factors, the declaration of the Nixon Doctrine toward the end of the Vietnam War compelled nations like Korea to reevaluate the development of a defense industry. Thus, in January 1973, President Park Chung-Hee urgently called for the development of the nation's heavy and chemical industries. Accordingly, from 1973 to early 1979 government protection was strengthened

Table 1. Foreign Debt in 1960–1996 (Millions of Dollar).

		Foreign Debt				Foreign Debt to GDP %	
Year	Total	Intermediate to Long Term U.S.$ (mil.)	Percentage	Short Term U.S.$ (mil.)	Percentage	Total	Net
1960	83	82	98.8	1	1.2	3.9	n.a.
1962	89	89	100	n.a.	n.a.	3.8	n.a.
1964	177	167	94.4	10	5.6	6.1	n.a.
1966	392	385	98.2	7	1.8	10.7	n.a
1968	1,199	1,110	92.6	89	7.4	22.9	n.a
1970	2,245	1,872	83.4	393	16.6	28.1	20.1
1972	3,587	2,947	82.2	640	17.8	33.9	n.a.
1974	5,933	4,694	79.1	1239	20.9	32	n.a.
1976	10,520	7,475	71.1	3,045	28.9	36.7	n.a.
1978	14,823	11,672	78.7	3,151	21.3	28.5	18.5
1980	27,170	17,794	65.5	9,376	34.5	44.4	32.4
1982	37,083	24,656	66.5	12,427	33.5	52	39.7
1984	43,053	31,628	73.5	11,425	26.5	49.5	37.9
1986	44,510	35,254	79.2	9,256	20.8	42.3	31.7
1988	31,150	21,370	68.6	9,780	31.4	18	4.2
1991	317,170	17,345	54.7	14,355	45.3	13.3	2
1992	62,900	259,000	41.2	37,000	58.8	20.4	8.4
1994	88,800	30,400	34.2	58,400	65.8	23.3	8
1996	157,500	57,500	36.5	100,000	63.5	32.4	7.2

n.a. = not available
Source: Bank of Korea.

and incentives expanded for so-called strategic industries (iron and steel, chemicals, nonferrous metals, machinery, electronics, and shipbuilding).

Since 1979, entry restrictions have been strengthened by limiting foreign ownership and imposing export requirements on FDIs. Interestingly enough, the number of manufacturing firms increased only marginally during the early 1980s, because governmental policy at the time encouraged the expansion of production capacity in existing firms.

When the Korean economy started to experience a construction boom in the Middle East in the late 1970s. the government also urged businessmen to accelerate the expansion of existing heavy and chemical industrial complexes, most of which later became the boundauons for large corporations called chaebols. Moreover, in some cases the government was directly involved in establishing large-heavy businesses such as the Pohang Iron and Steel Mill and the Korea Heavy Industry Company. For this purpose, incentives were provided to domestic producers in strategic industries in the form of preferential financing and tax exemptions or reductions. Long-term loans by public banking institutions like the Korea Development Bank were directed mainly toward these select industries, as was the allocation of commercial bank loans. Furthermore, the National Investment Fund was established in 1974 to expand the funds available to strategic industries, and these funds were provided at preferential interest rates until the late 1970s.

Thus, from 1973 to 1979, the government was deeply involved in the allocation of resources to promote the development of specific industries. The government policy of combining long-term economic planning with the aggressive entrepreneurship of the chaebol had transformed the Korean industrial structure and secured the dominant position of the chaebol, Table 2 shows the growth of the top six chaebols from 1973 to 1987, which clearly indicates the unprecedented level of expansion rarely witnessed in other countries.

In the course of economic development some controversial issues, have emerged. How much have the chaebols contributed to the Korean economy? What are the merits and problems of the chaebol system? Since the chaebols siphoned off the lion's share of Korea's financial resources, should they not be scrutinized in terms of cost efficiency? And if the system turned out to be the main problem, should all the chaebols be dismantled? In the aftermath of the Korean economic crisis, these and similar questions are being raised with increasing frequency. Entering 1998, it has become an unavoidable task for Korea as a nation to examine what role the chaebols has really played in the development of the Korean economy as a whole.

Table 2. Growth in Sales by the Six Largest Chaebols (Billion Won, %).

	1973	1980	1987
Gross Output	11,064	86,170	232,778
Growth rate		16.2a	15.3b
Sale of six chaebols	417	14,743	54,240
Growth rate		50.9a	18.6b
Sales/gross output	3.8	17.1	23.3

Source: Lee, Y. (1987). The Business Groups in Korea: Formation and Evaluation (in Korean) Seoul. Korea Economic Research Institute.

a: Average annual growth late between 1973 and 1980.

b: Average annual growth rate between 1980 and 1987.

c: The six chaebols are Hyundai, LG, Samsung, SK, Daewoo, and Sangyong.

Many Korean entrepreneurs have succeeded in building empires during the last three decades by entering into new industries and taking full advantage of the protection offered by the government. While some chaebols have diversified after thoughtful consideration and adequate preparation in terms of technology, capital, and market perspective, most chaebols have expanded their businesses recklessly. The expansion has shammed partly from the chaebols desire to internalize all elements of production up to the assembly of the final product. Instead of sourcing parts and components from SMEs (small- and medium-sized enterprises) outside the group, the chaebols established their own subsidiaries, crowding out existing SMEs operating in the same field. The reason behind such vertical integration was that it permitted them to economized on transaction costs and improve the quality of parts. This mode of behavior persisted and was not the object of severe criticism before the financial crisis took place.

The fierce competition among the chaebols has also stimulated expansion. They considered an increase of market share to be more important than the rate of profit. Two or more groups often have simultaneously made large-scale investments in the same industrial held, in order not to fall behind rivals. The chaebols were not willing to coordinate and cooperate in investment among themselves as well as others.

The chaebols also tend to expect the government to do whatever is needed to facilitate this expansion, since the government came to rescue them in the past. Such examples can be found in their pleas to the government to lower interest rates in order to reduce their debt service burden and lighten the cost of new investments. Yet at the same time they were loathed to see the same government hinder their expansion efforts.

While the chaebols accumulated problems during the economic development stage, in their view they also played a significant role in transforming the country from a rural backwater to an industrialized society. This was accomplished because they had easy access to financial resources that were in turn used by them as the means to invest in research and development. They were able to learn from imported technology, and in many cases they have succeeded in assimilating and improving the technology from abroad. Many efforts have also been made to advance technical levels in machinery, electronics, shipbuilding, and automobiles, and the chaebols even initially succeeded in their large-scale investments in the field of new technology-intensive industries.

In particular, throughout the 1970, and 1980, the chaebols have been a driving force in the economic development of Korea. The chaebol system worked well in allocating resources that fueled high rates of growth and bolstered worker productivity. Rising productivity in turn increased wages, although for some time it could be said that productivity outpaced wage growth by a small margin so that unit costs in Korea fell but standards of living did not keep pace.

However, the economic efficiency of the chaebols is put in doubt when viewed from the economy as a whole. Expansion is the major source of the chaebols' strength, but it is ironically the primary Source of weakness of the Korean economy. Some have even said that the strength of the chaebols is rooted in rents rather than profits and that the individual efficiency of a chaebol does not necessarily ensure efficiency for Korea.

In light of such problems, the government once tried to check the rising power of these groups. The Korean government passed a competition law and created the Korean Fair Trade Commission (KFTC) to enforce this law in 1980. However, the law in itself contained numerous exemptions to ensure that competition would remain subordinate to industrial policy goals. Until 1994 there were few actions in number that were taken against the chaebol by the KFTC, and those that were initiated seemed to be based more on political than economic considerations.

Such failures to check the chaebol resulted in fifty or so chaebols literally dominating the Korean economy. In 1993, the top five (Samsung, Hyundai, LG, Daewoo, and SK) accounted for more than half of Korea's Gross Domestic Product (GDP) and the top 30 of them for 75% of all economic activity in the country. They have also become major borrowers, with debt ratios in the manufacturing sector reaching 3.03, which are significantly higher than the 2.01 in Japan, 1.67 in the United States, and 0.87 in Taiwan.[3] Government and business circles have the attitude that allowing the chaebol to collapse would pose a systemic threat to the economy. The argument was that if they were to collapse, the whole economy might break down with them. They were indeed regarded as "too big to fail". This rationale was implicitly embedded in the government policy that the chaebols should be financed in their expansion whether it was justified or not. These problems placed a drag on the whole economic system.

3. FINANCING CHAEBOLS

In the early stages of economic development, the short-term goals of achieving growth targets were easily attained by funneling aid to the chaebols, and the government mobilized all available policy measures to support them. Industrial targeting policy in the early 1970s and the introduction of the general trading companies in 1975 gave a great stimulus to the growth and expansion of firms, and a group of "vertically integrated", family-owned conglomerates emerged in less than a decade. In many cases the government simply designated an individual to invest in a particular project, allocating to him designated amounts of bank loans and imported foreign capital.

In the case of banks, the policy loan was possible because bank managers had no autonomy either in the decision-making of bank operations or in determining interest rates. From the bankers' point of view, there was little incentive to review the proposed investment projects or provide advice and other assistance to loan recipients. Banks had only to follow government directions in making loans, and thus many bank loans could easily slip into investment projects with high default risks. Ideally speaking, banks should not be just passive financial intermediaries but also appraise their loans to see whether or not they are reasonable. The process is critical not only in the perspective of the asset portfolio but also of screening out unworthy investments. The banking system, which is deprived of this function made it easy for enterprises to access money and invest it in various fields without adequate scrutiny. This phenomenon inevitably resulted in producing many businesses wholly dependent upon government assistance, which could only be sustained by further bank loans. Consequently, although the banks realized that large borrowers such as the chaebols were on the verge of default, they could not even dare declare them as failing firms for fear of repercussions throughout the whole economy.

The government also created new financial institutions and implemented new practices to facilitate the financing of the development plans. Through these measures, the chaebols could have the increased means to finance their needs. Several specialized banks were established during the early 1960s to implement the First Five-Year Economic Development Plans When the Second Five-Year Economic Development Plans began in 1967, regional bank were established and foreign banks allowed to open branches in Korea. In 1972, when the Third Five-Year Economic Development Plans was launched, the Presidential Decree for Economic Stabilization and Growth was enforced. During this period, the National Investment Fund was created and new non-bank financial institutions were established: investment finance companies, merchant banking corporations, mutual savings and finance companies and credit unions. These non-bank financial institutions were designed to support short-term working capital to businesses.

The short-term financing companies were allowed to charge higher interest on their bill than regular banks so that they could increase the supply of loanable funds. The demand for short-term loans was also large because the banks had been unable to meet the demand for operating funds. In this way short-term financing companies were able to make sizable profits. In addition, the management of these companies was relatively free from government intervention compared with the banking institutions, which resulted in mushrooming of the former in the latter part of the 1980s.

The fast growth of non-banking financial institutions is a popular phenomenon throughout the world. It is natural to observe the same tendency in Korea, but what is unusual in Korea is the proliferation of investment and finance companies created to replace banks. However, even though they complement the banks, they did not substitute them adequately, since they perform different functions from those of banks. Investment and financial companies were intended to absorb underground funds in the curb market by issuing their own bonds and make short-term loans to businesses by discounting the firms' commercial paper at interest rates approaching those in the curb market. Because of higher rates, the savers purchases short-term securities issued by these companies, which attracted an enormous volume of funds from the curb market as well as banks. The investment and finance companies succeeded in many ways, but this success was achieved at the expense of banks. The companies could supply only very short-term operating funds for those firms that satisfied qualifications set by the government. In reality, the proliferation of the nonbank financial institutions weakened rather than strengthened the structure of corporate finance. Long-term investments are financed by rolling over short-term debts, and this has increased borrowers' risk and the cost of their investments.

To finance the nation's economic development, the government also heavily borrowed foreign capital in the 1960s and 1970s, and by the mid-1980. Korea became the fourth most heavily indebted nation in the developing world. The government's encouragement of foreign borrowing, particularly by giving government and bank guarantees, sometimes resulted in excessive inflows of foreign capital. Furthermore, since there was a wide disparity between domestic and foreign borrowing costs in the I960s and 1970, that favored borrowing, there naturally developed a trend to receive loans from foreign sources.

Consequently, the demand for foreign loans was high. It is difficult to judge how foreign borrowing has helped the Korean economy as a whole. There are many who believe that this excess demand did not substantially distort the allocation of resources in the 1960s, as most borrowing was allocated to labor-intensive export sectors, in keeping with the outward-looking development strategy.

However, in the current situation the short-term foreign loans have become a serious problem. Table 3 reports the composition of foreign indebtedness held

Table 3. Foreign Borrowing in Korea (Unit: Billion, U.S.$).

	1995	1996	1997			
			Jun.	Sept.	Nov.	Dec. 20
Long-term liability (%)	33.1	60.7	60.7	66.6	64.7	72.8
	(42.2)	(37.8)	(37.1)	(39.0)	(41.2)	(47.6)
Financial Institutions	NA	44.7	43.4	47.6	45.1	44.2
Farms	NA	13.6	15.1	16.9	17.6	17.6
Public	NA	2.4	2.2	2.1	2.0	11.0
Short-term liability (%)	45.3	100.0	102.8	104.0	92.2	80.2
	(57.8)	(62.2)	(62.9)	(61.0)	(58.8)	(52.4)
Financial Institutions	NA	78.0	77.7	78.3	66.4	54.6
Firms	NA	22.0	25.1	25.8	25.8	25.6
Total	78.4	160.7	163.5	170.6	156.9	153.0

Note: Values in the table are the end-of-period values. Data are from the Korea Institute of Finance, "The Korean Banking Sector in the Changing Financial Environment," paper presented at the annual Breakfast Meeting with Bank Managers, 1998.

by Korea. The short-term borrowings with maturities less than one year constitutes a major portion of Korea's foreign indebtedness. The share of short-term foreign debt, which amounted to 58% of the total external liability of Korea in 1995, again increased to 62% in 1996, edging up to U.S.$100 billion. Although the share of long-term foreign debt appears to have Increased over time in 1997, the short-term foreign debt still remains a major source of overseas borrowings. This phenomenon wreaked havoc last December, when many Korean financial institutions experienced difficulties in rolling over their short-term foreign debts, with their international credit lines suspended due to the crisis.

4. CURRENT FINANCIAL CRISIS IN KOREA

As 1996 drew to a close, the Korean economy was deteriorating rapidly. On 23 January 1997, leading Korean steelmaker Hanbo Steel Corp. defaulted on loans. The failure of company resulted in the bribery scandal related to the son of former President Kim to aggravate the already weak credibility of the financial sector. Starling with Hando, marginal companies began to suffer from slow sales, and banks, heavily exposed to Handio loans, have seen their credit ratings erode. Threatened by the unstable Korean domestic situation along with the overall crisis of other East Asian countries, the foreign investors started to leave Korea, further aggravating the already volatile Korean financial situation.

Table 4. Business Failures in Korea.

	1996	1997				
		1st Qr.	2nd Qr.	3rd Qr.	Oct.	Nov.
Number of business failures (number of firms)	11,589	3,443	3,790	3,834	1,432	1,469

Source: Korea Institute for Finance, "The Korean Banking Sector in the Changing Financial Environment," paper presented at the annual Breakfast Meeting with Bank Managers, 1998.

Since early 1997, the major companies of the country tried to expand their areas of business and thus failed to concentrate into industries where they could enjoy a comparative advantage. Two months later, on 19 March Sammi Steel Co., a unit of Korea's 26th largest conglomerate Sammi Group, defaulted. Junto, Korea's largest distillery, collapsed under mountain of debt and was placed under a bailout program by its creditor banks on 21 April. Kia Group, Korea's eighth largest conglomerate and its third largest automobile maker became a candidate for bank bailout after it failed to make some payments on its 10-trillion-won debt. The financial crisis made many other Korean corporations, including Ssangbang Wool and New Core Group, face debt defaults. Recently, Midopa, one of the nation's largest department stores, went bankrupt on 18 March 1998. Table 4 provides recent statistics on number of business failures, which can be used as indirect evidence that bank loans were extended to risky projects. The number of business failures in the first 11 months of 1997, already exceeded those recorded in the entire year of 1996 by more than 2,000 firms.

Alongside the bankruptcies, which were triggered by financial collapse in other Asian countries, Korea has had to deal with a financial and currency crisis of major proportions. The drop in the value of the won reflected concerns by investors and lenders over the viability of Korean financial institutions and the ability of the country to meet its short-term payments and other problems associated with its manufacturing industries. As a result of these concerns, investors pulled funds out of the Korean market. Consequently, the value of the won and the composite index of stock values plummeted drastically. As shown in Fig. 3, the value of the won had dropped from 841.3 in January to 1,577 in December. The index of values on the Seoul Stock Exchange also suffered heavy losses, and several Korean merchant banks edged toward bankruptcy (see Fig. 3).

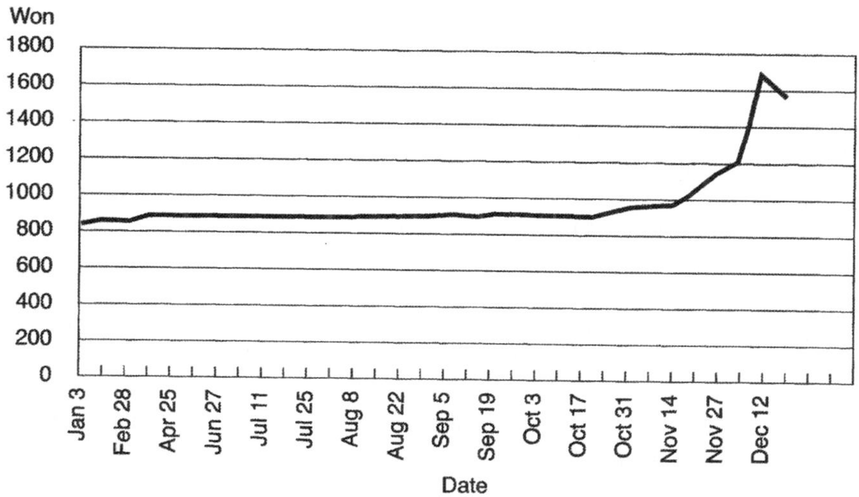

Fig. 3. Exchange Rate for Korea. Source: PACIFIC Exchange Rate Service.

The Hanbu loan scandal, the Asian financial crisis that originated in Thailand, the weakening Japanese yen, labor strikes, and financial problems have brought the Korean economy to a critical stage. On the whole, the long overlooked problems in the financial sector could he said to be the main culprit in the current financial crisis. According to Sachs et al.[4] a lending boom, especially when it is formed in a very short period of time, is likely to increase a bank's exposure to credit risk and raise the likelihood of the bank becoming insolvent. When the lending boom occurs, banks are unable to supervise and monitor all the credits they extend, therefore, many of the loans could easily be invested in projects of high default risk. In addition, financial liberalization may compel banks to invest in risky projects in order to offset the declining profitability caused by increased competition in the financial sector. Table 5 shows recent statistics on non-performing loans held by Korean commercial and merchant banks. According to this finding, non-performing loans held by commercial and merchant banks in Korea increased dramatically in a period of less than one year. The increase in non-performing loans is more obvious in the case of merchant banks. The ratio of non-performing loans to total loans made by merchant banks rose by 4.4 percentage points between December 1996 and November 1997, whereas a 2.1 percentage point increase to the non-performing ratio was recorded by commercial banks between December 1996

Table 5. Nonperforming Loans of Korea.

	Dec. 1996	Jun. 1997	Sept. 1997
Commercial banks	4.2	6.1	6.3
Non-performing loan ratio (%)			
Merchant banks	Dec.1996	Oct. 1997	Nov. 1997
Non-performing loan ratio(%)	1.6	4.5	6.0

Note: The non-performing loan ratio is the ratio of non-performing loans to total loans extended by each financial institution.
Source: Korea Institute for Finance, "The Korean Banking Sector in the Changing Financial Environment," paper presented at the annual Breakfast Meeting with Bank Managers, 1998.

and September 1997. The long-neglected and unproductive operating procedures of Korea's financial institutions forced two major commercial banks to face the risk of default in late 1997.

5. FUTURE CHALLENGES

In late 1997, Korea's usable foreign exchange reserves were around $10 billion. Facing the crisis of an impending moratorium, the Korean government scrambled to bolster confidence in its financial institutions and searched for additional foreign loans. Finally, on 3 December 1997, the International Monetary Fund (IMF) and Korea signed a bailout deal. Under the U.S.$57 billion agreement, IMF will provide a $21 billion standby credit, the World Bank $10 billion, the Asian Development Bank $4 billion, the United States $5 billion, and other countries about $I billion each. In return for its multibillion dollar loan, the IMF asked Korea to cut its growth forecast for world economic growth for 1998 from 4.3% to 3.5%, reduce the budget, keep interest rates high, and adjust its own growth target downward.

To meet the IMF request, the government drastically cut the budget for social overhead capital (SOC) projects. According to budget plans released by the Ministry of Finance and Economy, the SOC budget was set as 9.7 trillion won ($5.98 billion) for this year, down 13.1% from the original budget plan and 4% lower than the previous year. The reduction in the SOC budget was made to meet the conditions that the IMF attached to its rescue loans to the government. But the sharp decline in infrastructure investment will not help the nation's economy in the long run. This is because the reduction in infrastructure investment is estimated to raise the nation's ever-increasing burden for distribution and transportation costs thus weakening its international competitive edge.

In addition, while making additional loan agreements with the World Bank, the Korean government agreed to strengthen the accounting standards for local industrial and financial corporations. Under the financial restructuring reforms that were agreed upon with the World Bank, domestic commercial banks will be instructed to publicize quarterly account settlements starting from September 1998. The quarterly account-settlement system is aimed at upgrading timeliness and transparency of bank-management information.

Meanwhile, the government, moving to heighten the transparency of the corporate governance systems in line with the World Banks request, will take steps to strengthen the Fair Trade Commission (FTC) and provide it with the right to search and seize as well as investigate sources of corporate funding. The government also agreed to toughen fine-imposition rules to eliminate unfair corporate practices.

Since the nation requested the IMF's bailout loans on 3 December 1997, the effect and appropriateness of the IMF's program have raised controversies inside and outside the country. People have argued that the IMF's remedy programs for Korea are turning out to be too costly and painful and sometimes unsuitable for the nation's situation.

The IMF formula calling for retrenchment in state finance and low economic growth has led to growing unemployment, inflation, and reduction in overall economic volume. The IMF program of high interest rates and low economic growth has also driven local companies operating on heavy borrowing into back-to-back bankruptcies. In January alone, a record 3,323 businesses fell due to a lack of funds as interest rates hovered around 20%.

For Korea, the high growth rate was only one of many factors that contributed to the financial crisis that led the country to IMF bailout. The major source of the current financial crisis has been the chaebol's reckless expansionism, that was backed by an inefficient financial system, all supported by strong government intervention. Thus, the IMF rescue program should inevitably focus on the restructuring of financial institutions together with reforming corporate governance, with particular emphasis on reforming the chaebols. However there is a likelihood that the austerity demanded by the IMF would worsen the situation in macroeconomic management rather than help it.

It has now become obvious that the current economic crisis in Korea was caused by inefficiency in the government and financial sector together with reckless expansionism of the Korean chaebols of the last 47 years.

Nevertheless, no one could deny that the structure of the Korean economy viewed from the real sector remains quite sturdy and full of promise. For example:

(a) its industrial base is still supported by one of the most advanced structures in the developing nations, while Korea's automobile, electronics, semiconductor, chemical, shipbuilding, and steel industries belong to the group of top 10 in the world,
(b) its technological level is much higher than that of the average developing country, although it is somewhat behind the level of developed economics.
(c) the quality of its labor force is very high due to a high level of educational aspirations of Korean parents, and
(d) its quality of social overhead capital, such as telecommunications, highways, electricity supplies, and railway systems, is almost as good as those of the developed countries. In addition to these basic economic structures, Korea can also boast a stable democratic system, which is something rarely seen in other developing Asian countries.

Looking over the basic structure and the real-sector fundamentals, Korea still has a lot of growth potential. The core of the concern is how to overcome the current economic crisis without nullifying the past accomplishments as well as preserve the accumulated development experiences of the past four decades. Yet whatever the outcome and whichever course the nation may take, the Korean people should feel proud of their painstaking endeavor that raised the nation from absolute poverty to present affluence.

What should be done to ignite the new energy that would propel the economic dynamism in Korea one more time? First of all, the main task for rekindling the move toward a further development is to renovate the whole nation by introducing various new systems that as proven to be effective in most of the developed countries. This task requires the nation to reform the government, restructure businesses, and overhaul the outmoded banking sector.

First and foremost, the structure of government trust be re-examined. Its regulator functions should be reduced to the minimum level, and its service functions should facilitate voluntary participation and autonomous ruling of the private sector. The government should demonstrate in a definitive manner that the domestic market will operate based upon the "market principle" and "fair competition". This firm position of government should apply to all the facets of the economy, including the financial and goods and services markets.

Financial reform should be executed as quickly as possible. It is imperative to swiftly withdraw ailing and inefficient financial institutions from the market while the restructuring process takes place. These ailing institutions would have three options: finding foreign partners who can pitch in additional capital, receiving bailout funds from the government, or simply going out of business. All these options entail costs that would have to be covered eventually by their shareholders and depositors.

The reform program for the corporate sector, particularly the chaebols, should also be specified and is aimed at three main objective: regain financial soundness by selling of their assets, prohibit mutual payment warrantee systems within chaebols and introduce the consolidated financial bookkeeping practices.

Finally, labor stability is another essential factor that is needed to accomplish the reform programs and thus overcome the crisis. Labor unions should refrain from extreme resistance against the reform policies; at the same time, both government and corporate employer should try their best to work out measures to minimize unemployment during the reform period.

Every time Korea has been confronted by a national crisis, the nation as a whole has found a way out through its unique social cohesiveness and coherence. Considering its past history and current potential, one may predict that some time in the not-too-distant future the world will witness the emergence of a new nation with a surprisingly advanced economic system that is equipped with an efficiency and sound competitiveness comparable to the world's developed economics.

NOTES

1. In 1986 the trade surplus of the country was up to US$4.2 billion and reached US$11.4 billion in 1988. However, in 1990 the trade balance recorded a US$2 billion trade deficit.
2. Sakong, I. (1993). *Korea in the World Economy.* Washington, DC: Institute of International Economics.
3. Chia. L. L. *Korean Sovereign Credit.* Merrill Lynch & Co., Global Securities RC search & Economics Group. (1997, P. 4).
4. Sachs, J. D., Tornell, A., and Velasco, A. (1996). Financial crisis to emerging markets: The lessons from 1995. *Brooking Paper on Economic Activity,* 147-198.

REFERENCES

Chia, L. L. (1997). *Korean Sovereign Credit.* Merrill Lynch & Co., Global Securities RC Search & Economics Group.

Graham, E. M. (1998). Korea and IMF. *Cambridge Rev Intern Affairs.*

Han, S. T., & Nam. S. W. (1996). The development process of the Korean economy. *The Korean Economy: Current Status and Policy Directions.*

Korea Development Institute (1995). *Korean Economic Policies in the Past 500 years.* (In Korean).

Korea Institute for Finance (1998). The Korean Banking Sector in the Changing Financial Environment. Paper presented at the Annual Breakfast Meeting with Bank Managers.

Lee, J. Y. (1987). *The Business Groups in Korea: Formation and Evaluation.* Korea Economic Research Institute.

Nant, D. K., & Jones, V. C. (1997). South Korea's economy and 1997 financial crisis. *CRS report for Congress.*

Sachs. J. D., Tornell, A., & Velasco, A. (1996). Financial crises in emerging markets: The lessons from 1995. *Brooking paper on Economic Activity,* I147–198.

Sakong, I. (1993). *Korea in the World Economy.* Institute for International Economics.

Song, B. N. (1997). *The Rise of the Korean Economy.* Oxford, U.K.: Oxford University Press.

11. THE CURRENT ECONOMIC SITUATION AND TRENDS IN CHINA WITH A DISCUSSION ON EMPLOYMENT-RELATED ISSUES IN THE PROCESS OF INDUSTRIALIZATION

Fan-Zhang Huang

China's economy walked out of the two-year retrenchment in 1991 and entered into a super-high growth period thereafter. Between 1991 and 1995, the average annual rate of growth was 12%, with 1993 and 1994 hitting 14.2% and 13.5% respectively. The economy was heated up by the spree of investments, property developments, development zones and stock markets, which led to a drastic rise in the rate of inflation. The retail price for social commodities went up by 21.4% in 1994, with October of the year reaching a high of 27%. The Chinese government adopted austerity measures to curb the inflation since the second half of 1993. Starting from the middle of 1995, signs of the economy cooling down were seen and the economic operation was brought back to the track of stable growth. The GDP growth rate settled down at 10.5% in 1995 and the rate of increase in retail prices for social commodities was lowered to 14.8%. In 1996, GDP growth further stabilized at 9.6% and the rise in retail prices was correspondingly reduced to 6.1%. The Chinese government announced in 1996 that the tasks of "soft-landing" had been fulfilled.

Restructuring Asian Economics for the New Millennium, Volume 9A, pages 215–228.

ISBN: 0-7623-0709-9

It was anticipated that China's economy would enter a period of high growth and low inflation after its fulfillment of the economic "soft-landing". However, the economic trend had taken a path that runs counter to people's expectations. The rate of growth of the economy in the first of half of 1997 was lingering at a high of 9.5% and there was no obvious indication that there would be a slide. But, starting from the second half of 1997, product sales became stagnant and stocks piled up. A large number of enterprises were operating under capacity or came to halt or closed down. The economic growth took a plunge and the overall price level continued to decline. As a result, the annual growth rate of the economy in the whole of 1997 slowed down to 8.8%, the lowest in seven years since 1991. At the same time, the increase in retail prices precipitated to 0.8%, 5.3 percentage points lower than in 1996. All of these had put China's economy under the threat of deflation. How to stop and revert the sluggish trend of the economic growth had become a challenging task for China in the year of 1998.

1. PERFORMANCES OF THE MACRO-ECONOMY IN 1998

(1) The economy in the first half of 1998 continued to operate on the descending track, and started to pick up since the later part of the third quarter.

In the first half of 1998, GDP growth was 7%, 2.5 percentage points lower than the same period of the previous year. Meanwhile, the industrial growth rate was 7.9%, dropping by almost 3 percentage points in contrast with the corresponding period in 1997. Although negative economic growth did not occur and the rate of growth remained on the still high growth track of 7%, the speed of growth had slowed down and the economy was far weaker than it had been in 1997. Fortunately, however, because both production and investment had shown obvious signs of recovery in the third quarter of 1998, GDP for the first three quarters of this year registered a growth rate of 7.2% over the same period in the last year. The trend of recovery continued in October. In fact, the signs of recovery became perceivable in the second quarter in terms of the economic growth pattern. In June, for example, contrasting the persistent decline in the increase of industrial productions by state-owned and collective-owned enterprises was an increase of 12% in the industrial value-added of foreign-funded enterprises, private and individually owned enterprises, which was 1.2 percentage points higher than in May. Again in June, the production of key industrial products such as TV sets, computers, steel products, plate glass and fertilizers grew in the range of 8–16%. Judging from the basic trend of the current economic operation, 1998 GDP is expected to reach 8%, among which the value-added increase of the primary industry will be about 3.2%, the secondary industry around 9.2% and the tertiary industry 8.3%.

(2) The overall price level will continue to fall. From the third quarter of 1997 to the end of the third quarter in 1998, the retail price index for social commodities had experienced a consecutive 12-month negative growth. Despite the fact that the capital goods were revived as a result of the expansion of government investment and the prices for some products (such as building materials, timber and nonmetal minerals) began to rise since the second quarter, the general trend is that the prices will stay low. According to the statistics, the retail prices in the period between the first and the third quarter of 1998 had declined by 2.5% as compared with the same period in the last year. It is expected that rate of increase in the retail prices for the whole year will be about 1.5%.

(3) The growth of investment in fixed assets was accelerated, becoming a key factor that pulled the economy to swing up. In the first three quarters of 1998, the investment by state-owned entities totaled RMB 1086.4 billion yuan, an increase of 20% over the same period in the last year. The investment in capital construction, property development, renovation/transformation projects were up by 20.9%, 18.9% and 18.7% respectively as compared with the corresponding period of the last year. The financial input was tilted toward infrastructure and high-tech industries. In addition, the investment in the mid-west had significantly increased and the rate of increase in the first three quarters was 33.7%, which was faster than the increases in the eastern and middle areas of China by 15.7 and 20.2 percentage points respectively. In October, the investment in fixed assets by state-owned entities was up by 28.4% as against the same month in 1997. The investment in fixed assets by state-owned entities in January–October period of 1998 contrasted the same period in 1997 by an increase of 21.1% and the rate of increase was also 1.2 percentage points higher than the first three quarters of 1998.

(4) Foreign trade continued to grow while the utilization of foreign capital stayed at a desirable level. According to the statistics, the total import and export value had reached U.S.$232.9 billion in the first three quarters of 1998, an increase of 2.4% over the same period in the last year. China enjoyed a trade surplus of U.S.$35.3 billion, with the total value of export being U.S.$184.1 billion, an increase of 3.9%, and the total value of import U.S.$98.8 billion, an increase of 0.4%. In October, however, both import

and export experienced a drastic fall. The export fell by 17.2% while import by 9.1%, as compared with the same month in the previous year. As a consequence, the total export value for January-October period increased by only 1.3% and the import value in the same period fell by 0.7%, with the accumulated trade balance totaling U.S.$38.4 billion. Meanwhile, foreign direct investment in the first three quarters almost paralleled with the same period in the last year, at U.S.$31.4 billion. It is predicted that the total foreign trade value will see moderate growth while utilization of foreign direct investment will stay at about the same level as in the last year (about U.S.$45 billion).

(5) Execution of the fiscal budget remained fair. The statistics for the first three quarters of 1998 indicated that the fiscal revenue (excluding revenues from borrowings) was RMB 653.951 billion, which represented an increase of 10.4% over the same period of the last year. The fiscal expenditure (excluding expenditure on debt), on the other hand, increased by 17.5% to RMB 659.352 billion. What worth a mention here is that the declining trend of fiscal revenue had been reversed since June as a result of the effort to rectify tax collection and fight against smuggling. The fiscal revenue in June, July, August and September achieved increases of varying degrees, at 14.8%, 5.8%, 11.2% and 23% respectively. But, the expanded investment in infrastructure designed to stimulate the economy and pull the domestic demand and the increase in disaster-relief expenditures had quickened the increase in the total fiscal expenditure. The fiscal expenditure from the first quarter to the third quarter of 1998 had outgrown the fiscal revenue by 7.1 percentage points. It is predicted that the fiscal revenue and expenditure will continue to be contained in a rational and stable state.

(6) Foreign exchange reserve continued to increase while Renminbi remained stable. Although China's economy had experienced ups and downs such as over-heated development, soft-landing and slowdown, foreign exchange reserve was on continuous increase because of the surplus from both trade and non-trade balances, from U.S.$21.12 billion in 1993 to U.S.$51.6 billion in 1994, and then to U.S.$73.6 billion in 1995, U.S.$105 billion in 1996 and U.S.$139.9 billion in 1997. By the middle of 1998, China's foreign exchange reserve had reached U.S.$140.5 billion and is likely, as it is predicted, to hit a record of U.S.$150 billion. The huge foreign exchange reserve has given China the material strength to counteract the contagion of the financial turmoil and maintain the stability of its currency-Renminbi. With the lowering of the interest rates in the U.S. and U.K., in particular, the value of Renminbi will become even more stable.

(7) Unemployment has become an increasingly prominent issue. For a long time, the unemployment in China was chiefly manifested in the form of pervasive invisible unemployment in rural and urban areas, or, in other words, invisible employment in the form of underemployment. The deepening of enterprise reforms and the economic retrenchment have made the invisible unemployment prominently visible. Unemployment is not only a problem attributable to the imbalance of the macro economy, but also a problem of the economic structure and regime. This problem will be discussed in detail later in this article.

Generally, in view of the economic operation in 1998, China's economy is still operating on the track of high economic growth and low inflation. Production and investment began to pick up in the third quarter. GDP growth is expected to reach 8%. Even if a 7.5% growth rate is realized, the economy is still running on the high-speed track. If the turmoil in the external economic environment caused by the financial crisis in the Southeast Asia and the losses incurred in floods in the Yangtze River areas and Northeast of China are taken into account, it is indeed a commendable attainment to maintain a growth rate of 7–8%.

2. ANALYSIS OF THE MACROECONOMIC SITUATION

It had been widely believed that China's economy would enter a new round of economic growth cycle characterized by high growth (9–10%) and low inflation (below 6%) after the "soft-landing" was achieved in 1996. Contrary to people's expectations, however, was that the economy experienced continued slowdown and the economic scenario had been deflationary. The attributable reasons are: (a) the deflationary macroeconomic policy adopted since the second half of 1993 to achieve the objective of soft-landing has its inertial effects and the policy "time lag" will extend the effects beyond the targeted timeframe; (b) the implementation and deepening of the enterprise reforms has made the long-hidden contradictions to become evidently prominent, undisguising the problems of the once-invisible surplus production capacity and surplus labor; (c) the grain price increase is difficult because of the grain harvests and as a result, the growth in farmers' income has slowed down; (d) the promotion of the reform on social security systems has raised people's expectations that their own financial burdens will increase as a result of the new pension system, medical insurance system and unemployment insurance system, and consequently, people's propensity to save has increased, which unavoidably puts dampening effects on the growth in consumer demand; (e) the supply of consumer durable such as electrical appliances in urban areas has been largely in a state of saturation and there has been no new "hot spot" for

Table 1.

Year	GDP Growth Rate (%)	Growth Rate of Pr. Industry (%)	Growth Rate of Sec. Ind. (%)	Growth Rate of Ter. Ind. (%)	Rate of Inflation (%)	Registered Unempl. (%)	Fixes Asset Investment: Value (100 mn yn)	Fixes Asset Investment: Real Growth (%)	Investment Ratio (%)	Foreign Trade Value (U.S.$bn)	Export: Value US$ bn	Export: Rate (%)	Import: Value US$ bn	Import: Rate (%)	Used Foreign Dir. Investment USD bn
1978	11.7	4.1	15.0	13.7	0.7	5.3	899.0	19.5	24.8	20.64	9.75	14.4	10.89	51.0	
1979	7.6	6.1	8.2	7.8	2.0	5.4	977.0	4.7	24.2	29.33	13.66	33.1	15.67	44.0	
1980	7.8	–1.4	13.6	5.9	6.0	4.9	910.8	–8.5	20.2	38.14	18.12	33.4	20.02	27.7	
1981	5.3	7.0	1.9	10.4	2.4	3.8	961.0	2.9	29.8	44.02	22.01	21.5	22.01	10.1	1.16*
1982	9.0	11.5	5.5	13.0	1.9	3.2	1230.4	25.1	23.2	41.61	22.32	1.4	19.29	12.4	
1983	10.9	8.3	10.4	15.2	1.5	2.3	1430.1	13.3	24.1	43.62	22.23	–0.4	21.39	10.9	0.63
1984	15.2	13.0	14.5	19.4	2.8	1.9	1832.9	23.9	25.6	53.55	26.14	17.6	27.41	28.1	1.26
1985	13.5	1.8	18.6	18.3	8.8	1.8	2543..2	30.1	28.4	69.60	27.35	4.6	42.25	54.1	1.66
1986	8.9	3.3	10.3	12.1	6.0	2.0	3120.7	15.8	30.6	73.85	30.94	13.1	42.91	1.5	1.87
1987	11.6	4.7	13.7	14.4	7.3	2.0	3791.2	14.1	31.7	82.65	39.44	27.5	43.21	0.7	2.31
1988	11.3	2.5	14.5	13.2	18.5	2.0	4746.8	10.0	31.8	102.78	47.51	20.5	55.27	27.9	3.19
1989	4.1	3.1	3.8	5.4	17.8	2.6	4410.3	–12.9	26.1	111.68	52.54	10.6	59.14	7.0	3.39
1990	3.8	7.4	3.2	2.3	2.1	2.5	4517.6	–3.0	24.4	115.44	62.09	18.2	53.35	-9.5	3.48
1991	9.2	2.4	13.9	8.8	2.9	2.3	5194.6	15.0	25.9	135.63	71.84	15.7	63.79	19.1	4.36
1992	14.2	4.7	21.1	12.4	5.4	2.3	8080.4	25.3	30.3	165.53	84.94	18.2	80.59	26.3	11.01
1993	13.5	4.7	19.9	10.7	13.2	2.6	13072.3	27.8	37.7	195.70	91.74	8.0	103.96	29.0	27.51
1994	12.7	4.0	18.4	9.6	21.7	2.8	17042.9	18.1	36.4	236.62	121.01	31.9	115.61	11.2	33.76
1995	10.5	5.0	13.9	8.4	14.8	2.9	20019.3	10.9	34.2	280.86	148.78	22.9	132.08	14.2	37.52
1996	9.6	5.1	12.1	7.9	6.1	3.0	22913.8	10.4	33.9	289.88	151.05	1.5	138.83	5.0	41.72
1997	8.8	3.3	10.8	8.2	0.8	3.1	25300.0	10.1	33.9	325.06	182.70	20.9	142.36	2.5	45.25

* Total between 1978–1982.

consumption and investment, which contributed, to a great extent, to the stagnant situation of the market; (f) the growth of export, being affected by the financial crisis in the Southeast Asia, has dropped drastically, particularly, the shrinking markets in ASEAN countries, Korea and Japan as a result of the devaluation of Japanese Yen and other Southeast Asian currencies has brought about a drastic fall in China's export to these countries by 12.9%, 30.2% and 4.3% respectively.

The basic trend in the economic operation in recent years has indicated that the most-felt problem is undoubtedly the insufficient effective demand. Under the circumstance that China's export is facing difficulties, the key solution is to start up domestic demand to avert the economic slowdown and maintain a stable and fast economic growth. Since the second half of 1997, particularly from 1998 onward, the Chinese government has taken a series of economic measures, including monetary and fiscal measures, to stimulate the economy and boost economic growth.

(1) Adjusting the monetary policies on a continued basis so as to increase the stimulating impacts on the economy. The following measures have been taken in this regard:
  - January 1, 1998, the People's Bank of China (the central bank) announced the removal of the ceilings imposed on the size of loans by commercial banks, which was intended to give commercial banks full autonomy in their loan decisions, thereby changing the role of the central bank from direct control to indirect adjustment and control of the total monetary credit supply.
  - In late January of the same year, the People's Bank of China issued RMB 78 billion yuan to commercial banks for re-loaning purposes. Mobilizing huge amount of money for commercial bank re-loaning in a short period of time has never happened in China's history.
  - From the end of March, the People's Bank of China has decided to consolidate the legal deposit reserve account and the reserve-for-payment account into a unified account of deposit reserve and the reserve ratios were reduced from the high of 18–20% in the austerity period to a low of 8%.
  - On March 23 and July 1 of 1998, the People's Bank of China lowered the interest rate twice, the fourth and fifth rate cuts since 1996. The central bank's effort to lower the interest rates for five times had brought down the deposit rates by an average of 4.23 percentage points and the lending rates by an average of 5.17 percentage points. Preliminary statistics indicated that the five cuts in the interest rates had helped ease

enterprises' burden of interest payment by about RMB 240 billion yuan.

- In April, the People's Bank of China announced an addition of fix-asset loans in its directive loan program, which was up to RMB 370 billion from RMB 270 billion set out in the beginning of the year. The addition of fixed-asset loans took about 40% of the total RMB 900–1000 billion of the national directive loan portfolios by state banks, which was much higher than the 22% in 1997 and the highest in recent years.
- On May 26, the open-market operations for bond repurchase were resumed, thereby satisfying the needs for liquidity by other commercial banks. So far, the total liquidation was over RMB 50 billion.
- The People's Bank of China successively issued two circulars – *Directive Views on Improving Financial Services and Supporting the Development of the National Economy* and *The Views on Further Improving the Financial Services for Small- to Medium-sized Enterprises*. The former requires the commercial banks actively support the expansionary effort to boost consumer demand and expand the domestic market, while the latter calls for improved services for small- to medium-sized non-state-owned enterprises in a bid to better serve the objectives of economic growth and re-employment.

Since the beginning of this year, the monetary policy has been loosened time and again. According to statistics, the accumulated loans made by financial institutions in January–July period increased to RMB 452.9 billion yuan, up by RMB 3.8 billion yuan over the same period of last year. The end-of-July balances for various loans reached RMB 7956.3 billion yuan, an increase of 15.8%. The increase in money supply has helped alleviate the economic slowdown, however; the pull of the monetary policy at the economic growth was neither evident nor strong. This is not because the monetary policy is not appropriately employed; it is because the policy has some limitations. In the context of the current situation in China, quite a number of state-owned enterprises has long been loss making and given the loosening monetary policy by the People's Bank of China, the enterprises are afraid of committing to new burdens of interest payment and therefore, feel "reluctant" to borrow. The commercial banks, on the other hand, are afraid of taking on non-performing loans and feel "reluctant" to lend. All of these have contributed to the slow increase in the supply of M2. If the monetary policy is to be excessively used, the operating mechanism of the commercial banks will be jeopardized and the economy may be subjected to the danger of a new round of inflation. Aware of the danger, the Chinese government started to adopt policies more on the fiscal side to re-ignite the economy.

(2) Proactive fiscal policies have been adopted. The main measures taken include:
- To increase government spending with a view to improve the economic structure, the State Development Planning Commission announced six priority investment areas that included projects of irrigation and water conservancy, infrastructure and urban development. An initial calculation indicated that the investment by state-owned entities from the first to the third quarter was RMB 503.6 billion yuan, up by 28.2% as compared with the same period in the last year and also an increase of 14.4 percentage points over the first half of this year.
- In order to mitigate the burdens of enterprises, the Ministry of Finance and five other government departments jointly announced the removal of the third batch of "funds" (funding obligations allotted to enterprises under various names and reasons) and altogether 147 "fund" items were demolished, with an expected reduction of the enterprise burden by about RMB 12 billion yuan. In addition, the run-away exercises of fee collections at different levels were also rectified, saving the enterprises a total outlay of about RMB 15 billion yuan.
- The government allocated RMB 5.4 billion yuan from the fiscal budget to cover the outlays in relation to the effort to re-assign laid-off workers and to streamline over-manned enterprises to increase efficiency. At the same time, additional RMB 3 billion yuan was allocated to the mid-west and old industrial bases in a bid to help with the re-employment efforts for the laid-off workers from the state-owned enterprises.
- To strengthen the capital adequacies of the state-owned banks, the Ministry of Finance issued RMB 270 billion yuan of special bond, of which RMB 93.3 billion yuan went to the Agricultural Bank of China, RMB 42.5 billion to the Bank of China, RMB 43 billion to the Construction Bank and RMB 90.8 billion to the Industrial and Commercial Bank of China.
- On September 4, the Ministry of Finance once again issued 10-year RMB 100 billion yuan bond with the interest rate being 5.57%. The revenue from the bond issue will be used in the priority projects such as water conservancy, biological environment, communications and telecommunications, electric power grids in urban and rural areas, urban infrastructure development, and warehouses of state grain reserve and construction of low-cost residential apartments.

- Also in September, the State Development Planning Commission announced 601 urban development projects, which will entail a total investment of RMB 200 billion. 30% of the new RMB 100 billion bond issued will be used to finance these projects.
- The State Development Planning Commission recently announced a plan to invest a total of RMB 250 billion yuan over three years in the construction of electric power grids in both urban and rural areas.
- The range of export products that enjoy tax rebates was expanded, and the rates for certain sectors have been raised for a number of times. By the end of July, the total export tax rebate accumulated to RMB 20.47 billion yuan, an increase of 9.7% over the same period in the last year.
- Strong support is given to the development of small- to medium-sized enterprises, especially those set up by laid-off workers or individuals, who will be given preferential access to tax concessions and other assistance.

The above fiscal measures, which have achieved desirable effects in reining back the economic slowdown, combined with strong financial measures, have contributed to the resurgence of the economic growth since the third quarter. Due to the "time lag" effects, the pull of the policies (especially the fiscal policies) at the economy will manifest itself more evidently and strongly later.

In the effort to enlarge domestic demand and boost growth, the effects produced by fiscal policies are more distinct than monetary policies, though there are two aspects in which care should be taken.

Firstly, a great deal of effort has been made by the Chinese government to expand domestic demand and stimulate the growth. The slowdown has been held up and the economy has started to pick up. However, even if 8% growth is realized by the end of 1998, the aggregate demand is still less than being sufficient and there are still further growth potentials. 1999 growth rate will be targeting at 8.3% or 8.6% and there will be still stronger stimulations. But, after implementing the expansionary policies for over a year of time, the room for further stimulation by fiscal policies will be limited. Take the issue of treasury bond, for example, the level of dependency of the central government on debt has reached 58%, far higher than the internationally accepted standard of 25–30%. Cautions should be taken in order to prevent the economy from engaging into another round of inflation.

Secondly, the short-term adjustments in an effort to revert the economic slowdown will have to be implemented with considerations for the long-term development and the readjustment of the economic structure. Expanding domestic demand is of necessity, though, what is more so is that the structural readjustments should be embodied in the effort. In the process of expanding domestic demand, attention should be paid to the implementation of the industrial policies and improvement of the competitiveness by realizing industry upgrading and technological transformation. Maintaining production and employment should not be counted as an excuse not to close, suspend, merge or shift those enterprises suffering from technological backwardness, production of unmarketable products and prolonged loss making. Continued injection of large amount of money and materials into those enterprises will only result in accumulation of large inventories of "output". In face of the large amount of "laid-offs" by state-owned enterprises and increasing unemployment in both urban and rural areas, structural readjustments and technological transformation and upgrading are of special importance.

3. THE CURRENT UNEMPLOYMENT ISSUE IN CHINA

Why has the unemployment issue now become so prominent? This is because the unemployment issue is not only a short-term problem arising from insufficiencies of effective demand; it also has other deep-seated reasons related to both the economic structure and the economic regime.

(1) China has a huge population base of labor supply, which itself constitutes great pressure for employment. According to the estimates by the World Bank in 1997, China's population of labor supply (including the employed, unemployed and soldiers on active service) was 539 million in 1980 and 709 million in 1995, accounting for 26.4% and 26.3% of the world labor supply respectively. That is to say, China was using 3.4% of the world resources to create jobs for over one forth of the world labor supply. This pattern of employment situation cannot be changed within a short period of time, because the growth of China's labor population will not reach "zero growth" until the year of 2020. China will, therefore, as it can be predicted, face increasing pressure of employment from now on to 2020.

(2) Among the 1.2 billion population, about 70% is living in rural areas. In the past, agricultural sector had long been a main reservoir that absorbed new additions of the labor force. With the advance of industrialization process, the amount of new labors taken by agricultural sector is on the decrease and the trend that total labor force is decreasing in absolute terms has also emerged. It was revealed that the agricultural sector absorbed an average of about 4.03 million new labors per annum in the 1980s, about one third of the total annual addition of new labors. Changes had taken place in the 1990s, namely, the absolute amount of agricultural labors dropped sharply. Statistics also indicated that between 1994–1995, the total agricultural

labor force in China decreased by an average of 5.67 million. According to a research by Professor Zhou Qiren of Beijing University, in the 1990s, an annual average of 5 million farm laborers migrated to the cities for employment, which has exacerbated the competition for employment between agricultural sectors and non-agricultural sectors as well as between rural and urban areas.

(3) China's economy is experiencing extremely profound changes in its industrial structure. Some enterprises in the traditional industrial sectors have started to be closed, suspended, merged, shifted or go bankrupt, all of which have resulted in the drastic increase of laid-off workers. Some other enterprises in the traditional industrial sectors are suffering from surplus or underutilized capacities. The rates of capacity utilization for some products such as machine tools, automobiles, colored TV sets and washing machines are less than 50%. The production of air-conditioners, microwave ovens, VCRs is running under 40% of the capacity. Less than 70% of the production capacity is utilized in the light industries and textile industry. As a consequence, a large amount of industrial workers has been laid off in sectors such as textiles, garments, machinery, coal mining, metallurgy, forestry and defense. According to a sample survey, 39% of the workers in logging industry and 31% in textile industry were laid off.

(4) The state-owned enterprises and urban collectives have long been overmanned in the form of underemployment. The rigidity of flow had led to an accumulating number of redundant workers in these enterprises. According to statistics, the urban areas in China had 22 million redundant workers that need to be dispatched for re-placements, accounting for one fifth of the total work force (102.32 million) of the state-owned enterprises and collectives in the urban areas.

It is thus evident that, the deepening of the enterprise reform and the development of industrialization will result in a large number of workers disengaged from the enterprises operating in traditional and backward industries and a large number of farmers leaving the farmland or rural areas. They need to be soaked up by creating large quantities of job positions. This is not a short-terms problem arising from the temporary imbalances between the aggregate supply and aggregate demand. It is instead a problem brought about by the changes in the economic regime and industrial structure. When the economic performance is in good shape (GDP growth rate at about 9%), unemployment or employment problems will then be alleviated or even disguised. When the economy slows down, the problems will surface or become acute.

Then, what is the dimension of the current unemployment in China? Since 1993, the real unemployed population has been rapidly increasing, including the increases not only in the publicly admitted or registered unemployment, but also the unemployed among the laid-offs and various forms of invisible unemployment or underemployment. According to expert estimate, the urban unemployment population will reach 13–15 million in 1998, with the rate of unemployment being 7–8%. In addition, there are about 150–220 million surplus laborers in rural areas who need to be relocated.

In face of the severe employment or unemployment situation, the basic strategy China adopts should focus on boosting economic growth; developing diversified economies with the economy of public ownership remaining dominant; and striving for job creations. The concrete policy recommendations, among others, include:

- To maintain a high rate of growth so as to provide material backing for the continued effort to create employment opportunities. As determined by the growth pattern of the Chinese economy, GDP growth should not be lower than 8%.
- To implement Chinese-style "New Deal" policy. This "New Deal" encompasses two infrastructures, one of which is physical infrastructure, namely, large scale construction of infrastructure with the government spending on public projects such as roads, bridges, water conservancy, urban development, hospitals, schools and environment protection. The other is social infrastructure, striving for the establishment of social security system, medical insurance system, unemployment insurance system, private housing loan system, etc., and the setting up of labor markets and associated information networks and training centers for laborers' re-employment, all of which are designed to construct a good social safety net.
- To develop labor-intensive industries, and with a selective approach, to develop resource-intensive, capital-intensive and technology-intensive industries. Generally speaking, large enterprises, "backbone" enterprises and a considerable part of the medium-sized enterprises should concentrate on capital-intensive and technology-intensive operations, while small-sized enterprises and a part of the medium-sized enterprises should be focusing on labor-intensive technologies. In the circumstance that small enterprises and some medium-sized enterprises adopt labor-intensive technologies, attention should be paid to raising the level of processing and value-added. At the same time, attention should also be given to the appropriateness of the rise in salaries and welfare packages so as to prevent labor costs from increasing too fast.

- To encourage the development of the enterprises in small cities and towns and the township and village enterprises, thereby absorbing the surplus labor drifting out of the agricultural sector. The industries in small cities and towns should be centering on the productions of agriculture, forestry, husbandry, sideline products and fishery in the neighboring rural areas and rural household productions, while at the same time servicing the productions and livings in the large- to medium-sized cities. The laborers shifting out from agricultural sectors and rural areas will largely move to the nearby small cities and towns, with a small portion flowing into large- to medium-sized cities. In this manner, the advocate on rural laborers "leaving farmland without leaving homeland", as a guiding principle, can be implemented. This is a relatively wise path for China, a country with huge population, to realize industrialization.

12. THE EVOLUTION OF COMMERCIAL BANKING AND FINANCIAL MARKETS IN TAIWAN

Tzong-shian Yu

1. INTRODUCTION

Over the last five decades, Taiwan has achieved high economic growth with mild inflation (except for the periods of the two oil crises), low unemployment, and more equal income distribution. Correspondingly, Taiwan has gradually been transformed from an agriculture dominated economy to an industry-dominated one, and thence from an Industry-dominated economy to a finance-dominated one. Needless to say, the financial system has played a key role in the process of Taiwan's economic development. Through high economic growth, the financial system is gradually being transformed from a backward and controlled system to a modernized and liberalized one, although there are still many traditional obstacles to overcome.

Since the East Asian financial crisis began in Thailand on July 2, 1997, none of the countries in the region could completely avoid its impact. However, three countries, namely, Singapore, Taiwan, and mainland China, seem to have been less affected by the crisis and have maintained a moderate growth rate. Many people have become interested in Taiwan's financial system and its operation because it has weathered the financial crisis very well. Now is a good time to examine Taiwan's financial system.

Restructuring Asian Economics for the New Millennium, Volume 9A, pages 229–248.
2001 by Elsevier Science B.V.
ISBN: 0-7623-0709-9

The purpose of this paper is not to explain how Taiwan has coped with the financial crisis but to explore the evolution of the commercial banking and financial markets in Taiwan. In order to examine this issue, the paper analyzes the structure of the financial system and its evolution from a controlled system into a liberalized one, from a state banking system into a private banking one, and from an inward investment system into an outward investment one. We shall then discuss the impact of the Asian financial crisis on Taiwan's economy. Finally, some concluding remarks will be drawn from the analysis.

2. THE STRUCTURE OF TAIWAN'S FINANCIAL SYSTEM

Like many other developing countries, Taiwan has a dual financial system; i.e. a legal financial system and an illegal financial one. What we are more interested in here is the legal financial system, which covers the financial markets and financial institutions.[1]

2.1. Financial Markets

Prior to the 1960s, Taiwan's financial markets were rather incomplete and simple. Since that time, after rapid economic growth, Taiwan has gradually established a complete financial market to meet the demands of industrialization. The financial market includes the capital market, money market, and foreign exchange market. The capital market include stock markets, which include listed and over-the-counter; bond markets, which include listed and over-the-counter; and the futures market, which was set up early in 1998. The money market includes the interbank money market and the short-term bills financial market. As for the foreign exchange market, it includes the bank customer foreign trading market, the interbank foreign transaction market, and the foreign currency call loan market. All thee markets serve as the medium for transactions under rules set by the government.

2.2. Financial Institutions

Taiwan's financial institutions can be classified into two groups. One is established under banking laws and the other is controlled by other laws. If their setup is not based on laws, they are called underground financial institutions, shadow financial institutions, or illegal financial institutions.

2.2.1. Financial Institution Established Under Banking Laws:

(i) Commercial banks
  Domestic commercial banks
  Branches of foreign banks
(ii) Specialist banks
  Industrial banks
  Agricultural banks
  Land banks
  Export–import banks
  Medium business banks
(iii) Savings banks
  Savings department of banks
(iv) Trust and investment companies
  Specialist trust and investment companies
  Trust departments of banks

2.2.2. Financial Institutions Established Under Other Laws:

(i) The postal savings system
(ii) Cooperative banks
  Regional credit cooperatives
  Credit departments of farmers' associations
  Credit departments of fishermen's associations
(iii) Offshore financial institutions
  Offshore banking units
(iv) Insurance agencies
  General insurance
    Life insurance companies
    Property and causality insurance
  Special insurance
    Reinsurance corporations
    Central deposit insurance corporations
    Insurance department of the Ex-Im Bank (only this bank is allowed to set up a department in charge of insurance affairs)
(v) Specialist financial market agents
  Bill finance companies
  General securities
    Dealers
    Brokers
    Underwriters

Securities finance
Securities investment and trust corporations
Interbank money centers
Foreign currency call loan centers

The above outline is a conventional classification. In fact. Taiwan has no saving banks or trust banks; they exist as the departments of commercial banks. The postal savings system is located in post offices and is not an independent system. As of June 1998, the number of branches of financial institutions in Taiwan was as follows:

Domestic banks: 40 (1,900)
Branches of foreign banks: 45 (72)
Medium business banks: 7 (423)
Trust and investment companies: 5 (62)
Credit cooperative associations: 370 (1,453)
Life insurance companies: 31 (103)
Bills finance companies: 15 (38)
Security finance companies: 4 (2)
The postal savings system: (1,534 post offices)
Investment trust companies: 28
Securities firms: 220

2.3. The Illegal Financial System

It may be reasonable to say that the more backward a country's financial system, the more developed its underground financial system. Before 1990 all banks in Taiwan were either state-owned or state-dominated, making it almost impossible for small-scale enterprises and consumers to apply for loans if they had no mortgages. Under such conditions they had to obtain loans from underground financial agents; otherwise they could not survive. It is generally recognized that during the past 5 decades, small businesses, including small- and medium-sized enterprises, have made a great contribution to the expansion of foreign trade and the creation of job opportunities in Taiwan. Even after 1990, when many private commercial banks were set up, financial difficulties remained because mortgages were still required to borrow money from banks. However, underground financial agents could fill the hole, even though their interest rates are extremely high and the duration of their loans very short.

2.4. *The Financial Examination System*

A complete and efficient financial system must have a strong and fair examination system, not only in order to help financial institutions achieve sound development but also to prevent the need for risk-taking. The principal government agencies responsible for the supervision of financial institutions are the Central Bank of China, the Ministry of Finance, and the Central Deposit Insurance Corporation. The Central Bank is responsible for implementing monetary policies and handling a considerable amount of financial supervision, while the Ministry of finance is responsible for financial administration and the supervision and regulation of financial businesses. The authority of the Central Deposit Insurance Corporation is conferred by the Ministry of Finance.

The division of the functions of the financial examination bodies is as follows:

2.4.1. *Central Bank of China*

(1) Domestic banks
(2) Branches of foreign banks
(3) Postal savings bureau
(4) Bills finance companies

2.4.2. *Ministry of Finance*

(1) New private banks
(2) Branches of foreign banks
(3) Domestic banks
(4) Trust and investment corporations

2.4.3. *Cenntial Deposit Insurance Corporation*

(1) Domestic banks
(2) Trust and investment companies
(3) Credit cooperatives associations
(4) Credit departments of farmers' and fishermen's associations

It should be noted that before 1979, the Central Bank of China came under presidential authority. Since 1979 the Central Bank of China has been under the control of the Executive Yuan, which is now not as politically and economically independent as it was before.[2] The Central Deposit Insurance Corporation was set up in 1985. Before that time local financial institutions were examined by the provincial Cooperative Finance Bank, which ceased examination operations in 1996.

3. THE EVOLUTION OF TAIWAN'S FINANCIAL SYSTEM

When a country lacks natural resources, it is natural for its government to adopt a controlled financial system because its scarce resources is insufficient to meet demand. However, once the country has started to achieve rapid growth and has accumulated sufficient resources to meet demand, it is natural for its government to gradually adopt a more modern and open financial system. Taiwan is certainly a good example of this.

The evolution of Taiwan's financial system can be characterized by its gradual transformation from a controlled system into a liberalized one, from a state banking system into a private banking one, and from an inward investment system into an outward investment one. Such an evolution depends mainly on changes in external and internal economic conditions.

3.1. From a Controlled to a Liberalized System

In the early stages of Taiwan's economic development, the government controlled the financial system, including monetary instruments such as the determination of the interest rate, foreign exchange rate, etc., and it controlled the establishment of financial institutions such as banks and stock companies. In addition, the government took certain administrative measures to intervene in the operations of financial institutions to achieve stability, as mentioned above, through rapid economic growth the financial system has been changing, and concrete achievements can be seen in the liberalization of regulations on the interest rate, foreign exchange rate, and capital movement.

3.1.1. The Liberalization of Interest Rate Control

Before 1975, the interest rate was wholly determined by the Central Bank. All banks were either owned or partly owned by the government and had to accept the designated interest rate.

According to Paul C. H. Chiu, efforts to liberalize interest rate control in Taiwan since 1975 may be divided into three phases.[3]

(i) The first phase (1975–1981). During this period banks were allowed greater flexibility in loan pricing, and a money market was established as an additional conduit of funds for enterprises. Major development in the liberalization of interest rate regulation include (a) In July 1975, spread-hands

of 0.25% on unsecured loans and 0.75% on secured loans were introduced to provide flexibility for borrowers and lenders; (b) In May 1976, the short-term bills market was established. The instruments introduced included commercial paper and bankers' acceptance, which were guaranteed by banks and investment and trust companies, as well as treasury bills and negotiable certificates of deposit; (c) In November 1980, the "Essentials of Interest Rate Adjustment" was promulgated and implemented with the purpose of allowing banks more flexibility in determining interest rates during times of change in the money market situation. Three types of interest rate on bank deposits and loans were liberalized: negotiable certificates of deposit and bank debentures, discounts on bills of exchange, and on-lending foreign currency loans and foreign currency accommodation for negotiations on the issuance of letters of credit.

The interest rates on the first two types were liberalized with open-market financial instruments. Liberalization of the last type allowed foreign loans to compete with local funds.

(ii) The second phase (1982–1984). The prime mover of interest rate liberalization in this phase was the Central Bank's revision of the postal savings system redeposit scheme. From April 1982 on, new increments in postal savings deposits could be redeposited in four designated, specialized banks rather than in the Central Bank; this step reflects an important deregulation of the supply of marker funds. The increase in liquidity resulted in a downward slide in short-term money market rates, and the Central Bank responded to this by lowering the ceiling on I-year deposits still further (to 8% per annum in 1984). Correspondingly, loan rates were also adjusted downward.

(iii) The third phase (1985–7989). A loan rate liberalization policy was implemented in March 1985. Banks were required to set their own prime rates and use them to replace the minimum or maximum loan rates set by the Central Bank. The spread between the maximum long-term and minimum short-term loan rates was also broadened to allow for more flexible loan pricing and negotiation. During this phase, the Central Bank adopted several important measures to further lower the ceiling on deposit interest rates, such as in January 1986, when the Central Bank cut the number of ceding rate categories on deposits from 13 types to 4. These four are passbook demand deposits, passbook savings deposits, time deposits with maturity of one year or less, and time and savings deposits with maturity of over one year.

The effect of interest rate liberalization on the efficiency of the financial system can be seen from the movement of the interest rate spread between deposit and loan rates in response to changes in market rates. Adjustment in the deposit rates was very flexible with respect to changes in money market conditions.

3.1.2. The Liberalization of the Foreign Exchange Rate

Before the 1950s, Taiwan's financial situation was weak. Faced with a trade deficit and a shortage of foreign exchange, the government had to adopt foreign exchange control measures and a foreign exchange certificate system. Under this system, enterprises that received foreign exchange from exports had to surrender 20% of it to the Bank of Taiwan in exchange for New Taiwan dollars.[4] The remainder could be optionally exchanged for foreign exchange certificates equivalent to the original value, and the certificates could then be transferred, sold to the Bank of Taiwan, or used for imports.

From 1949 through 1978, the foreign exchange rate was fixed. In the beginning, one U.S. dollar could be exchanged for five New Taiwan dollars. This was supposed to make the New Taiwan dollar overvalued, i.e. not good for exports but good for imports. In order to reduce the trade deficit and save limited foreign exchange, the government adopted a foreign exchange censor system

In view of the need for foreign exchange for different purposes, the government adopted of multiple foreign exchange rate system on April 4, 1951. However, due to its disadvantages, the government abandoned the foreign exchange certificate system in 1958; due to its shortcomings, the government also abandoned the multiple foreign exchange rate system and adopted a single foreign exchange rate system instead. During the 1950s and 1960s, long procedures in applying for foreign exchange for individual use were imposed to reduce the demand for foreign exchange.

In 1978, the government abandoned the fixed foreign exchange rate system and adopted managed foreign exchange. Under the new system, the foreign exchange rate was determined not by the Central Bank but by the Center of Foreign Exchange, which was composed of five large Taiwanese banks, namely the Bank of Taiwan, the Chinese International Commercial Bank, the First Commercial Bank, Hua-Nan Commercial Bank, and Chang-Hua Commercial Bank. Daily adjustment to the foreign exchange rate could not go beyond 2.25% of the foreign exchange rate of the previous day: it was hoped this would prevent drastic fluctuations in the foreign exchange market.

During the 1980s Taiwan's trade surplus increased substantially, and foreign exchange was no longer a scarce resource. The government was confident enough to release foreign exchange controls and in 1987 substituted the floating foreign exchange rate system for a managed foreign exchange rate system. Under this new system, the foreign exchange rate was determined by market forces. Since then, the Central Bank has been able to make use of open-market operations to adjust for imbalances in the foreign exchange rate.

From 1985 to 1992, the New Taiwan dollar appreciated from 39.80:1 to 25.40:1 against the U.S. dollar, which means that the New Taiwan dollar appreciated by 56.7%. However, due to the collapse of Taiwan's bubble economy in 1990, the growth rate of Taiwan's economy has tended to decline since 1993, the trade surplus started to decrease, and the New Taiwan dollar has started to depreciate against the U.S. dollar. Meanwhile, from 1995 to 1998, two important events caused the New Taiwan dollar to further depreciate: the first was mainland China's sabre-rattling military exercises conducted in the waters around Taiwan in 1995: the second was the East Asian financial crisis that broke in Thailand and was threatening Taiwan's economy by late 1997. In these two specific years there was a rush among people in Taiwan to change their New Taiwan dollars to U.S. dollars, and these U.S. dollars were either remitted abroad or deposited in the foreign exchange departments of commercial banks in Taiwan, as can be seen from Table 1.

In the domestic market, as people rushed to exchange New Taiwan dollars for U.S. dollars, the New Taiwan dollar depreciated further. In 1995, the missile tests and other military activities conducted by mainland China in the Taiwan straits caused an outflow of about U.S.$7 billion from Taiwan. In response to

Table 1. U.S. Dollars Deposited in Taiwan Banks, 1990–1998.

Year	Foreign exchange rate (New Taiwan $/U.S.$)*	Growth rate of U.S.$ deposits in Taiwan banks	Significant events
1990	27.11	76.8	Stock market collapse
1991	25.75	–23.9	Collapse of bubble economy
1992	25.40	47.0	
1993	26.63	34.3	New Taiwan dollar about to depreciate
1994	26.24	14.5	
1995	27.27	67.5	Threat of military exercises
1996	27.49	43.8	
1997	32.64	90.4	East Asian financial crisis
1998	32.22	86.1	

* By the end of December, not an average.
Source: Data prepared by the Council for Economic Planning and Development.

this situation, the Central Bank released more U.S. dollars into the foreign exchange market to stabilize the fluctuations. In October 1997, the New Taiwan dollar continued to depreciate even though Taiwan was still enjoying a trade surplus and had foreign exchange reserves to the tune of around U.S.$84 billion. The government at this time released more U.S. dollar in order to increase the supply of foreign exchange in the hope of stemming the further depreciation of the New Taiwan dollar. Unfortunately, the result did not meet expectations.

3.1.3. The Liberalization of Capital Movement

Before 1978, capital investment was strictly controlled by the government. Once the trade surplus started to increase and foreign exchange reserves to accumulate, the government started to relax its controls on capital movement, particularly after 1985, when the government announced its official policies of economic liberalization and internationalization. Restrictions on trade-related capital transactions and controls on the inward remittance of foreign exchange were dropped; however, non-trade-related outward remittance by local residents remained, subject to a strict ceiling imposed by the Central Bank. Both outward find inward remittances for direct investment were subject to approval by the Investment Commission of the Ministry of Economic Affairs.

However, when the economy continued to show rapid growth in the late 1980s, huge increases in both the current account surplus and short-term capital inflow caused the government to take steps toward foreign exchange liberalization. In particular, as the price index of stocks skyrocketed and the New Taiwan dollar continued to appreciate, speculative capital inflow flooded the financial markets, accelerating the formation and eventual collapse of Taiwan's bubble economy in late 1990.

It can be argued that the foreign exchange market in Taiwan today is still only partially liberalized, since the Central Bank has often intervened in the spot exchange rate when it rises too rapidly or drops too quickly, rendering it unable to fully or instantaneously fluctuate in response to market forces. For example, in the 1986–1988 period the Central Bank often Intervened in the foreign exchange market, generating an obvious gap between market expectation of the future rate of the New Taiwan dollar and the market spot rate. This in turn induced huge capital inflow into Taiwan and made adjustment of the current account surplus even harder.[5]

Because of the abundance of inward capital, each person or company was allowed to make outward remittance of up to U.S.$5 million a year, while inward remittance was decreased to U.S.$50,000 a year to reduce the supply of foreign exchange in Taiwan. Restrictions in capital movement for direct investment, except in the case of direct investment in mainland China, were also abolished.[6] However, these ceilings have repeatedly been changed since led. In 1990 both were adjusted to U.S.$3 million.

To date the current account has been completely liberalized, while the capital account has not yet been fully deregulated in the sense that the amount of foreign investment permitted in the stock market is still subject to some restrictions.[7]

3.2. From a Public Banking System to a Private Banking One

Historically speaking, there are three kinds of banking establishment in Taiwan's banking system: the first kind was set up during the period of Japanese occupation and left to the Chinese government in 1945, when Japan surrendered to the Chinese government; the second kind constitutes the renewal of the banks originally set up in mainland China; and the third kind is the set of bunks established since 1945.

3.2.1. Bank Setup in the Japanese Occupation Period

There were six banks and four departments of savings and credit in associations and cooperatives in Taiwan under the Japanese. All these banks have been under the control of Taiwan's local government since the Republic of China moved there in 1945.

3.2.2. Renewal of the Banks Originally Set Up in Mainland China

Four important banks and two savings and credit bureaus were renewed in Taiwan. Then are owned by the central government, although one was privatized in 1971.

3.2.3. Banks Established Since the 1950s

By 1991 there were 20 commercial banks, 38 foreign bank branches, and numerous trust and investment companies and credit departments of farmers' and fishermen's associations in Taiwan. Since 1991 the establishment of sixteen private commercial banks has been permitted, and in Taiwan several private banks have already been set up.

As mentioned earlier, since 1985, when the government decided to adopt economic liberalization and internationalization, the privatization of financial institutions has become a big issue and has attracted much public attention. Before the 1990s, all banks in Taiwan were either owned or partially owned by the government. In the provincial government three large commercial banks played a significant role in promoting industrial development, and many local

Table 2. Change in the Number of Public and Private Banks.

	Number of Banks	
	Before 1991	By the end of 1998
Public banks	13	9
Banks owned by central government	4	4
Banks owned by provincial government	7	3
Banks owned by municipal government	2	2
Private banks	4	31
Total	17	40

financial institutions assisted farmers and small businesses a great deal. Meanwhile, in the central government, many banks played a key role in promoting economic development during the last five decades.

Although the Taiwan government has expressed its intention to liberalize many banks, the process of liberalization has been very slow because many groups with vested interests have tried to hold back the process. The reason for this is simple. If all state-owned banks are privatized, the groups with vested interests will no longer be able to take advantage of examining the budget of the privatized banks and thus have no right to borrow money from the banks without any mortgage, or be unable to intervene in the personnel arrangements of the banks.[8] In any case, thanks to tremendous changes in political and economic conditions in Taiwan, the process of privatization has been brought to the forefront of government policy since 1997.

Table 2 provides a clear picture of the results of privatization in Taiwan.

As of 1998, only four banks formerly owned by the provincial government had been privatized. As for the banks owned by the central government, so far no definite time schedule for privatizing them has been announced. However, it should be pointed out that since 1991, 27 private commercial banks have been established.

3.3. *From an Inward Investment System to an Outward Investment One*

Taiwan started out as a poor country with scant capital. In 1952 the gross savings ratio was about 15%, which was much less than the gross investment ratio.[9] Before 1971 Taiwan had seen a trade deficit every year, so no significant foreign exchange reserves had been accumulated. Under these conditions, what Taiwan needed was foreign investment.

According to the official registration from 1952 to 1997 (see Table 3). foreign direct investment (FDI) in Taiwan amounted to U.S.$28.988 million.

Table 3. Foreign Direct Investment in Taiwan.

Year	Amount	Year	Unit: U.S.$ million Amount
1952	1	1987	1,419
1955	5	1988	1,183
1960	15	1989	2,418
1965	42	1990	2,302
1970	139	1991	1,778
1975	118	1992	1,401
1980	466	1993	1,213
1985	702	1994	1,613
1986	770	1995	2,925
		1996	2,641
		1997	4,267
		1952–1997	28,988

Source: The Ministry of Economic Affair, Republic of China.

Before 1965 FDI was not very significant, and U.S. aid was imperative in promoting Taiwan's economic takeoff. During the 1970s, FDI became very important to the expansion of exports. The establishment of export processing zones supported mainly by FDI made substantial contributions to job creation and the transfer of foreign management skills. After the 1990s, FDI became less important because Taiwan started to enjoy a huge trade surplus and burgeoning foreign exchange reserves; moreover. the gross savings ratio increased to 38% in the post-1980s period. However, FDI has not been increasing at a steady rate.

As for outward investment before 1970, no records of overseas direct investment exist. However, after the 1970s, Taiwan started to make overseas direct investment in countries like Brazil and South Vietnam. In the second half of the 1980s, due to the shortage of labor on the island, the sharp rise in land prices, and the appreciation of the New Taiwan dollar, many labor-intensive enterprises found they could not survive in Taiwan and had to relocate their factories in many ASEAN countries, where they could make use of cheap labor and abundant natural resources and thereby maintain their competitiveness in the world market. Furthermore, in the wake of the abolition of martial law in Taiwan, people were permitted to visit their relatives in mainland China. In addition, many small businesses started to look to mainland China for development opportunities, some making direct investments there. It can be seen from Table 4 that the total amount of Taiwan's overseas direct investment from 1991 to 1997 amounted to U.S.$51,426 million. which was much larger than the total amount of FDI made from 1952 to 1997.

Of the total amount of overseas direct investment made during the 1991–1997 period, direct investment in ASEAN countries accounted for 49.8%, while that in mainland China accounted for 34.2%. In 1991, direct investment in ASEAN countries accounted for 66% of total overseas direct investment, while that going to mainland China accounted for only 16.3%. However, in 1997 these respective figures had changed to 50.5% and 26%, meaning that the ASEAN bloc was still the most important area for Taiwanese direct investment.

4. THE IMPACT OF THE EAST ASIAN FINANCIAL CRISIS

The incipient financial crisis first hit the region on July 2, 1997, and swept through East Asia like a tornado, engulfing Malaysia, Indonesia, the

Table 4. Overseas Direct Investment (1991–1997).

	1991	1992	1993	1994	1995	1996	1997	Unit: U.S.$ million Total
ASEAN	3,490	1,985	1,148	4,953	4,216	4,496	4,836	25,126
(%)	66.0	54.9	21.2	51.8	50.1	47.1	50.5	49.8
Mainland China	863	1,050	3,139	3,391	3,162	3,475	2,490	17,570
(%)	16.3	29.1	57.9	35.5	37.6	36.4	26.0	34.2
Other Asian countries	210	61	230	162	142	74	177	1056
(%)	4.0	1.7	4.2	1.7	1.7	0.8	1.8	2.1
U.S. and Canada	312	194	529	145	248	272	563	2263
(%)	5.9	5.4	9.8	1.5	2.9	2.8	5.9	4.4
Other Americas	347	255	111	843	539	1,171	1,353	4,619
(%)	0.6	7.1	2.0	8.8	6.4	12.3	14.1	9.0
Europe	60	46	256	22	60	12	59	515
(%)	1.1	1.3	4.7	0.2	0.7	0.1	0.6	1.0
Other	7	22	10	47	42	48	101	277
(%)	0.1	0.6	0.2	0.5	0.5	0.5	1.1	0.5
Total	5,289	3,613	5,423	9,563	8,409	9,548	9,581	51,426
(%)	100.0	100.0	100.0	100.0	100.0	100.0	100.0	100.0

Source: Liu, T. N. (1998). Reconsideration of overseas investment after the Asian financial crisis. *Quarterly Deposit Insurance Information 12*(1), 99–105

Philippines, and Singapore and encroaching upon Hong Kong, Taiwan, Korea, Japan, and China. No one country in the region managed to completely evade its impact. All the countries that succumbed to the crisis found themselves facing a sharp depreciation in their currencies and stock market collapse: these consequences in turn resulted in a decline in exports, a slowdown in economic growth, and a rise in unemployment.

Taiwan suffered much less from the crisis than other East Asian countries. This was not because the Taiwan government had implemented effective preventive measures but because Taiwan has extremely sound economic fundamentals, which are outlined as follows: (a) up until 1998 Taiwan had enjoyed a large trade surplus for years, and it has abundant foreign exchange reserves; (b) foreign debt is negligible; (c) national competitiveness is comparatively high; (d) the unemployment rate is low: and (e) the economic growth rate has consistently topped 5% over the last 5 years. As a result, Taiwan suffered less from the financial crisis in the period from the end of June, 1997 to the end of June, 1998. For instance, in the wake of rile crisis, the New Taiwan dollar depreciated by only 19%, much less than the currencies of Indonesia, Thailand, Malaysia, the Philippines, and Korea, and its stock price dropped by only 16.4%, again much less than that of all the other East Asian countries (see Table 5).

In October 1997, as the New Taiwan dollar was continuing its decline against the U.S. dollar, the Central Rank adopted open-market operations; that is, the

Table 5. The Impact of the Financial Crisis in East Asia.

Country	Currency depreciation	Drop in Stock prices	Unit: Percent decline in growth rate		
			1996	1997	1998*
Indonesia	83.6	38.5	8.0	4.6	–13.4
Thailand	38.7	49.3	6.7	–1.5	–6.4
Malaysia	39.0	57.7	8.2	7.8	–1.7
The Philippines	37.3	40.9	6.8	5.8	1.8
Singapore	15.4	46.3	7.0	7.8	1.2
Hong Kong	0.0	43.8	5.0	5.3	0.9
Korea	35.3	60.0	7.1	5.5	–3.8
Taiwan	19.0	16.4	5.7	6.8	5.2
China	0.0	—	9.5	8.8	7.8
Japan	18.3	23.2	3.5	0.9	–0.5

* Forecast.

Source: prepared by the Council for Economic Planning and Development.

Central Bank released more foreign exchange. However, its effect was not as significant as expected. Since October 1998, two bills of finance companies and several manufacturing corporations have been swallowed up by the crisis, causing stock prices to drop sharply. In response to this sluggish economic performance, the government adopted many measures. Of these, the most important are (1) persuading banks to put back debt repayment deadlines by 6 months; (2) persuading bank to take over the management of financially troubled corporations through new investments in those corporations; (3) temporarily pausing the transaction of the shares issued by the companies in financial crisis, which is subject to official approval; and (4) setting up a stabilization fund for purchasing the shares from the stock market if their price index drops to 7,000 points. Whether or not these measures are effective depends mainly on whether the crisis-hit corporations can survive. In addition, the government set out to persuade big mutual funds such as the labor retirement fund, labor insurance fund, government employee insurance fund, and postal savings fund to purchase shares in the stock market to increase the demand for shares and thereby offset the fall in stock prices. One may speculate on whether this measure is really effective, it could be called "government intervention", it is appropriate or inappropriate, and it is legal or illegal. If stock prices continue to drop still further, who will be responsible?

5. CONCLUDING REMARKS

Up till now, the financial crisis in Asia has not yet died out. Now is the time for all affected countries to scrutinize their own financial systems and financial management methods. In the case of Taiwan, I am very happy to have had such a good opportunity to review the evolution of Taiwan's commercial banking and financial systems.

As the times and socioeconomic conditions change, a country's economic system should strive to cope with these changes by changing itself. During the last five decades, Taiwan's economy has undergone huge transitions. As an integral part of the economic system, the financial system has also changed. However, change in the financial system resulted mainly from the operation of two forces: (1) the necessity of promoting economic development and (2) external pressure.

5.1. The Necessity of Promoting Economic Development

The scarcity of resources in Taiwan is regarded as the primary factor in the government's takeover of full control of economic development. In the early stages of Taiwan's development, due to the shortage of everyday necessities, the government controlled their prices; due to the scarcity of foreign exchange, the government controlled foreign exchange. In order to maintain the stability of general price levels, the government used to intervene in the operation of the market. In fact, these measures were either only minimally effective or missed the target completely. As the economy continued its rapid growth, the government, under pressure from internal public opinion and academic policy recommendations, gradually relaxed its market control measures and adopted a "hands off" policy toward market operation. In 1985, official approval of economic liberalization and internationalization policy became a significant milestone in the evolution of Taiwan's financial system.

5.2. External Pressure

External pressure has played an important role in the evolution of the financial system in Taiwan, especially that from the U.S. government. Since the late 1960s, Taiwan has enjoyed a huge trade surplus with the United States, which made a great contribution to the rapid accumulation of foreign exchange reserves in the second half of the 1980s. In order to eliminate this huge trade deficit with Taiwan, the U.S. government urged the Taiwan government to open its doors completely to U.S. goods. Although the Taiwan government had formally adopted a policy of economic liberalization, it was still something of a hollow slogan at that time, and no effective steps had been taken. Up until 1990, when the bubble economy burst, the government had already carried out some financial reform and allowed 16 private banks to be set up. Also, by that time the interest rate and foreign exchange rate were being completely determined by market forces.

Despite the fact that Taiwan has weathered the Asian financial crisis very well, it has provided us with many important lessons on how to improve our financial system so that it will be in an even better position to withstand future challenges. The most important lessons learned from the crisis is summarized below.

Finance has gradually come to dominate the global economy. Through financial liberalization and internationalization, many countries of the world are now more closely linked together. Nations must not only maintain sound economic conditions but also modernize their financial systems. Otherwise, a sudden financial crisis may turn a prosperous country into a poor one.

Today there are many mutual funds in the world, and some of them have rapidly expanded and become nightmares for stock markets because of the high rate of return on investments. Since mutual fund managers have plenty of information on each country's financial and economic conditions, they take advantage of their monopolistic power to control stock markets. The question of how to evade the encroachment of international speculators should be an issue of international concern. The solution to this problem lies not in controlling a country's capital movement but establishing a mechanism restricting such speculative activities.

In newly industrialized countries, including Taiwan, politicians and business persons usually maintain a linked, dependent, and coexistent relationship. Enterprises in newly industrialized countries arc generally weak in combating external challenges. Tackling such crony capitalism is of *paramount* importance.

Recently, several large business groups in the region have become entangled in the financial crisis. mainly because they had been using shares as mortgages in order to borrow large sums of money from banks to invest in the stock market. When stock prices drop, these business groups have insufficient money to pay the interest on their loans, causing an increase in the bad debt ratio of the bank on the one hand and a decrease to the supply of funds to support its own activities on the other. Therefore, a sound loan inspection system is urgently recalled, and transparent financial statements from any corporation being issued stocks should also be made a condition of stock purchase.

Small businesses are considered the pillar of Taiwan's economic development. Unfortunately, their many financial problems are still neglected by the government, despite the fact that the government has set up many small-business banks to serve them. In fact, since many small enterprises have no mortgages to use as financial collateral, it is still extremely difficult for them to borrow money from hanks. One possible way to solve this problem is for the government to serve as a guarantor on their behalf, since we have considered that when they borrowed money from the underground financial agents, the bad debt ratio of these agents was very low.

NOTES

1. Illegal or underground financial institutions have played an important role in financing small businesses and consumers in Taiwan. Since many borrowers cannot obtain money from legal financial institutions without having mortgages, they have to borrow money from underground financial institutions, instead, even though they must pay an extremely high rate of interest. This Demand–Supply relationship is the main reason for the existence of underground financial institutions.

2. Whether the Central Bank should be under presidential control or under the control of the Executive Yuan is a question that was long debated in the years before 1979. In Taiwan, the Executive Yuan is the highest administrative institution in the government. However, many scholars were worried about the independence of the

Central Bank under the Executive Yuan and strongly opposed the change. Finally, the central government decided in favor of the new arrangement.

3. Refer to Paul C. H. Chiu, "Money and Financial Markets: The Domestic Perspective," in Taiwan: From Developing to Mature Economy, edited by Gustav Rants, Westview Press, 1992. In his essay. Dr. Chiu, who served at the Central Bank for many years. gives a comprehensive description of the liberalization of interest rate and capital movement regulation.

4. During the period 1949–1961, Taiwan had no formal Central Bank. The functions of the old Central Bank of China, established in 1929, were suspended and not reactivated until July 1, 1961. In the interim, the Bank of Taiwan performed most of the functions of a Central Bank, e.g., issuing notes, handling the banking business of the government and the treasuries of the municipal and local governments, and maintaining custody of foreign exchange assets.

5. Refer to Christina Y. Liu, "Money and Financial Markets: The International Perspective", in Taiwan: From Developing to Mature Economy, edited by Gustav, Rains, 1992.

6. In 1987, Taiwan's small-business persons started to make direct investment in mainland China. From 1987 through 1997, total Taiwan investment in the mainland amounted to U.S.$20 billion. However, since 1995 and mainland China's missile shooting exercises in waters around Taiwan, a thinly veiled threat against the island, the Taiwan government has restricted direct investment in China to not more than U.S.$50 million per case.

7. There is a ceiling on the amount foreign nationals can invest in the stock market: for an individual foreign national purchasing shares in a company, the amount of shares purchased cannot exceed 25% of the total shares of that company; meanwhile, overall foreign investment in a company cannot exceed 30% of the total shares of that company.

8. The gross savings ratio is gross savings divided by the gross domestic product; the gross investment ratio is gross investment divided by the gross domestic product.

9. From 1950 to 1965, the total amount of U.S. aid to Taiwan amounted to U.S.$1.5 billion. This sum contributed much toward economic reconstruction in Taiwan.

REFERENCES

Chiu, P. C. H. (1992). Money and financial markets: the domestic perspective. In: G. Ranis (Ed.), *Taiwan: From Developing to Mature Economy* (pp. 121–194), edited by Westview Press.

GrlllI, V., Masciandaro, D., & Tabellini, G. (1991). Political and monetary institutions and public finance policies in the industrial countries. *Economic Inquiry, 12*, 341–392.

Huang, R.-R., Yang, K. L., Ro, K. H., & Huang, P. Y. (1983). Underground Financial Problems (in Chinese). Nantou: Center of Local Financial Research and Training.

Lai, Y. C. (1997). Retrospect and Forecast of Taiwan's Financial Institutions (in Chinese). Taipei: The United Publisher Co.

Lau, L. J. (1997). Is independence of the central bank desirable? In the economic essays in Memory of Professor Liang Kuo-shu, the Department of Economics, National Taiwan University, pp. 15–38.

Lee, Y. S., & Chen, S. C. (1984). Retrospect and Forecast of Taiwan's Financial Institutions (in Chinese), The Iinstitute of Economics, Academicia Sinica.

Liang, K. S., & Hou, K. Y. (1990). Financial liberalization and internationaliation (in Chinese). In: Jian-dong Shea (Ed.), *Collected Works on Taiwan's Money and Finance*. United Publisher Co.

Liang, K. S. (1992). The banking system and financial reform in Taiwan. *Industry of Free China, LXXVIII*(3), 25–46.

Lin, W.-Y. (1998). The structure and functions of the deposit insurance system – The performance of the CDIC. *Quarterly Deposit Insurance Information, 12*(1), 1–64.

Liu, C. Y. (1992). Monday and financial markets: the international perspective. In: G. Ganis (Ed.), *Taiwan: From Developing to Mature Economy.* Westview Press.

Liu, T. N. (1998). Reconsideration of overseas investment since the Asian financial crisis. *Quarterly Deposit Insurance Information, 12*(1), 99–105.

Shea, J.-d., & Yang, Y.-h. (1996). Taiwan's financial system and economic development (in Chinese). In: T. S. Yu & J. C. Lee (Eds), *Economic Policy and Economy Development – Evaluation of Taiwan's Economic Development*, edited by Chung-hua Institution of Economic Research.

Yu, K. H. (1982). How to propel the liberalization of interest rates (in Chinese). *The Quarterly Central Bank of China, 4*(2), 7–18.

13. A LONG AND BUMPY ROAD TO ECONOMIC RECOVERY: THE CASE OF INDONESIA

Anwar Nasution

INTRODUCTION

Politic of destruction that has been widely practiced since the past two years is tearing the social fabric and cohesion of Indonesia. Such a process of internal destruction is likely to continue with aggravating force when we have general elections this June and the presidential election later this year. There is no assurance whether the coming elections can produce responsible, innovative and creative leaders to reunite the already torn nation, restore civil order, and reestablish a new social contract which address systematic inequities and concentration of economic and political powers and is inclusive of all members of society. Decision-making process is likely to become more difficult with the growing number of political parties and eroding dominance of Golkar, particularly when the elections produce a coalition government. The way we do politic is also changing, from command and instruction to suasion. As of now, none of the existing (old and new) political parties has a platform or clear program on how to solve the current parallel and independent tracks of economic and social crises. In reality the 'parties' are only election clubs without strong ideology and ideas how to rebuild the social system of Indonesia.

Restructuring Asian Economics for the New Millennium, Volume 9A, pages 249–255.

ISBN: 0-7623-0709-9

Slow Progress of Economic Recovery

The uncertain political and security conditions have slowed down the progress towards economic stabilization and recovery and pushed Indonesia deeper into international debt trap. External value of the rupiah has depreciated again from previously nearly Rp7,000 to a U.S. dollar to presently around Rp9,000. The depreciation of the rupiah raises prices of traded goods and push inflation rate up. It also makes more difficult for Bank Indonesia to ease monetary policy and lower interest rates.

The combination of sharp devaluation of the rupiah, high interest rates and economic contraction has caused corporate bankruptcies as it sharply increased liabilities of borrowing firms and reduced their sales. The problem with their customers and high negative spread between deposit and lending interest rates have further contributed to the problems of the already fragile financial system. In 1998, the economy (as measured by GDP – gross domestic product) was contracted by minus 13.5%. This is lower than the original prediction at minus 20%, yet one of the worst in modern history. The unemployment rate and the number of population lives under the poverty line rose sharply while the non performing loan ratio of the banking system is estimated to hover between 60 and 70%.

Monetary Policy

Monetary policy in Indonesia has swung back and forth from defending exchange reserve position of the central bank to lowering interest rates. In the beginning of the crisis, Bank Indonesia provided massive liquidity credit to financially distressed banks. This resulted in doubling monetary base between November 1997 and March 1998. The rapid expansion of money supply was partly sterilized by Bank Indonesia intervention in foreign exchange market, selling its own certificates of deposits and shifting the public sector deposits from state-owned banks to the central bank. The policy shift to defend the central bank's foreign exchange reserves begun in April 1998 and Bank Indonesia allows interest rate to adjust freely.

The rupiah appreciation since November of last year had been mainly because of the availability of massive financial supports from international community for Indonesia's economic stabilization and recovery program. The financial assistance helps finance the twin deficits of the balance of external payments and the public budget. Foreign donors pledge a $42.5 billion new loan to Indonesia following the IMF standby arrangement of 24 June 1998, the fourth agreements since October 1997. Because of the social unrest, private capital flows have not returned. Demand for foreign exchange is rising because of the panic-induced capital flight and the requirement for repayment of public and private sectors' external debts. Meanwhile, the large devaluation of the rupiah has helped compress imports but not powerful enough to boost non-oil exports.

Fiscal Policy

The economic crisis forced the government to redefine the scope of government change the role of public budget from, previously, a tool of resource allocation to a principal instrument of macroeconomic policy. The rise in both domestic and external debt burden requires the government to cut non-debt expenditures. The use of fiscal policy as an 'automatic stabilizer' is limited due to the rigidities of external assistance from official sources. As the inflexible tax system cannot be expected to pursue equity objective, the expenditure side of the government budget is also expected to do this job. For these, the public budget had been revised four times in fiscal year 1998/99 to have deficit amounted to over 6% of annual GDP, lower than the original prediction at 8.5% of GDP. Moreover, the non-debt public budget expenditures of the public budget have been spent more on the non-traded sector of the economy. This sector is labor intensive and requires minimal imported inputs. Such expenditures include primary school, extension and outreach services, public health care program and public work programs in rural areas.

Meeting the targets for the next fiscal year of 1999/2000, starting in April, will face with four main obstacles. The first two obstacles come from domestic recession that erodes fiscal efforts in two ways. On revenue side, the economy's downturn hurting both tax and non-tax revenues. The real value of government tax revenue is eroding because of the relatively long collection lag and high inflation and interest rates (Tanzi effects). On the expenditure side, there are growing demands for public resources that potentially undermine support for spending austerity. The rising demand, among others, comes from the following budget items: subsidies on state-vended products, repayment of external public debt, social safety net program, interest expenses of government bonds issued for financing the restructuring and capitalization programs of the banking system (including Bank Indonesia, the central bank). The third obstacle to fiscal policy is the presently weak price of oil in international markets, the lowest in the past decade. The United States and United Kingdom military attack on Iraq last December had hardly affected the price of oil mainly because the target of the attack is only weapon and not oil.

Other obstacle comes from the popular demand for decentralization of power to local governments to implement the principle of *Bhineka Tunggal Ika*, the system of government that combines the unity with diversity. Decentralization reduces monopoly power of the government on the main sources of revenues: rents on exploitation of natural resources (oil, mining, timber and marine resources), and taxes on income, internal and external transactions and consumption. The eroding revenue base of the central government also reduces its the capacity to correct horizontal imbalances among local governments through fiscal transfers. The horizontal imbalances is inevitable because of differences in fiscal capabilities of local governments due to variation in their revenue base and administrative capacities, together with the costs of providing government services.

The alternative methods of delegating authority over delivery of public services should be carefully studied to be able to deliver the private merit goods on competitive markets to the target groups at least costs. In the presently adopted top down centralized system, the authority is delegated to bureaucrats by the central government that has limited ability to monitor their performance with respect to either service delivery or cost control. Because of this, the system is prone to corruption and lack of accountability as shown by the bad performance of the present social safety net programs. As Indonesia has no modern social safety net system, the authorities try to supplement the falling primary income of the poor (i.e. income derived from labor and assets), among others, by raising their social income through provision of basic necessities at subsidized prices. The traditional informal arrangements based on familial and communal supports (*'gotong royong'*) have been weakened along with the urbanization, migration, and specialization during the process of development. The program requires high public budget subsidy, but the products do not necessarily reach the needy. At the same time, the subsidies create large wedge between domestic and international prices of the state-vended products. The price differential provides incentive for smuggling of the products to neighboring countries.

In theory, elected local governments in the decentralized system should have better information about local needs and costs of specific communities. In such a system the authority is given to elected local governments or client groups which may be subject to capture by local elites. Nevertheless, inter-community targeting is also ambiguous under this system. Because of this, if it is not carefully prepared, the decentralized system is also prone to corruption and lack of accountability.

Public budget must adjust to the new reality of eroding capacity of government to raise tax revenues and to borrow both from domestic and foreign sources. The scope of the government in the economy has to be redefined. The role of the budget has to be shifted, from a tool of resource allocation to a principal instrument of macroeconomic policy as the vehicle for providing public goods.

The fiscal policy must be directed for stabilizing the public sector external debt (as a share of annual GDP). High reliance on external debt for financing the public budget deficit is not sustainable. The budget deficit for fiscal year of 1999/2000 is expected to amount to around 6% of annual GDP. To reduce the need for external financing, budget requires even larger primary surplus or revenues from privatization of state-owned enterprises and quick sale of assets auction of the financially distressed banks with high recovery. As of now, only 41 of the existing 160 state-owned enterprises are financial sound and a small number of them are eligible for public offering.[1] Market values of the rest are very small due to weak structural problems.

The fiscal distressed requires restructuring of public sector external debt. It is true that the bulk of the public sector external debt is long-term in nature. However, the associated principal and interest servicing cost still run high, around $9 billion a year. To provide a temporary relieve for pressures on budgetary and balance of external payment, Indonesia need a more comprehensive rescheduling of its public external debt.

Deregulation Program and Program to Rebuild Economic Institutions

Aside from expansionary fiscal expenditures, economic recovery can be promoted by accelerating growth of non-oil exports and by stimulating of private sector capital inflows. The large devaluation of the rupiah has led to substantial improvement in Indonesia's external competitiveness. Nevertheless, more need to be done to improve business and investment climate by eliminating economic distortions, by removing administrative restraints, by supplying adequate working capital with reasonable interest rates and by improving market transparency to cut transaction costs. All of these require forceful implementation of the medium-run deregulation program and consistent reforms of the social system, particularly legal and accounting systems, to improve market infrastructure. The deregulation program will improve efficiency and productivity. Improvements in market infrastructure reduce transaction costs.

To summarize, the present crisis is not only about economic contraction, high inflation and interest rates, high unemployment rate and compression in purchasing power. The present crisis is also characterized by melting down of economic institutions, bad market infrastructure and torn social cohesion. As a result, beyond economic and political stability, there are three priorities to restore economic growth. First, to restructure the rotten banking system, including Bank Indonesia, the central bank. It is clear that the public funds will be needed to restructure the system particularly as the publicly owned banks (Bank Indonesia, state-owned banks and regional development banks) are the weakest parts of the system. The most urgent need is to re-capitalize the viable banks so that normal banking activities can be resumed as soon as possible. The second priority is to restructure the corporate sector (state-owned, private owned and cooperatives) and solving their external debt overhang.

Solving those two problems require rebuilding of market infrastructure, particularly legal and accounting systems. This is the third priority that needed to address the problems in an open and transparent framework, with due consideration to the principle of justice and equitable distribution. The issues to be solved include dealing with the financial distressed banks and non-bank companies, non-performing loans, asset management of owners of the banks and companies and repayment of their debts to the government. Solving these problems require an effective bankruptcy procedures, external assistance and helping hands from the public sector to provide funds and guarantee for alleviating risks.

Need Responsible, Innovative and Creative Statesmen

All of the above programs require a lot of efforts and energy. More importantly they require visionary, innovative, creative and responsible political leaders who can take necessary bold and daring measures. For sure, cutting (consumption, investment and public budget) expenditures and raising the prices of state-vended products that required by short-run stabilization program is politically unpopular. As has been discussed early, the structure of the cuts is equally important as it affects distribution of the adjustment costs and potential growth for the future. Some segments of the society are hurt by deregulation program (such as holders of monopoly and oligopoly rights). Rebuilding the social system requires elimination of *KKN* (corruption, collusion and nepotism).

To get out from the present economic and social crises we should stop killing each other and restore law and order. The competing political parties must adopt politic of civility (*madani*) and reach a common ground and consensus for giving top priority to solve the economic crisis. We have to restore peace and political stability and show strong will to solve our own problems of our own makings. Without this, no donor country or international institution will extent helping hands to us that badly needed for providing food and medicine for the needy, for stabilizing the exchange rate, creating job opportunity and for reducing inflation and interest rates.

The government and political parties should strongly resist the temptation for adopting dangerous populist policy, 'Robinhood economics' and money politic for short-run political objective. Asset redistribution and resource reallocation through non-market mechanism that favors one ethnic group over others will not work and will only worsen the matters. Such arbitrary non-market mechanism will waste the scarce resources, raise inflation rate and disturb the required deregulation program to improve efficiency and investment climate that conducive for exploiting natural-resource base and labor intensive sectors where lies Indonesia's comparative advantage.

NOTE

1. In fiscal years 1998–1999 and 1999–2000 equities of 13 state-owned enterprises (SOEs) are to be partly sold to private sector. These companies are PT Telkom (domestic telecommunications), PT Indosat (satellite and international communications), PT Aneka Tambang (mining), PT Tambang Timah (tin mining), PT Semen Gresik (portland cement), PT Tambang Batubara Bukit Asam (coal mining), PT Pelindo II and PT Pelindo III (sea ports), PT Angkasa Pura II (airport), PT Jasa Marga (toll road), PT Krakatau Steel (steel mills), PT Perkebunan Nusantara IV (plantation). Part of the equities of the first five SOEs have been sold in local and international stock exchanges.

14. ECONOMICS FOR THE MILLENNIUM

Pasuk Phongpaichit

The topic for the panel discussion is Asian Economies in the New Millennium. But I'm afraid I want to wander off the subject just a little. I find it difficult to talk about the 21st century when we are having such difficulty at the end of the 20th. Besides, if you want to know the future, I suggest you consult astrologers.

We have a huge crisis of the economy. But, even more, we have a crisis of economics.

What roles have professional economists played here or made major contributions to getting the country over this crisis?

I think it is quite symbolic that the most important figure in the Thai economy in this crisis is not an economist at all. Tarrin Nimmanhaeminda, our Minister of Finance, was not trained as an economist. By profession, he's a banker. This is not in any way a criticism.

The fact is, during this economic crisis, we professional economists have been largely powerless spectators.

Moreover, how well do we think our students can understand this crisis? I have some students who don't even seem to know there is a crisis, but that is another matter. What I really mean is: how much do the courses we teach enable our students to understand real-world economics today?

This is not just a problem for us here. It's not just a problem for Thai universities. It's a problem for economics faculties in universities everywhere around the world.

Restructuring Asian Economies for the New Millennium, Volume 9A, pages 257–259.

ISBN: 0-7623-0709-9

Recently, Paul Krugman said the American people might be horrified to learn that if they opened the best-selling U.S. economics textbook, they would find nothing to help them understand this crisis. He could guarantee this, because he had co-authored the textbook.

I'm not a great Krugman fan, but I think this is a very brave and very liberating statement.

In the west, there are some economists (like Paul Ormerod) who recognize that the economics of the real world have changed so much that the textbooks are becoming more and more irrelevant. The whole "institutional" school represents one attempt to bring the subject of economics a bit closer to reality.

For developing countries, 1 think the problem is even more complex. Old-style "development economics" did not die, but simply faded away. Since then, the rise of neo-liberalism has told us that we don't need anything to put in its place. We can sit back and leave everything to the market.

1 attended a conference recently. Two groups of our most eminent economists reported back on a year-long project to find ways to help Thailand to move beyond this crisis. They produced nothing new. They *admitted* that they produced nothing new. Their message seemed to be: sit tight, do what the foreigners tell us, and hope. Above all else, hope.

I don't think I was the only one who found this hugely depressing. Here we are in a crisis which we all know is totally new, totally unprecedented, and very dangerous. And our most eminent economists cannot come up with anything to match the novelty and scale of the event.

Now, I'm not blaming anyone; I count myself among the guilty. But I would like to offer something about the 21st century which is somewhere between a wish and a prediction.

I hope this crisis will prompt a significant new look at the subject of economics, especially the part we used to call "development economics", and especially the sub-set of that which is relevant to Asia. I doubt very much whether this new look will come from universities and economists in the west. I think it is much more likely to come from within the region.

I hope this new look will take its starting point not as "micro" or "macro" but as "globo" – the understanding of the modem world economy, the role of firms and financial flows, the position of the modem state within this context, and the issue of regional and global governance.

I hope also that this approach will concentrate on how, within this international context, we can still strive to improve the wealth and well-being of all members of the society.

This is a critical point. At the moment, we seem to have two paradigms which simply don't talk to one another. On one side, the neo-liberal view says the market must rule. But this view has no answer to the proven facts that market liberalization leads to greater social division and social exclusion.

On the other side, we have the self-reliance paradigm, which urges individuals and communities to protect themselves as best they can. But this view has no macro or "global" theory. Hence, it is bound to be limited.

In my view, this crisis shows that we have to protect ourselves from the dangers in the global economy (we have to reform too, but that won't be enough); we have to be more self-reliant; and we have to pay much more attention to the management of our internal economy at the macro economic level, particularly to the distribution of assets and wealth.

We have been so committed to export-led growth (not just in the industrial phase of the last 15 years, but in the agricultural phase before that) that all our economic management is geared to the external economy.

Now we find that we cannot depend on the external economy as before. Suddenly, we need to think how we can generate and manage internal demand. But we don't have the tools. Our leaders don't seem able to make the theoretical leap. And we are not teaching our students how to confront this problem either. Last week, 1 asked my class what could be done to stimulate internal demand. The question simply floored them.

The big risk of this crisis is that we stumble through it and emerge with a renovated banking system but little else. If we economists could get angry enough about our own apparent irrelevance during this crisis, then I think we might be able to create something which would give us a bit more hope for the 21st century.

15. CHALLENGES TO HUMAN RESOURCE DEVELOPMENT IN ASIA

M. G. Quibria[1]

INTRODUCTION

The last thirty years saw the dramatic transformation of developing Asia. This dramatic transformation of the Asian economy, now widely referred to as the Asian miracle, reflects a constellation of conducive factors such as stable macroeconomic policies, openness to trade, high saving rates, generally sound institutional frameworks, high literacy rates, and favorable demographic characteristics. However, not all of developing Asia share these characteristics, nor have all of developing Asia been part of the Asian miracle in the same manner as East Asia has been. Indeed, the miracle largely bypassed some parts of Asia, such as South Asia and the Philippines. However, since 1990 these countries have undertaken significant reforms of their policy frameworks and laid strong economic foundations for growth.

Nevertheless, despite improvements in policies and institutions, many of Asia's high-performing economies seem to be facing a serious challenge to their economic growth as a result of the ongoing financial turmoil. To most observers, this turmoil reflects structural problems such as institutional frailties in the financial sector and weaknesses in its systems of corporate governance. It is believed that once these issues begin to be dealt with, growth will resume.

Restructuring Asian Economics for the New Millennium, Volume 9A, pages 261–277.

ISBN: 0-7623-0709-9

However, the challenge to a resumption of "miracle" growth rates may appear from another source: gaps in human skills and in technological capabilities. The main precursor to the financial crisis was the export deceleration of 1996 (Asian Development Bank, 1998). Much of that deceleration in exports may well have been cyclical, for example, caused by the downturn in demand for electronic products. But for some of the Southeast Asian tigers, the lack of an adequate supply of well-trained technicians, engineers, and scientists constrained the ability of these economies to move from simple assembly-line operations in foreign-built plants toward designing and developing products in the face of competition from lower-wage economies such as the People's Republic of China (PRC) and Viet Nam. For the other more advanced economies of East Asia, it has been asserted that they are fast approaching the end of growth attainable through input mobilization (Krugman, 1994). Further acceleration of growth would require technological progress attainable through new innovations that stem from a more sophisticated endowment of human capital than is available in these economies.

There is another reason why the returns from investing in human resources will be critical for future prosperity even if they have not always been so in the past. Globalization has increased technology flows from developed to developing countries. It is the well educated who are in a position to gain the most from the introduction of new technology.

HOW IMPORTANT IS HUMAN RESOURCE DEVELOPMENT TO GROWTH?

In order to improve its standard of living, an economy must produce greater amounts of goods and services per worker. This may be achieved by equipping workers with more tools, that is, investing in physical capital, but may also be achieved by investing in *human* capital to improve the quality of human resources. In fact, a strategy based only on accumulation of physical capital will soon run into diminishing returns (Krugman, 1994). It is only technological progress that can allow sustained growth of output per worker. Additionally, contrary to what many early economic growth models assumed for simplicity (for example, Solow, 1956, 1957), technological progress does not fall like "manna from heaven." Instead, it is the outcome of human endeavor and depends critically on the quality of human resources. Even when technology does appear to fall like manna from heaven, as may be the case with technological latecomers, adopting successfully the new technology developed elsewhere is no easy task. It is ultimately tied to the quality of human resources once again.

Investment in human capital to augment the quality of human resources can take many forms. Perhaps the most basic of these are investments in people's health. Research has confirmed that the adverse effects on productivity of poor health and nutrition – as reflected in inadequate calorie intake, low weight-for-height, or low body mass – are considerable. Moreover, it is not just the ability to carry out strenuous work that suffers. Inadequate nutrition, especially when measured in terms of low height-for-age among children, has long-term and lasting adverse effects on cognitive development and schooling achievements. These are likely to lead to lower labor productivity in adulthood.

While the productivity effects of nutrition and health are strongest at low levels of nutrition and income, they level off at higher income and nutrition levels (Strauss, 1986). Further improvements in productivity can only be sustained if workers possess the knowledge and skills required not only to utilize given tools and technology effectively, but also to generate and manage new tools and technology. Workers accumulate this technological capability in myriad ways. On-the-job training and learning are important, as are research and development (R&D) activities that are undertaken precisely for the purpose of developing new technologies or utilizing effectively those developed elsewhere. However, it is no exaggeration to say that formal education is the key to technological capability.

To what extent is worker productivity associated with education? It is difficult to deny that schooling yields important pecuniary returns to individuals in the form of higher earnings. These returns can be relatively large in developing countries and some estimates indicate private rates of return to schooling in Asia to be as high as 31% (Psacharopoulos, 1988). It is, of course, possible that only a small part of these returns reflects the productivity-enhancing effect of education. The rest could reflect the importance of credentials in higher paying jobs or education as a screening device that allows employers to distinguish between high-ability and low-ability individuals, because high-ability individuals will go through school more easily than low-ability individuals (Spence, 1976).

However, the effect of schooling on wages works primarily through its positive impact on cognitive achievements. In addition to enabling workers to do everyday tasks more efficiently, education, even of the basic kind, is found to facilitate the adoption of new tools and technologies. However, the contribution of education varies with the context and would be higher where technology changes rapidly.

A recent study uses India's experience with the green revolution to shed light on this issue (Foster & Rosenzweig, 1996). While farmers with a primary education were in general more productive than their uneducated counterparts, the productivity differential was greatest in those regions that were especially conducive to the cultivation of the new high-yielding varieties of seeds. In these regions educated farmers' profits grew to be as much as 46% more than those of uneducated farmers. The findings strongly suggest that it was the educated who responded best to the new technology.

Supportive evidence comes from research using industrial data from three developing economies including Taipei, China (Tan & Batra, 1997). While firms' investments in activities to enhance workers' knowledge, such as training programs, lead to higher wages and productivity for both skilled (typically well-educated) and unskilled labor, the gains made by skilled workers are much larger. In addition to this micro-level evidence, some recent evidence using cross-country data also points to the dependence of returns to education on an economy's context. In particular, a recent study by Mingat and Tan (1996), which separates countries into three groups: low, middle, and high income, suggests that the benefits by level of education are sensitive to countries' economic circumstances. In particular, whereas the typical low-income country would benefit most by expanding its coverage of primary education, middle and high-income countries would benefit most by expanding their coverage of secondary and tertiary education, respectively. The simplicity of production technology, economic processes, and commercial and legal institutions, including the limited scope for specialization of labor in low-income countries relative to higher-income ones, could explain why expanding the supply of highly educated people in these economies may not have too much impact on economic growth. By contrast, for more developed countries that are at or near the frontiers of technology, sustaining growth comes from developing new technologies and refining older ones. And in this it is practically impossible to deny the key role higher education plays in enabling activities, such as R&D, which are undertaken precisely to push forward the frontiers of technology and knowledge.

However, higher education and the technological capability it instills in a workforce can be of benefit to developing economies also. This can be seen by examining the returns to activities such as R&D, which are intensive in their use of highly educated individuals. In general, investigators have looked most closely at the economic benefits to developing countries from R&D activities in the field of agriculture. In this context, the evidence from public sector agricultural research programs in Asia suggests that the internal rates of return to R&D activity have been on average 56% (Evenson & Westphal, 1995). While few estimates of the rate of return for industrial R&D in the developing world exist, they suggest high returns.

HOW MUCH IS DEVELOPING ASIA INVESTING IN EDUCATION AND TECHNOLOGY?

Literacy, Education, and Technology Development

One rough but ready measure of the stock of human capital is the literacy of the adult population. As an examination of adult literacy rates across selected ADEs reveals, typically more than 80% of the adult populations of East and Southeast Asian economies are literate. This is in sharp contrast to many South Asian economies that still have literacy rates below 50, if not 40%! Of course, it is true that these South Asian economies (Bangladesh, India, Nepal, and Pakistan) have shown much improvement over the last 30 years. However, some of the most impressive improvements in adult literacy have taken place in Southeast Asia. For example, whereas Indonesia had a literacy rate that was around 11 percentage points higher than that of India in 1961, the gap between the two had increased to about 32 percentage points by 1995.

Over time and with the increasing coverage of basic education in all of these economies, the fraction of the adult populations that are illiterate will decline substantially. However, the large illiterate populations in South Asia are likely to continue to act as a drag on their economies for some time to come unless programs designed to target adult illiteracy are introduced in a big way.

What of Asia's investments in human capital? Despite a number of problems, enrollment rates[2] at various levels of schooling – i.e. the number of students enrolled in education at a particular level divided by the population in the age range corresponding to that educational level – do provide a measure of the investments being made in human capital. They can also be used as an approximate measure of education outcomes and can be useful in highlighting various broad features regarding education. Indeed, a comparison of gross enrollment rates across the selected ADEs reveals a number of interesting features.

Consider enrollment rates for primary education. Data reveals that many East and Southeast Asian economies have been quite successful historically in the provision of primary education especially in comparison to perhaps all South Asian economies with the exception of Sri Lanka. However, there has been a rapid improvement in recent years in the provision of primary education in South Asian economies (plus such Southeast Asian laggards such as Laos). Therefore, the dispersion of enrollment rates among Asian economies is much less today than it was three decades ago.

Nevertheless, a more careful look beyond the enrollment rates reveals that access to a decent primary education is a serious problem in the low-income economies of Asia, especially in South Asia and some economies of Indo-China. First, access by children across income groups and gender in these economies can vary greatly. For instance, in Cambodia and Laos, the gross primary school enrollment rate for children from the wealthiest 20% of families is, respectively, 35 and 43% greater than that for the poorest 20% (Deolalikar, 1997). Similarly, in India the gross primary school enrollment rate of girls is only 80% that of boys, while in Pakistan the figure is only 50% (Deolalikar, 1997).

Second, enrollment rates do not tell us too much about the quality of education being provided. Direct measures of quality of education, such as performance on standardized test scores, are hard to come by. Nevertheless, there are a number of indirect measures of the quality of education. One manner in which low quality can manifest itself is through high rates of dropout and grade repetition. Although high rates of dropout could reflect the perceived lack of relevance of education, especially in rural areas, there is evidence to suggest that poor quality of schooling can lead students (prompted by their parents) to drop out (Dreze & Gazdar, 1997; Hanushek, 1995). The problem of dropping out seems particularly acute for South Asian economies and for some of the Indo-China countries. For example, of the cohort of children who enter primary school in low-income ADEs such as Bangladesh, Laos, Nepal, and Pakistan, only about half complete sixth grade (Deolalikar, 1997). Moreover, in some of these countries, about a quarter of all children repeat a grade and, consequently, some children spend two to four years longer in primary school than normal.

Consider next enrollment rates across the ADEs at the secondary level of education. As is the case for primary education, these enrollment rates do not capture issues relating to quality. Nor do they describe to us the related problems of high dropout rates, grade repetition, and gender and income biases. Nevertheless, as rough measures of the investments in education, these do reveal some interesting features. First, the cross-country variation in enrollment rates is much greater at the secondary level than at the primary level. For example, while Bangladesh had a gross secondary enrollment rate of 20% in 1992, Korea's enrollment rate was 90%. Second, the enrollment rates in secondary education in some of Asia's higher-performing economies seem inadequate, particularly given their higher income levels.

This can be seen most clearly by comparing actual enrollment data at the secondary level alongside those that predicted by each country's level of development as proxied by their income levels.[3] As may be expected, countries with higher per capita incomes also tend to have higher enrollment rates although, as the figures make clear, the relationship is by no means a watertight one. Such calculations indicate that there are large and negative differences between actual and predicted enrollment rates for such high-performing economies as Hong Kong, China, Malaysia, Singapore, and Thailand at the secondary level. Whereas Hong Kong, China and Singapore may be classified as high-income economies, Malaysia and Thailand can be classified as middle-income economies.

Finally, consider enrollment rates at the tertiary level of education. As may be expected, the enrollment rates for the more developed newly industrialized economies or NIEs (Hong Kong, China, Korea, Singapore, and Taipei, China) are higher than those of other ADEs with the exception of the Philippines. Again, comparisons of actual and expected enrollment rates at the tertiary level indicate large and negative differences in such economies as Hong Kong, China and Singapore. However, given that the future growth prospect of the NIEs is probably most critically dependent on education at the highest levels, the large and negative differences between actual and predicted enrollment rates for Hong Kong, China and Singapore do not bode well.[4]

Unlike basic education, which entails development of basic skills relating to literacy and numeracy, higher education, especially tertiary education, prepares students for more specialized occupations and thereby varies by field of study. For the purposes of developing technological capability, higher education in the natural and applied scientific fields is an obvious necessity. However, an examination of the enrollments in these fields as a proportion of total tertiary enrollment indicates that this proportion across the more dynamic Southeast Asian economies and the lagging economies of South Asia is very similar. Indeed the relatively low proportion in the scientific fields in these economies relative to the dynamic NIEs is consistent with the finding that Korea and Singapore have around ten times as many R&D scientists and technicians per capita as the other countries in the region and are comparable to the industrial economies in this regard.

Similarly, an examination of R&D spending as a percentage of GNP for the same cross section of countries reveals that the more dynamic Southeast Asian economies do not invest as much in technology-related activities as would be suggested by their higher incomes. For example, R&D expenditures in Indonesia, Malaysia, Philippines, and Thailand are much less than those of India or Pakistan.

The Importance of Developing Technological Capabilities

While it is true that foreign technology represents an important opportunity that the ADEs must exploit, it is equally true that the ADEs ignore the development of local technological capabilities at their own peril. Indeed, recent trends in some Southeast Asian economies suggest how important local technological capability can be in sustaining growth.

The economic growth experience of Asia's high-performing economies has often been described in terms of the metaphor of the "flying geese." According to this metaphor, as Japan, the most technologically advanced country in Asia, moved away from being an exporter of labor-intensive manufactured goods such as textiles to more high-technology products, the NIEs took its place. Similarly, as the latter set of countries developed more skill intensive exports, other Asian countries, essentially Southeast Asian ones, stepped in to fill the vacancy.

At first glance, an economy such as Thailand would seem to be a good example of this process, but a more careful examination suggests that there is cause for concern about its future exports, and hence, growth performance. Certainly, Thailand follows the pattern in that it has shown a marked increase in the proportion of its total manufactured exports, from 32% in 1980 to 80% in 1995, with a consequent decline in the share of agricultural exports. Moreover, manufactured exports have shifted away from labor-intensive products to products classified as medium to high technology, so that by 1993 the latter exceeded the former. In 1995–1996 the growth of exports fell well below the average for the preceding four years, but the medium to high-technology products performed relatively less badly than the other categories.

Why then is there cause for concern? The problem is that the decline in the growth of Thailand's low-technology exports has come from increasing competition from the low-wage economies of Bangladesh, PRC, India, and Viet Nam. As the examination of wage rates across ADEs in Table 1 suggests (adapted from Lall, 1998), economies such as Bangladesh, PRC, India, and Sri Lanka have an advantage over such ADEs as Malaysia, Philippines, and Thailand in the manufacture of unskilled and semiskilled labor-intensive products. This makes it less likely that the latter group of countries can continue to specialize in traditional manufactured exports in which the primary unit cost advantage stems from low wages, and which have grown as rapidly as they have in the last 10 to 15 years.

All of this would not be serious if it were more than compensated for by sustained growth in the medium to high-technology exports. Indeed, it could be said to be highly desirable if it reflected increasing sophistication of Thailand's industrial base. However, the crux of the problem is that Thailand is essentially acting as an assembly point for these medium- to high-technology exports (Lall, 1998). Thus while these goods, such as electronics and cars, are classified as medium to high-technology products, they do not involve a highly skilled labor force for that part of their production that actually takes place in Thailand.

Table 1. Annual Wages in Manufacturing (U.S.$), Selected Asian Economies and Years.

Economy	1970	1994
NIEs		
Hong Kong, China	n.a.	15,160
Korea, Rep. of	502	14,295
Singapore	n.a.	17,794
Taipei,China	n.a.	14,469
PRC		340
ASEAN-4		
Indonesia	165	1,001
Malaysia	707	4,555
Philippines	651	2,857
Thailand	n.a.	4,917
Cambodia	249	n.a.
South Asia		
Bangladesh	315	1,016
India	445	1,269
Pakistan	453	n.a.
Sri Lanka	n.a.	837

Source: UNIDO Industrial Statistics Database.

Hence, Thailand's production of these exports is likely to come under increasing pressure from the low-wage and unskilled labor economies.

A similar pattern is discernible in the Philippines, where during 1991–1996, electrical and electronic exports increased by 37% per year while exports of textiles and garments increased by merely 8.2%, the slowest growth rate among all ADEs with the exception of Korea and Taipei, China (World Bank, 1997). Unfortunately, electrical and electronics exports of the Philippines have very low value added – a reflection of the fact that the local industry is primarily engaged in the simplest types of assembly and testing activities. For example, the average local content in semiconductors – which accounted for 77% of total electronics exports in 1995 – is only about 20%. By comparison, Taipei, China has achieved an average local content of about 75% in the production of semiconductors (World Bank, 1997). Moreover, the local content in the electrical and electronics industry has not increased appreciably in the Philippines during the past two decades, indicating low development of technology and technological capability.

If the process of the hollowing-out of the industrial and export base continues, it will spell new difficulties for these economies. Of course, the steep currency devaluations in these economies have given them some breathing space. However, the difficulty is that at the moment, the human resource base to produce those exports where skills are important and that provide a measure of protection from competition from the low-wage, unskilled economies does not exist. While Thailand enjoys almost universal enrollment rates in primary education, its enrollment rates in secondary education remain low and its tertiary education appears to be somewhat biased against basic and applied science. Estimates suggest that Thailand is producing less than two thirds the number of engineers and scientists with undergraduate and graduate degrees that it requires (Lall, 1998), and Thailand has a lower proportion of tertiary-level students enrolled in natural and applied scientific fields than such low-income ADEs as Bangladesh and India.

For economies that have lost their competitiveness in the production of semi-skilled labor-intensive goods, regaining competitiveness in the absence of cuts in wage rates will involve either improving productivity in the production of the same goods or moving into the production of new and improved goods. And the latter will entail a greater emphasis on the design, development, and marketing of these goods as opposed to assembly-line operations. However, both options require workforces to upgrade their technological capabilities.

HOW TO IMPROVE HUMAN RESOURCES?

Allocation of Resources to Education and Technology Development

The case for public financing of the various human resource development activities must depend on two grounds. The first is the difference between social and private returns across various activities. When this difference is positive and large for any particular activity, the decisions of private individuals will lead naturally to an underinvestment in that activity from the point of view of a social optimum. Education generally falls into this category as it is believed to bring substantial benefits to society as a whole, and not just to those individuals who receive it. That is, education has a significant positive externality.

While activities geared toward the creation of new technologies such as R&D can reap substantial benefits to those conducting them – and this is precisely why profit-seeking firms conduct R&D – such activities can nonetheless benefit others as well. That is, the social returns can exceed the private returns. This happens because the knowledge developed as a result of

R&D activities is difficult to appropriate completely by those carrying it out. Others can also derive benefits by virtue of imitation and reverse engineering. To the extent that this takes place, the market will undersupply them relative to the social optimum. Additionally, relying excessively on imports of new technology can reinforce the tendency for undersupply of R&D. Importing technology represents a very important opportunity for the ADEs. However, to some extent, such imports will substitute for local efforts and over time this can have adverse consequences for domestic technological capability.

The second justification for public financing is on grounds of equity and income distribution. If education is left entirely to private financing, the poor and disadvantaged are likely to be denied access to it. Even though poor people may wish to partake of education in the knowledge that this would greatly enhance their future earnings, they may have no way to pay for it. Typically, future earnings cannot be used as collateral for a loan, and so they cannot borrow against the stream of future incomes to pay for the education. The situation is similar with R&D. R&D activities are typically quite costly, and generally involve substantial outlays up front (for example, in setting up a well-equipped laboratory). While the costs of carrying out R&D may be sufficiently recouped through higher profits over time, obtaining credit can be a problem, especially for smaller firms. Effectively, these firms are denied the opportunity of conducting R&D.

In principle, therefore, there is a case for public financing of both education and technology development. However, in the face of limited public sector budgets, choices have to be made regarding which activities deserve the support of public finances and to what extent. A widely held view is that there is ample scope for reallocating public finances toward primary education and away from higher education in many developing economies, particularly those where access to primary education remains a serious problem. First, despite the many difficulties in computing social returns and, therefore, the difference between social and private returns, many believe that these are greater for primary education as opposed to higher education. This view is perhaps best characterized by an illustration in Psacharopoulos (1996): the probability that higher education leads through research to a life-saving breakthrough is very small. On the other hand, lack of schooling guarantees illiteracy, which in turn is very likely to result in high transactions costs for the rest of the economy. Second, students enrolled in higher education can typically afford to pay for it.

These arguments suggest that it would seem to make eminent sense for the low-income ADEs to focus on their provision of primary education. The primary sectors of the many low-income ADEs, especially those in South Asia, suffer from a number of serious problems of access and quality that could, in principle, be tackled through the allocation of more resources towards them. Yet, many of these economies tend to underemphasize their primary education sectors relative to their secondary and tertiary sectors. The ratio of spending per student in primary versus secondary education, as Mingat (1995) notes, for four South Asian countries (Bangladesh, India, Pakistan, and Sri Lanka) and the PRC is on average 0.40. In contrast, the ratio for both the ASEAN-4 and the NIEs is on average 0.70 and 0.71, respectively. Similarly, the pupil-teacher ratios across levels and ADEs confirm the low relative importance of primary education in low-income ADEs. As computed by Mingat (1995), the "Asian means" for pupil-teacher ratios were 31, 23, and 18 at the primary, secondary, and tertiary levels respectively.[5] The largest deviations from these means obtain in primary education for three South Asian economies, Bangladesh, India, and Pakistan. For example, whereas the pupil-teacher ratios in Bangladesh at the secondary and tertiary levels were 27 and 19, respectively, and therefore not too far from the "Asian means," it was 64 at the primary level – more than double the mean pupil-teacher ratio.

Thus, in comparison with other Asian economies, especially the better performing ones of East and Southeast Asia, South Asian countries seem to devote fewer resources to the primary sector relative to higher levels. At the same time, they also have the poorest performance in terms of enrollments and student retention rates, suggesting strongly that relative neglect of primary education is responsible for its inadequacy.

At the same time, the higher-income economies of Asia that have been able to provide greater access to primary education than their South Asian neighbors cannot afford to rest on their laurels. As their economies have developed, sustaining future growth requires greater skills and knowledge. Yet, particularly for the high-performing Southeast Asian economies, the current systems of education have not delivered the required skills and knowledge.

This is not to suggest that public resources be used to provide blanket subsidization for higher education. Not all of higher education involves research and research-based training, activities for which there could be important externalities for society. In fact, much of higher education involves the training of students for relatively well paid professional functions. The appropriate role of the government insofar as the latter group of functions of higher education is concerned is to ensure an effective loans and scholarship program to complement user fees. Direct subsidies should essentially be reserved for research and research-based training that have important externalities. Moreover, these subsidies should be channeled in the form of research grants to departments active in research.

In addition to subsidizing research activities in institutes of higher learning, it is possible for governments to promote research activities by supporting special institutions that facilitate the development of technologies that the private sector can use. This is because in addition to the externality and credit constraints noted above, less developed economies lack the critical mass of researchers that individual firms would require for successful technology development. Thus, in Korea, for example, government research institutes took the lead in technological activities initially, with the private industry taking a supporting role. By the end of the 1970s, Korea had at least 16 R&D institutions (many of which were later consolidated under the Ministry of Science and Technology). Similarly in Taipei, China, government research institutes initiated a system of extension and contract research to provide technological support for private industry.

In their efforts to develop local technological capability, R&D institutions in both economies worked closely with the private sectors and were not meant to substitute private efforts but rather nurture them. Thus, whereas the Korean government accounted for nearly three quarters of the national R&D expenditures in the early 1970s, 80% of R&D expenditures was borne by the private sector by the early 1990s. Similar technology and R&D institutions in India have not worked as well, because they did not foster close partnerships with industry. As a result, these institutions have produced research and technologies that have not found many applications in industry.

In addition to fostering fruitful partnerships between the public and private sectors, the technology policies followed in both Korea and Taipei, China simultaneously encouraged the import of technology and developed local technology and technological capability. By contrast, India adopted a strict policy of import substitution with respect to technology (whether it is import of physical inputs such as equipment embodying foreign technology or licensed transfer of technological know-how). The policy was implemented by an across-the-board attempt at developing locally what could be imported by installing a stringent import licensing regime that worked in tandem with some of the highest tariff rates in the world. By disregarding comparative advantage to such an extent and spreading meager research resources too thinly, India did not develop export competitiveness in virtually any knowledge-based industry. Interestingly, India's very recent success in software exports – exports in 1995–1996 were $2.5 billion – has come about alongside greater market orientation in its economy and facilitative government policy toward this niche market.

In short, expansion of primary education makes greatest sense for the low-income economies striving to make a breakthrough in manufacturing and improve agricultural productivity. On the other hand, the higher-income economies simply cannot afford to neglect higher levels of education and research activities related to technology development as some seem to have done.

Reforming the Education Sector

Making appropriate investments in human resources is not a question of simply allocating more resources to the appropriate levels of education. Indeed, there is much to be gained through partnerships between the public and private sectors in research activities, where striking the right balance can lead to high payoffs. In fact, it has been argued that a fundamental change in the institutional setting, including the incentive structures, is needed.

The education sector has expanded tremendously in most ADEs over the last three decades. Much of this expansion, particularly at the lower levels of education, has taken place under government finance and provision. Nevertheless, in many circumstances the quality of education being delivered leaves much to be desired. On one hand, the problems of quality have to do with curricula. Especially at higher levels of education such as the senior secondary and tertiary levels, there is concern in many ADEs that current curricula do not emphasize scientific knowledge enough or are not developing the skills valued by the private sector. On the other hand, where there seems to be consensus on what constitutes an appropriate curriculum, there is concern that schools are unable to deliver it.

One answer to these problems is greater decentralization and market orientation of education. In addition to alleviating strained public sector resources, greater decentralization and market orientation may be needed to overhaul the incentive structures of educational institutions and for teachers, to make the education sector more effective. In particular, when combined with a student loan and scholarship program to mitigate tendencies toward inequity, increasing the extent of decentralized management and finance (including private finance) in education can provide greater incentives for utilizing resources more efficiently at the level of the individual educational institutions and for delivering better quality. Additionally, they can create incentives for ensuring that curriculums suit the needs of labor markets.

However, decentralization and market orientation are not panaceas. First, decentralization of finance and market orientation may well lead to growing inequities unless they are combined with effective student loan and scholarship programs for the poor. As is well known, administering these efficiently is no easy task. Second, decentralization of management may simply shift "the same old problems" to levels less capable of resolving them. Clearly local capabilities, including those of local administrators and headmasters, will be an important deciding factor in the success of decentralization. Where local capabilities at monitoring schools are scarce, appropriate incentive systems for teachers could be instituted. For example, teacher motivation is frequently assumed to be a critical factor in determining schooling outcomes; thus, it is often suggested that a portion of teachers' salaries be associated with student performance. In practice, unfortunately, such schemes are rarely successful.

These caveats do not mean that decentralization and market orientation of the education sector are not the appropriate direction for the ADEs. In fact, virtually all ADEs are embracing these to some degree. However, these caveats must be kept in mind by policymakers if they are to seriously address the myriad problems of the education sector.

CONCLUSION

A rapid demographic transition, especially in the NIEs, has led to a bulge in the proportion of young people in the population that is working its way through the age structure. The NIEs' economic success is attributable in part to an increase in the proportion of workers in the population and to the increased level of savings that resulted from the fast economic growth rates (Asian Development Bank, 1998). These factors mutually reinforce each other: high growth leads to high savings, and hence to investment, which in turn further stimulates growth. More recently, Southeast Asia has begun to experience a bulge in its working-age population, and might also benefit from this demographic bonus. Eventually, the same may hold true for South Asia.

However, an increase in the labor force is not by itself a prerequisite for economic growth. It could merely result in higher unemployment and greater poverty. What is crucial in this context is how this resource is enhanced. One of the key elements behind the economic miracle, and one that the relatively less developed Asian countries ignore at their peril, has been the effective use and development of the most fundamental of all resources, the people.

Obviously a single set of policies will not be appropriate for every economy. For low-income ADEs such as Cambodia, Lao PDR, Viet Nam, and those in South Asia, improving the access to and quality of basic education should be the main focus. To the extent that adult illiteracy remains a major problem in these economies, basic literacy programs targeted at illiterate adults will also be needed.

For the middle-income ADEs such as the high-performing economies of Southeast Asia, the focus will have to be on improving the access to and quality of higher education, particularly that relating to the scientific and technical fields. Moreover, this focus will need to be supplemented by greater efforts at developing local technological capability.

For the NIEs, their success at transforming their economies has naturally led them to now aspire to a position of leadership in scientific innovation and technological advances, rather than remain imitators. Hong Kong, China and Singapore, which provide high-technology services in finance, trade, and transport, will require more sophisticated information infrastructure and basic scientific and technological development. To achieve all this calls for further upgrading of tertiary education in general, and of science and technology in particular. This would entail not only making greater investments in research infrastructure in universities, but also establishing an environment to foster creativity through greater administrative flexibility and unfaltering commitment to academic and research excellence.

NOTES

1. M. G. Quibria is the assistant chief economist of the Economic Analysis and Research Division of the Asian Development Bank. This paper draws heavily on the special chapter, "Population and Human Resources," in the *Asian Development Outlook*, 1998.

2. For example, enrollment rates are typically based on annual enrollment surveys conducted at the beginning of the school year. If dropping out of school in the course of the academic year is significant, then the reported enrollment rates will overstate effective enrollment rates. See Behrman and Schneider (1994) for a more complete discussion of the limitations of enrollment data.

3. The predicted values were obtained from cross-country regressions based on 121 countries. The dependent variable is gross enrollment rate at the secondary level and the independent variables include an intercept and a polynomial in per capita income. Data is from *World Development Indicators* (World Bank, 1998).

4. The predicted values were obtained from cross-country regressions based on 109 countries. The dependent variable is gross enrollment rate at the tertiary level and the independent variables include an intercept and a polynomial in per capita income. Data is from *World Development Indicators* (World Bank, 1998).

5. The countries included were: ASEAN-4, NIEs, Japan, PRC, and four South Asian economies (Bangladesh, India, Pakistan, and Sri Lanka).

REFERENCES

Asian Development Bank (1998). *Asian Development Outlook.* Hong Kong: Oxford University Press.

Behrman, J., & Schneider, R. (1994). An International Perspective on Schooling Investments in the Last Quarter Century in Some Fast-Growing East and Southeast Asian Countries. *Asian Development Review, 12*(2), 1–50.

Deolalikar, A. (1997). Human Resource Development and International Competitiveness in Asia. Paper prepared for the Asian Development Bank, Manila.

Dreze, J., & Gazdar, H. (1997). Uttar Pradesh: The Burden of Inertia. Unpublished.

Evenson, R., & Westphal, L. (1995). Technological Change and Technology Strategy. In: J. Behrman and T. N. Srinivasan (Eds), *Handbook of Development Economics*, Volume III. Amsterdam: Elsevier Science.

Foster, A. D., & Rosenzweig, M. R. (1996). Technical Change and Human Capital Returns and Investments: Evidence from the Green Revolution. *American Economic Review, 86*(4), 931–953.

Hanushek, E. (1995). Interpreting Recent Research on Schooling in Developing Countries. *World Bank Research Observer, 10*(2), 227–246.

Krugman, P. (1994). The Myth of Asia's Miracle. *Foreign Affairs*, November/December, 62–78.

Lall, S. (1998). Thailand's Manufacturing Competitiveness: A Preliminary Overview. Paper prepared for the Conference on Thailand's Dynamic Economic Recovery and Competitiveness, May, UNCC, Bangkok.

Mingat, A. (1995). The Strategy Used by High-Performing Asian Economies in Education: Some Lessons for Developing Countries. Paper written for the Asian Development Bank, Manila.

Mingat, A., & Tan, J. P. (1996). The Full Social Returns to Education. Human Capital Working Papers, World Bank.

Psacharopoulos, G. (1988). Education and Development: A Review. *World Bank Research Observer, 3*(1), 99–116.

Psacharopoulos, G. (1996). Public Spending on Higher Education in Developing Countries: Too Much Rather than Too Little. *Economics of Education Review, 15*(4), 421–422.

Solow, R. (1956). A Contribution to the Theory of Economic Growth. *Quarterly Journal of Economics, 70*, 65–94.

Solow, R. (1957). Technical Change and the Aggregate Production Function. *Review of Economics and Statistics, 39*, 312–320.

Spence, M. (1976). Competition in Salaries, Credentials, and Signaling Prerequisites for Jobs. *Quarterly Journal of Economics, 90*(1), 51–74.

Strauss, J. (1986). Does Better Nutrition Raise Farm Productivity? *Journal of Political Economy, 94*(2), 297–320.

Tan, H., & Batra, G. (1997). Technology and Firm Size-Wage Differentials in Colombia, Mexico, and Taiwan (China). *World Bank Economic Review, 11*(1), 59–83.

UNIDO (various years). *Industrial Statistics Database*. Geneva.

World Bank (1997). *Philippines: Managing Global Integration, Volume 2 (Background Papers)*. Draft Confidential.

World Bank (various years). *World Development Indicators*. Washington, D.C.

SECTION THREE

16. EQUILIBRIUM REAL EXCHANGE RATE, REAL EXCHANGE RATE MISALIGNMENT, AND THE CURRENT ACCOUNT: AN ECONOMETRIC ANALYSIS OF THAILAND'S EXPERIENCES

Shuji Kasajima[1] and Sukanda Lewis[2]

1. INTRODUCTION

In this paper, we study the dynamics of real effective exchange rates and current account of Thailand during the 1980s and 1990s. This paper's focuses are first to examine whether the Thai authority's nominal foreign exchange rate policy, adopted during the period, led to a sort of misalignment of the real effective exchange rate. Second, we examine whether there were endogenous fundamental factors leading to the establishment of the equilibrium real exchange rates ensuring the internal and external balance, even when the misalignment of real effective exchange rates resulted. Third, we study how such real exchange rate misalignment would be associated with deviation of the observed trade-weighted real effective exchange rate from the equilibrium rate. Lastly, we analyze whether the real exchange rate misalignment is statistically related to the evolution of large current account deficit experienced since the mid-1980s.

Restructuring Asian Economics for the New Millennium, Volume 9A, pages 281–322.
2001 by Elsevier Science B.V.
ISBN: 0-7623-0709-9

Thailand experienced rapid economic growth since the mid-1980s, owing largely to capital accumulation and export growth associated with massive foreign direct investment and other private capital inflows. Rapid export growth was partly accelerated by depreciation of real effective exchange rate after 1985. Trade-weighted real effective exchange rate depreciated 7.4% on average during 1986 and 1990 and 9.2% on average during 1991 and 1995, compared to the level in late 1985. Despite the rapid export growth, however, the current account deficit showed a secular worsening trend, reaching a historically high level of 124 billion Baht deficit in the second quarter of 1996.

In spite of the observed real depreciation of Thai Baht over the period, the nominal exchange rate of Baht, essentially pegged to the U.S. dollar, appeared overvalued. The nominal overvaluation was revealed by the sharp depreciation of Baht against the U.S. dollar, once the exchange rate system turned into a floating one in July 1997. This implies that the real effective exchange rate should have depreciated further than the observed real depreciation. It is sometimes argued that the Thai authority's foreign exchange policy, namely "the Basket method" of bilateral exchange rate determination, might be responsible for the overvaluation of real effective exchange rate, albeit its actual depreciation trend since the mid 1980s. It is pointed out that the Thai authority assigned a much larger weight to the U.S. dollar than the actual trade-weight with the USA in deciding the Basket-based bilateral exchange rates, to avoid fluctuations of the Baht-dollar rate.[3] The acclaimed reasons for such policy are, first, the dominance of the U.S. dollar as the international currency for transaction settlement. The second reason is the potential capital loss, caused by fluctuation of the Baht-dollar rate, on the part of borrowers of foreign capital denominated in the U.S. dollars. We may call this exchange rate policy "currency-weighted basket system," in comparison to the standard "trade-weighted basket system." Essentially, the "currency-weighted basket system" may be interpreted as the U.S. dollar-pegged system

A hypothetical estimate of real effective exchange rate (REER) based on "currency-weighted basket method," in fact, shows an appreciation trend since 1986 (Fig. 1). From Fig. 1, we see that although the observed trade-weighted real effective exchange rate depreciated since 1985, the currency-weighted real effective exchange rate traced a slightly appreciating trend. This may be easily understood. Because inflation has crept into the Thai economy as the economy grows, fixing the Baht-dollar nominal rate within a narrow adjustable range, while assigning a large weight to the U.S. dollar, has led to the appreciation of real effective exchange rate.[4] This, in turn, implies that if the Thai authority had strictly followed the trade-weighted basket method in determining the bilateral exchange rates, the actual trade-weighted real effective exchange rate might

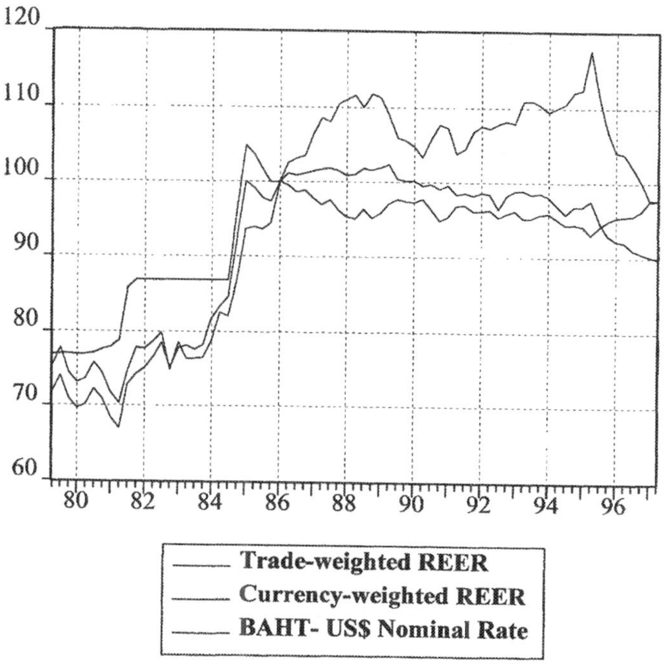

Fig. 1. Trade-weighted REER vs. Currency-Weighted REER (Index: 1986/IQ = 100).

have depreciated by a larger margin than the observed real depreciation. We interpret this exchange rate policy as a version of real exchange rate misalignment policy.

In the following analyses, we examine, by econometric methods, the time evolving dynamics of the observed trade-weighted real effective exchange rate ($REER^{TW}$), and the potential consequences of the real exchange rate misalignment on the current account. This paper consists of two parts. The first part studies the relationship between the equilibrium real effective exchange rate and the actual trade-weighted real effective exchange rate, and the effects of the real exchange rate misalignment policies on the evolution of the actual real exchange rate. We first examine the relationship between the hypothetical real effective exchange rate based on the "currency-weighted basket method" ($REER^{CW}$), and the realized trade-weighted real effective exchange rate ($REER^{TW}$). We define the discrepancy between these two REER a proxy measure of real exchange rate misalignment.[5] Second, we investigate the relationship between the actual trade-weighted real effective exchange rate

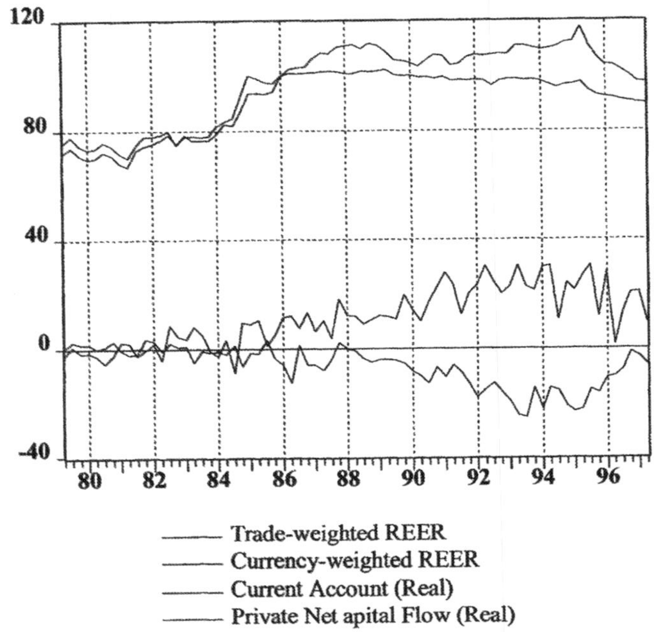

Fig. 2. Real Exchange Rate Misalignment and Current Account.

($REER^{TW}$) and unknown equilibrium (trade-weighted) real effective exchange rate, the latter of which is determined by fundamental, economic factors ensuring the internal and external equilibrium. We are particularly interested in two points. The first point is to find out whether there was an autonomous mechanism of reverting to the equilibrium real rate, when policy disturbances perturbed the actual real exchange rates away from the equilibrium rate. Inconsistent government's policies may work as the factors driving the actual real exchange rate to deviate from the equilibrium rate. The second point is to investigate how the deviation of the actual trade-weighted REER from the equilibrium real rate is attributed to the real exchange rate misalignment policy associated with the currency-weighted basket method.

The second part of this paper examines the potential adverse consequence of the real exchange rate misalignment policy associated with the currency-weighted basket method on the evolution of the current account deficit. We first estimate theoretically predicted current account, using a stochastic intertemporal model of current account. The stochastic intertemporal current account model draws on the method of the present value approach of the permanent income hypothesis of consumption due to Campbell (1987). The predicted current account fits the actual current account reasonably well. The model's prediction precision appears to decline since the mid 1980s, however, when large current account deficit, financed by massive capital inflow, emerged. Therefore, we attempt to associate this prediction error with the real exchange rate misalignment due to the Thai authority's foreign exchange rate policy.

This paper is organized as follows. In section 2, we briefly summarize the foreign exchange rate policy adopted by the Thai authority. We also present our measure of the currency-weighted real effective exchange rate ($REER^{CW}$), which is computed in reference to the foreign exchange rate policy. In section 3, we estimate the determinant of the equilibrium real exchange rate, the adjustment mechanism of the actual real rate to the equilibrium rate, and the effects of the government's policies on the evolution of the actual real effective rate. Section 4 estimates the stochastic current account model and examines the relationship between the exchange rate policy and the prediction error of the current account. Section 5 draws a conclusion.

2. THE FOREIGN EXCHANGE RATE POLICY

The foreign exchange rate system of Thailand underwent several major revisions during the past thirty years.[6] It is classified into five phases. In the first phase between the year, 1963 and 1978, which includes the period of the Bretton-Woods system, Thai Baht was fixed to the U.S. dollar and gold with several episodes of devaluation. In March 1978, the fixed link to the U.S. dollar was abolished, and Baht was pegged to a basket of major currencies (second phase). This basket method of exchange rate had a short life; in November 1978, the exchange rate began to be determined under a managed floating system. Under this managed floating system (phase 3), the Baht-U.S. dollar rate was determined between the Exchange Rate Equalization Fund and representatives of major commercial banks. In July 1981, the exchange rate system returned to the U.S. dollar-pegged system (phase 4). This U.S. dollar-pegged system lasted until November 1984. Because the U.S. dollar was overvalued during this period due to the "Reaganomics" policy, Thai Baht was also overvalued against major foreign currencies. Consequently, trade account worsened and international reserves approached the critical minimum level. The Thai government was forced to launch stringent government expenditure policies as well as measures to increase the revenue. The foreign exchange rate policy was another major policy issue to improve the current account. With this aim, the bilateral exchange rates of Baht began to be determined by the "basket method" from November 1984, in which the bilateral exchange rates of Baht were determined in reference to trade weights of the major trading partners (phase 5 until July 1997).

The basket method of exchange rates is, in its formal definition, a trade-weighted average of market exchange rates with foreign currencies. Therefore, if the trade-weight with the USA absolutely dominates the total volume of trade, the official Baht-U.S.$ rate will mimic its market rate of Baht-U.S. dollar rate. Although the exact weight assigned to each major foreign currency was not publicized by the Thai authority, an estimate shows that about 80–82% weight was given to the U.S. dollars in 1992. The remaining weights are estimated to be assigned to the Japanese Yen (10 ~ 13%), German Mark (6 ~ 8%), Singapore dollar, Hong Kong dollar, and other currencies of major trading partner countries of Thailand.[7] On the other hand, the trade weights of the major trading partners are 16.5% with the USA, 24.1% with Japan, 4.9% with Germany and 54.5% with the other countries in 1992. Apparently, an excessively large currency weight was given to the U.S. dollars, while the Japanese Yen received a much smaller currency weight compared to the trade weight. The plausible reasons for such U.S. dollar-dominated policy of basket-weight exchange rate determination are two-fold. The first is the dominance of the U.S. dollar as the international, transaction settlement currency. Tasaka (1995) shows, from the Bank of Thailand's statistics, that 91% of total export and 75% of total import were settled by the U.S. dollar in 1992, respectively. The second reason is the potential capital loss on the part of borrowers of foreign capital. Since most of the foreign capital inflow was made in terms of the U.S. dollars, swings of the Baht-U.S. dollar rate might discourage foreign capital inflows.

Given the fact that "the U.S. dollar-dominated basket method" determined bilateral exchange rates, we can construct a measure of real effective exchange rate misalignment by comparing the realized trade-weighted real effective rate ($REER^{TW}$) with the policy-determined, currency-weighted real effective rate ($REER^{CW}$). Suppose that the variable q_t represents the realized trade-weighted real effective exchange rate ($REER^{TW}$) which exercises an influence on the determination of trade account. The realized trade-weighted real effective exchange rate is computed as a weighted-sum of bilateral nominal exchange rates, adjusted for relative inflation rates between Thailand and major trading partner countries, with the actual value of trade used as the weights. The currency-weighted real effective exchange rate q_t^{CW}, on the other hand, uses policy-determined currency weights to compute real effective exchange rate, with a massive weight given to the U.S. dollars. Because the bilateral nominal exchange rates are determined based on these currency weights, we may be able to interpret the currency-weighted real effective rate representing the real value of Baht. We compute the difference in these two real effective exchange rates, $RERDIF_t = q_t - q_t^{CW}$, as a proxy measure of the real exchange rate misalignment. This is a crude measure because the realized trade-weighted REER, q_t, is computed with the data of nominal bilateral exchange rates that are determined basically by the currency-weight method. As the nominal Baht-U.S. dollar exchange rate depreciates, the currency-weighted REER depreciates more than the trade-weighted real effective rate does, making the discrepancy, $RERDIF_t$, smaller. On the other hand, when the nominal Baht-U.S. dollar rate appreciates, the currency-weighted REER appreciates more than the trade-weighted real effective rate does, making the discrepancy larger. This implies that the correlation between a change in q_t^{CW} and a change in $RERDIF_t$ is negative, $Corr(\Delta q_t^{CW}, \Delta RERDIF_t) < 0$. For quarterly data between the 4th quarter of 1981 and the 4th quarter of 1996, the contemporaneous correlation is found to be –0.1822, while one-period lagged (on Δq_t^{CW}) correlation is –0.058.[8]

Relative movements of the trade-weighted REER, q_t, and the currency-weighted REER, q_t^{CW}, over the sample period are plotted in Fig. 1. We find that the currency-weighted REER exhibited a secular appreciation trend since early 1985: the basket method was introduced in November 1984, and nominal Baht-U.S. dollar was depreciated by 17.4% in the same month. On the other hand, the trade-weighted REER soared between 1985 and 1988, and stayed high during 1990s until the middle of 1995. The figure shows that the discrepancy $RERDIF_t = q_t - q_t^{CW}$ follows a widening trend. Thai Baht appears to have been overvalued since the mid-1980s.

3. EQUILIBRIUM REAL EXCHANGE RATE AND REAL EXCHANGE RATE MISALIGNMENT

3.1. The Model of Equilibrium Real Exchange Rate

In this section, we set up a model of equilibrium real exchange rate determination, drawing on the intertemporal approach by Edwards (1989, 1994). Our empirical interest is to find an autonomous mechanism that equilibrated the deviation of actual real exchange rate from the equilibrium rate, when disturbances perturbed the actual real exchange rate. As will be shown, the vector error correction model is used to find the adjustment mechanism toward the long-run equilibrium real exchange rate.

We postulate first that the long-run equilibrium real exchange rate be determined by real economic fundamental variables. On the other hand, nominal and monetary factors associated with the government's policies may exercise disturbances to the equilibrating movement toward the equilibrium real rate, causing a discrepancy between the actual real rate and the equilibrium rate.

Let q_t be real exchange rate at time t. Theoretically, real exchange rate is defined as the price of tradable goods relative to the price of nontradable goods. As an empirical matter, it is usually defined as the nominal exchange rate (price of foreign currency in terms of home currency) adjusted for relative prices of trading partner countries. The latter definition is the one we use in the empirical section. In both definitions, an increase in real exchange rate means depreciation of the real exchange rate of the home currency. For analytical purpose shown below, however, we use the definition of the relative price of tradables and nontradables in this section.

The equilibrium real exchange rate, q_t^*, is the one that is consistent with both the internal and external balance. It may be thought of as determined by a vector of fundamental real variables Z;

$$q_t^* = q^*(Z_t), \qquad (1)$$

where the vector Z_t includes the following real variables.

1. TOT_t: External terms of trade, defined as relative international price of export goods to import goods (P_X^*/P_M^*)
2. TFP_t: Technological progress, measured by growth rate of total factor productivity of traded good sector
3. $KGROW_t$: Capital accumulation in the country
4. $GCONSNT_t$: Government's consumption expenditure on nontradable goods
5. $TRDCONTR_t$: Trade control measures: import tariff and non-tariff trade control measures
6. $CFLCONTR_t$: Controls on exchange and capital flows

The terms of Trade, TOT_t, is assumed to have a negative association with real exchange rate. As the terms of trade worsens due to, for example, decline of international prices of primary commodities that are main export products of a developing country, real exchange rate of such developing country will deteriorate. Usual argument on this negative relationship between the terms of trade and real exchange rate is based on income effect of worsening terms of trade on demand for nontradable goods. Because of adverse income effect of worsening terms of trade, demand for nontradable goods will decline, which pushes the price of nontradable goods downward. This, in turn, results in requiring depreciation of real exchange rate to restore the internal equilibrium by stimulating foreign demand for home tradable goods.

The technology progress variable TFP_t represents the Balassa–Samuelson effect of technology progress on real appreciation of exchange rate (Balassa, 1964, Samuelson 1964). The Balassa–Samuelson effect refers to the theoretical account to explain the empirical regularities that a country with higher productivity growth in the tradable sector relative to nontradable sector tends to have higher relative price of nontradables. That is, the country with higher technology progress tends to experience real exchange rate appreciation. Since the productivity growth in nontradable goods sector may be similar across countries, the difference in productivity in tradable goods sector will result in real appreciation of the currency of the more productive country.

Trade control measures, $TRDCONTR_t$, is assumed to work as a negative factor in determining equilibrium real exchange rate. The reason is that when tariff rates are reduced (decline in $TRDCONTR_t$), import of tradable goods will increase, requiring real exchange rate depreciation to maintain external balance. As Balassa (1982) puts it, "Eliminating protective measures would necessitate a devaluation in order to offset the resulting deficit in the balance of payment." Edwards (1989) showed, in an intertemporal utility optimization context of a small open economy, that an anticipated future tariff reduction would also result in depreciation of both current and future real exchange rates.

The measure of exchange and capital flow controls, $CFLCONTR_t$, will also be negatively correlated with the real exchange rate. When the controls on capital account are relaxed (increase in $CFLCONTR_t$), real exchange rate tends to appreciate. Traditional argument on this positive correlation between capital controls and real exchange rate such as Corden (1981) is as follows. When foreign capital flows in the country as a result of relaxed capital controls, total expenditures will exceed total income, which generates excess demand for nontradable goods. In order to restore the internal equilibrium, the price of nontradable goods has to rise relative to tradable goods, resulting in real appreciation of home currency.

Edwards (1989, 1994) extended the traditional argument in an intertemporal dynamic context. When the restrictions on capital flows are lifted in the current period, the cost of foreign borrowings declines, approaching downward toward the world interest rate. This will cause the home interest rates to decline too. Then, future consumption becomes more expensive relative to current consumption, inducing the people to substitute future consumption for current consumption. This intertemporal substitution in favor of current consumption will put an upward pressure on the price of nontraded goods in the current period, resulting in real appreciation of the exchange rate. Current account will worsen too due to an increase in consumption of foreign goods, which will be financed by increased foreign borrowing.

The variable $GCONSNT_t$ refers to the magnitude of the government's expenditures spent on nontraded goods, relative to traded goods. Given the equilibrium path of government's spending and its debt financing constrained by government's intertemporal budget constraint, an increase in government's spending on nontraded goods in current period, financed by debt, will exercise an upward pressure on the price of nontraded goods. On the other hand, the consumers have to bear a higher tax burden in the future to pay off the government's debt, which will reduce their demand for nontraded goods in both future period and in current period by adverse income effect. It may be likely, however, that net demand for nontraded goods will increase in current period, causing the real exchange rate to appreciate.

The variable $KGROW_t$ refers to the growth rate of capital stock. If capital accumulation advances more in traded goods sector, relative to nontraded goods sector, productivity growth will be larger in traded goods sector, which will lead to appreciation of the real exchange rate. The logic will be similar to the Samuelson–Balassa effect of productivity growth on real exchange rate.

Given the above definition of each real factor, a linear regression to represent determination of the equilibrium real exchange rate may be modeled as follows.

$$q_t^* = \alpha_0 + \alpha_1 \cdot TOT_t + \alpha_2 \cdot TFP_t + \alpha_3 \cdot KGROW_t + \alpha_4 \cdot GCONSNT_t + \alpha_5 \cdot TRDCONTR_t + \alpha_6 \cdot CFLCONTR_t + u_t \quad (2)$$

where u_t is assumed to be white noise process.

Equation (2) cannot be estimated, however, because the equilibrium real exchange rate q_t^* is not observable by an econometrician. We need to formulate an estimable equation without recourse to the equilibrium real exchange rate.

One method of doing so is to impose an assumption of partial adjustment mechanism, sometimes called stock adjustment mechanism, of actual real exchange rate toward its equilibrium value. This is the method used by Edwards (1989). By assuming a partial adjustment mechanism, we are interested in how actual real exchange is associated with the equilibrium real exchange rate. While the equilibrium real exchange rate is the long-run equilibrium rate that reflects real economic fundamentals, short-run nominal disturbances might disturb the realization of the equilibrium rate. We want to examine the mechanism on how the actual real exchange rate will tend to revert to its equilibrium rate, once short-run nominal disturbance would make the real exchange rate deviate from its long-run equilibrium rate.

Edwards (1989) suggested the following partial adjustment mechanism.

$$q_t - q_{t-1} = \theta(q_t^* - q_{t-1}) + \gamma(\Omega - \Omega^*) + u_t. \tag{3}$$

If the equilibrium real exchange rate at time t, q_t^*, turns out to be larger than the real exchange rate realized in the previous period q_{t-1}, which implies potential overvaluation of real exchange rate at time $t-1$, the actual real exchange rate at time t tends to correct the exchange rate misalignment by pushing current exchange rate q_t upward (real depreciation). The parameter θ, the speed of autonomous adjustment, is assumed to take value between zero and one. Yet, there are other factors that might influence the evolution of actual real exchange rate q_t over time. The vector Ω represents the government's macroeconomic policies that may be responsible for real exchange rate misalignment. Suppose that the vector Ω^* is a vector of macroeconomic policies consistent with maintaining equilibrium real exchange rate. Actual macroeconomic policies Ω may deviate from the optimal policies Ω^*, pushing the actual real exchange rate downward (overvaluation); γ is assumed to be negative. Such macroeconomic policies may include both expansionary fiscal policy financed by money creation and expansionary monetary policy.

In addition to the conventional macroeconomic policies associated with real exchange rate misalignment (Ω) studied by Edwards (1989), we include the difference between trade-weighted real effective rate and currency-weighted real rate as a measure of direct real exchange rate misalignment policy.

$$q_t - q_{t-1} = \theta \cdot (q_t^* - q_{t-1}) + \gamma \cdot (\Omega - \Omega^*) + \phi \cdot RERDIF_{t-1} + u_t. \tag{4}$$

The variable $RERDIF_{t-1}$ is defined as $q_{t-1} - q_{t-1}^{CW}$, where as shown before q_{t-1} is trade-weighted real effective exchange rate, and q_{t-1}^{CW} is currency-weighted real effective exchange rate in period $t-1$. The difference between trade-weighted real exchange rate and currency-weighted real exchange rate, $q_{t-1} - q_{t-1}^{CW}$, measures the magnitude of real exchange rate misalignment induced by the government's policy. The parameter ϕ is the dynamic adjustment parameter. It measures the degree of adjustment of the trade-weighted real exchange rate in period t, q_t, when there was discrepancy between the trade-weighted real exchange rate and the currency-weighted real exchange rate in the previous period. Instead of $RERDIF_{t-1} \equiv q_{t-1} - q_{t-1}^{CW}$, we may also use $\Delta RERDIF_t \equiv \Delta(q_t - q_t^{CW})$, where the symbol Δ is the first difference operator. Using this variable, we can examine how the change in the magnitude of real exchange rate misalignment in current period, $\Delta(q_t - q_t^{CW})$, will be related to the realized value of trade-weighted real effective exchange rate. Although there may be an endogeneity problem between q_t and $\Delta RERDIF_t$, we think that the difference formula would make $\Delta RERDIF_t$ be exogenous to the dependent variable q_t.[9]

Substituting equilibrium real exchange rate equation (2) into the adjustment equation (4), we obtain the regression equation that may be estimated by OLS.

$$q_t = \beta_0 + \beta_1 \cdot TOT_t + \beta_2 \cdot TFP_t + \beta_3 \cdot KGROW_t + \beta_4 \cdot GCONSNT_t$$
$$+ \beta_5 \cdot TRDCONTR_t + \beta_6 \cdot CFLCONTR_t + (1 - \theta) \cdot q_{t-1} + \gamma_1 \cdot BGTDFCT_t$$
$$+ \gamma_2 \cdot EXDMCRDT_t + \phi \cdot RERDIF_{t-1} + u_t. \quad (5)$$

In equation (5), we specified the macroeconomic policy misalignment $\Omega - \Omega^*$ as

$$[\Omega - \Omega^*] = \begin{bmatrix} EXDMCRDT_t \\ BGTDFCT_t \end{bmatrix},$$

where $EXDMCRDT_t$ and $BGTDFCT_t$ are measures of inconsistent monetary policy, and inconsistent fiscal policy, respectively, and each is defined as follows.

$EXDMCRDT_t$: Inconsistent monetary policy measured by growth rate of domestic credit minus one-period lagged growth rate of real GDP. We assume that over-supply of domestic credit in excess of growth of GDP will create an inflationary pressure on the economy, causing the real exchange rate to appreciate.

$BGTDFCT_t$ Inconsistent fiscal policy measured by the ratio of budget deficit to monetary base. This captures monetization of fiscal deficit, which often leads to inflationary process, resulting in overvaluation of real exchange rate.

In estimating the real exchange rate equation (6), the coefficients of our major interests are those on the terms related to reversion to the equilibrium exchange rate, ($\beta_1, \beta_2, \beta_3, \beta_4, \beta_5, \beta_6$), and the coefficient ($\phi$) on the real exchange rate misalignment variable. The expected signs of each coefficient in equation (5) are all negative;

$$\beta_1(TOT_t) < 0, \ \beta_2(TFP_t) < 0, \quad \beta_3(KGROW_t) < 0,$$
$$\beta_4(GCONSNT_t) < 0, \quad \beta_5(TRDCONTR_t) < 0, \quad \beta_6(CFLCONTR_t) < 0.$$
$$\gamma_1(BGTDFCT_t) < 0, \quad \gamma_2(EXDMCRDT_t) < 0.$$

The sign on $RERDIF_{t-1}$, ϕ, is not predetermined. Following the discussion in the previous section, however, it may be likely that ϕ is negative since $Corr(\Delta q_t, \Delta RERDIF_t) < 0$.

The problem associated with estimation of equation (5) is that most of the variables are nonstationary, containing a unit root, as show in Table 2. Estimation of equation (5) by ordinary least square method, therefore, may lead to a *spurious regression result*.[10] To avoid the non-stationarity problem of the data, we estimate equation (5) by the vector error correction (VEC) model.

3.2. The Vector Error Correction Model and Long-run Equilibrium

The vector error correction model (VEC) is essentially a vector autoregression model (VAR) in the first difference of endogenous variables, augmented by cointegrating equations of endogenous variables. Given a group of non-stationary series, we want to determine whether the series are cointegrated, and if they are, to identify the cointegrating (long-run equilibrium) relationships.

Consider a reduced form VAR of order p,

$$y_t = A_1 y_{t-1} + A_2 y_{t-2} + A_3 y_{t-3} + - - - - + A_p y_{t-p} + \sum_{j=0}^{q} B_j x_{t-j} + \varepsilon_t \qquad (6)$$

where y_t is $(k \times 1)$ vector of non-stationary $I(1)$ variables. x_t is a vector of exogenous variables. ε_t is a $(k \times 1)$ vector of white noise, each of which may be contemporaneously correlated. In our empirical work, the vector y_t is

$$y_t' = [q_t, TOT_t, TFP_t, GCONSNT_t, TRDCONTR_t, CFLCONTR_t, KGROW_t]',$$

and the vector x_t is

$$x_t' = [BGTDFCT_t, EXDMCRDT_t, RERDIF_{t-1}]'.$$

We can rewrite the VAR (6) in first difference series as

$$\Delta y_t = \Pi y_{t-1} + \sum_{j=1}^{p-1} \Gamma_j \Delta y_{t-j} + \sum_{j=0}^{q} B_j x_{t-j} + \varepsilon_t \qquad (7)$$

where $\Pi = \Sigma_{j=1}^{p} A_j - I$, $\Gamma_j = -\Sigma_{i=j+1}^{p} A_i$, and I is an identity matrix.

By Granger's representation theorem, if the coefficient matrix Π has reduced rank $r<k$, then there exist $(k \times r)$ matrices φ such that $\Pi = \psi \cdot \varphi'$, and $\varphi' \cdot y_{t-1}$ are stationary processes. Then, all endogenous variables in equation (7) become stationary, and therefore we can estimate each equation by OLS, which will yield consistent estimators. Equation (7) is called vector error correction model. The stationary processes $\varphi' \cdot y_{t-1}$ are called cointegrating equations, and each column of φ a cointegrating vector. The number r is the number of cointegrating relations among endogenous variables, which can be determined by Johansen's (1991) cointegration test. The elements of ψ are interpreted as

the adjustment parameters in the vector error correction model. When there is an equilibrium error such that the actual series $\vartheta' \cdot y_{t-1}$ deviates from the equilibrium relation $\varphi' \cdot y_{t-1}$ in period $t-1$, $\mu_{t-1} = \vartheta' y_{t-1} - \varphi' y_{t-1}$, the series y_t tends to correct the equilibrium error in the next period t at the speed ψ.

The vector error correction model (7) will have r cointegrating equations. In line with our theory of equilibrium real exchange rate, we are most interested in the following equation, if it turns out to be a legitimate cointegrating equation by the Johansen's test.

$$\begin{aligned}\mu_{t-1} = q_{t-1} &+ \gamma_1 \cdot TOT_{t-1} + \gamma_2 \cdot TFP_{t-1} + \gamma_3 \cdot KGROW_{t-1} \\ &+ \gamma_4 \cdot TRDCONTR_{t-1} + \gamma_5 \cdot CFLCONTR_{t-1} \\ &+ \gamma_6 \cdot GCONSNT_{t-1} + C + \delta \cdot T \end{aligned} \tag{8}$$

or,

$$\begin{aligned}q_{t-1} = &- \gamma_1 \cdot TOT_{t-1} - \gamma_2 \cdot TFP_{t-1} - \gamma_3 \cdot KGROW_{t-1} \\ &- \gamma_4 \cdot TRDCONTR_{t-1} - \gamma_5 \cdot CFLCONTR_{t-1} \\ &- \gamma_6 \cdot GCONSNT_{t-1} - C - \delta \cdot T + \mu_{t-1}. \end{aligned} \tag{9}$$

In equation (8), C is constant term and T is trend term, both of which may not be included in the cointegration equation. The cointegration equation μ_{t-1} is viewed as an equilibrium error that represents deviation of actual real effective exchange rate from the equilibrium real effective exchange rate in period $t-1$. The equilibrium error μ_{t-1} is assumed to be white noise process.

With this cointegrating equation, the first row of vector error correction model (7) may be written as

$$\begin{aligned}\Delta q_t = \psi \cdot \mu_{t-1} &+ \sum_{j=1}^{p} \alpha_{1,j} \cdot \Delta q_{t-j} + \sum_{j=1}^{p} \alpha_{2,j} \cdot \Delta TOT_{t-j} + \sum_{j=1}^{p} \alpha_{3,j} \cdot \Delta TFP_{t-j} \\ &+ \sum_{j=1}^{p} \alpha_{4,j} \cdot \Delta KGROW_{t-j} + \sum_{j=1}^{p} \alpha_{5,j} \cdot \Delta TRDCONTR_{t-j} \\ &+ \sum_{j=1}^{p} \alpha_{6,j} \cdot \Delta CFLCONTR_{t-j} + \sum_{j=1}^{p} \alpha_{7,j} \cdot \Delta GCONSNT_{t-j} \\ &+ \sum_{j=1}^{q} \beta_{1,j} \cdot BGTDFCT_{t-j} + \sum_{j=1}^{q} \beta_{2,j} \cdot EXDMCRDT_{t-j} \\ &+ \sum_{j=1}^{q} \beta_{3,j} \cdot RERDIF_{t-j-1} + \varepsilon_t. \end{aligned} \tag{10}$$

The variables q_{t-j}, TOT_{t-j}, TFP_{t-j}, $GCONSNT_{t-j}$, $TRDCONTR_{t-j}$, $CFLCONTR_{t-j}$, and $KGROW_{t-j}$ are endogenous variables, which correspond to the elements in the equilibrium real exchange rate equation (2). On the other hand, the variables $BGTDFCT_{t-j}$, $EXDMCRDT_{t-j}$, and $RERDIF_{t-j-1}$ are exogenous variables. Note that we use one-period lagged value of $RERDIF_{t-j}$ to avoid potential endogeneity problem between Δq_t and $RERDIF_t$. We estimated vector error correction model (7) in the following.

3.3. Data and the Unit Root Tests

The construction method of each variable and the sources of data are presented in Table 1. Almost all data are taken from the Quarterly Bulletin of Bank of Thailand, and International Financial Statistics of IMF.

Table 2 presents the results of Augmented Dickey-Fuller (ADF) test of unit root on each variable. The ADF test is carried out by running the following regression by OLS.

$$\Delta y_t \equiv y_t - y_{t-1} = c + \alpha \cdot t + (\beta - 1) \cdot y_{t-1} + \sum_{j=1}^{p} \gamma_j \Delta y_{t-j} + \varepsilon_t \tag{11}$$

Null hypothesis is unit root: $\delta \equiv \beta - 1 = 0$.

Examining the Akaike Information Criterion (AIC), Schwartz Bayesian Criterion (SBC), Durbin-Watson *d*-statistics, and significance level of each estimated coefficient, we decide the number of lags p on the first difference terms and inclusion of constant (C) and trend term (t). Visual inspection of evolution of each series on a plot graph is also employed. For all the variables, optimal number of lags is found to be at most two.

Among the endogenous variables, the null hypothesis of unit root cannot be rejected even at 10% significance level for the variables REER, TFP, GCONSNT, TRDCONTR, CFLCONTR, KGROW, and (as an alternative variable to replace KGROW) INVGDP. On the other hand, the ADF test reject the null for the endogenous variable TOT at 5% level, but cannot reject it at 10% level.[11] First difference of each endogenous variable is stationary (no unit root), so that we conclude that all endogenous variables are $I(1)$. Note that for the endogenous variables to be cointegrated, every endogenous variable has to be integrated of the same order.

Among the three exogenous variables, the variable RERDIF contains a unit root, as the null hypothesis of unit root cannot be rejected even at 10% significance level. The null for EXDMCRDT is rejected at 1% level, but cannot be rejected at 5% level. The variable BGTDFCT is a stationary process because the null of unit root is rejected at any significance level.

Table 1. Definition of Variables and Their Sources.

Variable	Definition	Data source
REER (q_t)	• Trade-weighted real effective exchange rate $REER = (\text{Trade-weighted NEER}) \times \frac{\text{Trade-weighted foreign price}}{\text{Thai price}}$ NEER is nominal effective exchange rate.	Quarterly Bulletin, Bank of Thailand International Financial Statistics, IMF
RERDIF	• Trade-weighted REER – currency-weighted REER To compute currency-weighted REER, we assign 75% to U.S.$, and use trade-weight for other major currencies.	Quarterly Bulletin, Bank of Thailand International Financial Statistics, IMF
TOT	• Terms of Trade defined as $\frac{p_x}{p_m}$, where p_x is average price of Thai export, and p_m is average price of Thai import.	Quarterly Bulletin, Bank of Thailand
TFP	• Total factor productivity growth We use Solow's growth accounting method to compute TFP.	National Income Account, NESDB Estimate of Capital Stock in Thailand: 1970–1996, NESDB Tinakorn and Sussangkarn (1996).
GCONSNT	• Government's expenditure on nontradable goods Because no data is available for expenditure in nontradables, we use government's total consumption expenditure. Definition: Government's total consumption expenditure divided by nominal GDP	Quarterly Bulletin, Bank of Thailand
TRDCONTR	• Trade control measures Because we do not have data on non-tariff barriers, we use only import tariff. Definition: the ratio of import tariff to Import value	Quarterly Bulletin, Bank of Thailand
CFLCONTR	• Controls on foreign exchange and capital flows As a proxy, we use the ratio of net capital flows to nominal GDP	Quarterly Bulletin, Bank of Thailand
INVGDP	• Investment/GDP ratio: a proxy for capital accumulation variable	National Income Account, NESDB
KGROW	• Growth rate of real capital stock	Estimate of Capital Stock in Thailand: 1970–1996, NESDB Tinakorn and Sussangkarn (1996) Quarterly Bulletin, Bank of Thailand
BGTDFCT	• Expansionary fiscal policy Ration of budget deficit (surplus) to nominal GDP	Quarterly Bulletin, Bank of Thailand
$EXDMCRD^T$	• Expansionary monetary policy Ratio of growth of domestic credit to growth of monetary base	Quarterly Bulletin, Bank of Thailand

Table 2. Augmented Dickey – Fuller Unit Root Test.

- Equation: $\Delta y_t = y_t - y_{t-1} = c + \alpha \cdot t + (\beta - 1) \cdot y_{t-1} + \sum_{j=1}^{p} \gamma_j \Delta y_{t-j} + \varepsilon_t$
- Null hypothesis: Unit root $\delta = \beta - 1 = 0$

Variable	No. of lags p for Δy_{t-j}	δ (τ)	c	α	γ_1	γ_2	Adjusted R^2	D-W statistics
REER	1	−0.1689 (−2.8834) <−3.171>	12.7487 (2.8391)	0.1587 (3.1994)	0.1492 (1.1927)		0.1230	2.0573
RERDIF	1	−0.0612 (−2.3320) <−2.592>	5.9727 (2.3941)		0.2669 (2.1981)		0.1715	1.936
TOT	1	−0.1079** (−2.7854) <−2.589>	11.051 (2.6722)		0.2239 (2.0046)		0.1098	2.084
TFP	2	−0.0416 (−1.8248) <−2.593>	0.0444 (1.5051)		0.6750 (5.1376)	0.0847 (0.6117)	0.4825	1.988
TRDCONTR	1	−0.1668 (−2.2552) <−3.171>	2.2649 (2.418)	−0.0207 (−2.7484)	−0.1392 (−1.0774)		0.1104	2.0465
GCONSNT	1	−0.2581 (−2.885) <−3.171>	4.2629 (2.9305)	−0.0305 (−2.977)	−0.3844 (−3.3549)		0.3022	2.044
CFLCONTR	1	−0.2453 (−2.304) <−2.592>	1.735 (2.015)		−0.3353 (−2.7121)		0.2515	2.079
INVGDP	0	−0.2297 (−2.8421) <−3.171>	4.3074 (2.733)	0.1109 (2.8217)			0.099	2.047
KGROW	2	−0.4259 (−2.323) <−2.593>	1.0909 (2.3555)		−0.6079 (−3.5379)	−0.2549 (−2.010)	0.5587	2.034
BGTDFCT	1	−0.5.41*** (−5.406) <−3.171>	−3.1725 (−3.3359)	0.1260 (3.9743)	0.2466 (1.9425)		0.3238	1.760
EXDMCRDT	1	−0.311** (−3.1121) <−2.593>	2.6275 (2.8130)		0.5969 (5.5171)		0.3607	2.136

Notes:
1. Sample period: 1982/1st quarter–1996/4th quarter
2. The number in the parenthesis < > is the ADF τ test statistics at 10% significance level.
3. Asterisks ***, ** and * stand for τ statistics being significant at 1%, 5% and 10%, respectively.

Table 3. Johansen's Cointegration Test Result: Trace Test.

Eigenvalue	Likelihood Ratio (Trace statistics)	(5% critical value for LR)	(1% critical value for LR)	Null hypothesis: Number of cointegrating equations (r)
0.7863	276.935	146.76	158.49	None**
0.6671	184.334	114.90	124.75	at most 1**
0.4989	118.344	87.31	96.58	at most 2**
0.3953	76.889	62.99	70.05	at most 3**
0.3689	46.706	42.44	48.45	at most 4*
0.2147	19.085	25.32	30.45	at most 5
0.0736	4.586	12.25	16.26	at most 6

Notes:
1. Trace statistics is $Q_T = -T \cdot \Sigma_{i=r+1}^{k} \log(1-\lambda_i)$, for $r = 0, 1, 2, ---, k-1$. λ_i is i-th largest eigenvalue of the $(k \times k)$ matrix Π, k is the number of endogenous variables, and r is the number of linearly independent rows in matrix Π (or, called the number of cointegrating rank of matrix Π.) T is the number of observations.
2. The null hypothesis is that the number of cointegrating rank is r, against the alternative hypothesis of more than r. Under the null hypothesis, $\lambda_i = 0$, for $i = r+1, r+2, --, k$, so that $Q_T = 0$. *The critical values of* Q_T is given by Osterward-Lenum (1992).
3. The asterisks ** and * denote rejection of the hypothesis at 5%, and 1% significance level, respectively.
4. Log likelihood of the first cointegrating equation, the one shown in Table 4, is -404.8891.

Although both RERDIF and EXDMCRDT show possibility of random walk, we include them as the level variable in the estimation, because they are exogenous variables to the VEC system.

3.4. Estimation Results

Table 3 presents the Johansen's (1991) test results on cointegrating relations for the seven endogenous variables. The trace-statistics using eigenvalue of coefficient matrix Πgives us the number of cointegrating equations (cointegrating rank, r). The test results reveal that there are three cointegrating equations at 5% significance level on χ^2 as tabulated by Osterward-Lenum (1992). Among the three cointegrating equations, we focus on the following equation, since this exactly corresponds to the equilibrium real exchange rate equation in equation (2).

$$\begin{aligned} \mu_{t-1} = q_{t-1} &+ \gamma_1 \cdot TOT_{t-1} + \gamma_2 \cdot TFP_{t-1} + \gamma_3 \cdot KGROW_{t-1} \\ &+ \gamma_4 \cdot TRDCONTR_{t-1} + \gamma_5 \cdot CFLCONTR_{t-1} \\ &+ \gamma_6 \cdot GCONSNT_{t-1} + C + \delta \cdot T. \end{aligned} \tag{8}$$

Table 4. The Cointegrating Equation.

Variables	Normalized Cointegrating Coefficient
REER	1.0000
TOT	0.7076
	(0.2348)
	[3.013]
TFP	6.5691
	(3.5330)
	[1.8594]
KGROW	7.7619
	(2.1259)
	[3.6511]
GCONSNT	9.7236
	(2.9938)
	[3.2479]
TRDCONTR	–13.4113
	(2.9706)
	[–4.5146]
CFLCONTR	–2.1148
	(0.8378)
	[–2.5244]
TREND	–0.7948
	(0.0643)
	[–12.3470]
CONSTANT	–133.3689
Log likelihood	–404.89
AIC	20.0630
SBC	26.939

Notes:
1. The original cointegration coefficients are normalized by setting the coefficient on REER to be unity. This allows us to find the speed of adjustment parameter in VEC model (10).
2. The figure in parenthesis (.) is standard errors, and the figure in [.] is t-statistics.
3. When we evaluate the sign of the coefficient in the equilibrium real exchange rate equation, we need to change the sign in the above cointegrating equation.

We included constant and trend term in the cointegrating equation. The estimation result is shown in Table 4.

Since the cointegrating equation is expressed by the equilibrium error term μ_{t-1}, we need to reverse the sign on each coefficient in Table 4, to evaluate the equilibrium condition of the real effective exchange rate determination. Among the six endogenous variables, terms of trade (TOT), technology progress (TFP), capital accumulation (KGROW), and the proxy for government's expenditure on nontraded goods (GCONSNT) have expected negative sign, and all of them are statistically significant at least 10% level. On the other hand, the variables related to the government's trade and capital controls, namely TRDCONTR and CFLCONTR, have wrong (positive) sign. They are statistically significant. The trade control variable, TRDCONTR, is measured by import tariff divided by import value. Therefore, decline in TRDCONTR means trade liberalization. The theory of equilibrium real exchange rate implies that trade liberalization would lead to real appreciation of exchange rate by the demand effect on nontradables. As Edwards (1989) notes, however, if income effect from lowering tariff would dominate substitution effects between tradables and nontradables, it may be possible for the real exchange rate to depreciate under trade liberalization. The foreign exchange control variable CFLCONTR is measured by the ratio of net private capital flow to nominal GDP. Therefore, an increase in CFLCONTR implies liberalization on foreign exchange. The equilibrium theory suggests that foreign capital liberalization would result in real appreciation of the currency by demand effect.

The VEC estimation results shown in Table 5, however, indicate that lagged trade liberalization and foreign exchange liberalization would lead to real appreciation of Baht. In particular, the coefficients on CFLCONTR are significant for both one-period lagged and two-period lagged values. Given the opposite sign in the cointegrating equation, this result poses a puzzle.

Despite these problems, we may be able to say that the equilibrium condition estimated in the cointegrating equation appears to be essentially consistent with the equilibrium theory of real exchange rate as presented in equation (2)

We next investigate the real exchange rate equation of the VEC model. Using the information of AIC and SBC, we select three lags for endogenous variables, and use current value of exogenous variables except for the variable $RERDIF_{t-1}$. The estimated model is as follows. We include a dummy variable $DUMMY84$, to control for nominal devaluation in November 1984.

$$\Delta q_t = \psi \cdot \mu_{t-1} + \sum_{j=1}^{3} \alpha_{1,j} \cdot \Delta q_{t-j} + \sum_{j=1}^{3} \alpha_{2,j} \cdot \Delta TOT_{t-j} + \sum_{j=1}^{3} \alpha_{3,j} \cdot \Delta TFP_{t-j}$$

$$+ \sum_{j=1}^{3} \alpha_{4,j} \cdot \Delta KGROW_{t-j} + \sum_{j=1}^{3} \alpha_{5,j} \cdot \Delta TRDCONTR_{t-j}$$

$$+ \sum_{j=1}^{3} \alpha_{6,j} \cdot \Delta CFLCONTR_{t-j} + \sum_{j=1}^{3} \alpha_{7,j} \cdot \Delta GCONDNT_{t-j}$$

Table 5. Vector Error Correction Model Estimation Results.

Explanatory variable	Lag 1	Lag 2	Lag 3
Cointegration equation:	–0.10559		
μ_{t-1}	(0.0555)		
	[–1.9037]		
Δq_{t-j}	0.0734	–0.1395	0.0615
	(0.1561)	(0.1502)	(0.1553)
ΔTOT_{t-j}	0.1254	–0.0687	0.1513
	(0.1348)	(0.1197)	(0.1088)
ΔTFP_{t-j}	9.1483	–4.9977	7.6669
	(4.3191)	(4.8256)	(3.8850)
$\Delta KGROW_{t-j}$	0.4536	0.04926	–0.0299
	(0.3984)	(0.4096)	(0.3405)
$\Delta TRDCONTR_{t-j}$	–0.7031	–0.8763	–0.5085
	(0.7046)	(0.6408)	(0.5700)
$\Delta CFLCONTR_{t-j}$	–0.2587	–0.2835	–0.1696
	(0.1350)	(0.1321)	(0.1273)
$\Delta GCONSNT_{t-j}$	0.7252	0.8956	0.3996
	(0.8187)	(0.8157)	(0.6176)
$BGTDFCT_t$	0.02551		
	(0.1308)		
	[0.1950]		
$EXDMCRDT_t$	–0.02101		
	(0.06322)		
	[–0.3323]		
$DUMMY84$	3.0556		
	(3.1923)		
	[0.9572]		
$RERDIF_{t-1}$	–0.1690		
	(0.07314)		
	[–2.3111]		
R^2	0.4917		
AIC (Akaike criterion)	4.8255		
SBC (Schwarz criterion)	5.7679		

Notes:
1. Dependent variable is Δq_t.
2. The figure in parenthesis (.) is standard errors, and the figure in [.] is *t*-statistics.

$$+\beta_1 \cdot BGTDFCT_t + \beta_2 \cdot EXDMCRDT_t$$
$$+\beta_3 \cdot RERDIF_{t-1} + \delta \cdot DUMMY84 + \varepsilon_t.$$

The estimation result is shown in Table 5. Our major interest in this equation is the coefficient on the cointegrating equation μ_{t-1}, and the coefficient β_3 on the real exchange misalignment variable $RERDIF_{t-1}$. As in the case of VAR, the coefficients on the lagged value of endogenous variables are difficult to interpret, and may be of little use.

The speed of adjustment parameter on μ_{t-1} has negative sign and is significant at 10% level. It implies that if the actual trade-weighted REER deviated from the equilibrium real rate, in previous period, there will be an autonomous mechanism working toward correcting the disequilibrium in current period. As an example, if the actual REER depreciated more than the equilibrium theory predicted in last period, that equilibrium error will be corrected by pushing the current REER downward in current period.

Another main interest is the coefficient on the real exchange rate misalignment variable $RERDIF_{t-1}$. It has negative sign and is highly significant. It appears to suggest that when real overvaluation is high in previous period, actual trade-weighted REER tends to correct for the over-valuation by appreciating in current period. This logic seems to be consistent with the logic of dynamic adjustment of equilibrium error as discussed above. This interpretation, however, may need further consideration, because currency-weighted REER will also appreciate as trade-weighted REER appreciates. In fact, the contemporaneous correlation between $\Delta RERDIF_t$ and trade-weighted REER, q_t, is 0.2131, and one-period lagged (on $\Delta RERDIF_t$) correlation $Corr(q_t, \Delta RERDIF_{t-1})$ is still 0.2017. The lagged correlation dies out slowly, becoming below 0.10 after 8th lags (2 years). This suggests another possibility that when real overvaluation remains, subsequent movement of trade-weighted REER may not direct toward the correction of the over-valuation.[12]

Let us turn to other exogenous variables. Expansionary fiscal and monetary policy variables, BGTDFCT (budget deficit/monetary base) and EXDMCRDT (growth rate of domestic credit minus growth rate of real GDP), do not have any explanatory power in determining the change in real exchange rate. Although each of them has correct negative sign, the estimated coefficients are too imprecise to draw any conclusions.

Although the estimation results are not fully compatible with the equilibrium theory of real exchange rate, we state some major findings.

- The equilibrium real effective exchange rate of Thailand, though not observable, appears to have been determined by real economic factors that have bearings on REER determination. In particular, such fundamental variables as terms of trade, technological progress and capital accumulation seem to be important factors for determination of equilibrium rate.

- When actual REER tends to deviate from the equilibrium rate, an autonomous mechanism seems to work toward correcting the disequilibrium.
- Considering that trade-weighted REER may be biased downward (appreciation) by currency-weighted basket method, we may obtain some different results on the effects of endogenous and exogenous variables on equilibrium real exchange rate if such bias can be removed.
- Time series properties of the data suggest that potential overvaluation of REER due to currency-weighted basket method seems to have long-lasting impact on subsequent currency overvaluation.

4. REAL EXCHANGE RATE MISALIGNMENT AND THE CURRENT ACCOUNT

In this section, we investigate the potential consequence of the real exchange rate misalignment on evolution of the current account over time. The current account exhibited large swings between the latter half of 1980s and 1990s. The expanding current account deficit during this period was financed by precipitous inflow of foreign capital. Increase in net foreign capital flows was made possible by high domestic interest rates policy, managed, stable Baht-U.S. dollar exchange rate, foreign exchange deregulation accelerated since early 1990s, and inward foreign direct investment. Phenomenal increase in private foreign capital inflows, particularly short-term capital flows, was observed with the start of Bangkok International Banking Facility (BIBF), the offshore market, in early 1993.

Our interest is whether the dynamic movement of the current account in the 1980s and 1990s is related to the foreign exchange rate policy. To this end, we first need to set up a model of current account in a general equilibrium framework, and examine whether such an equilibrium model could mimic the actual evolution of current account. As might be expected, the theoretical model will not be able to explain perfectly the actual movement of current account. Therefore, as the next step, we test whether the deviation of actual current account from the equilibrium value is associated with the real effective exchange misalignment associated with the exchange rate policy.

To formulate an equilibrium model of current account, we postulate that the current account will evolve as an outcome of a representative consumer's intertemporal utility maximization in a small open economy, following the literature of the intertemporal approach to the current account by Sheffrin and Woo (1990), Ghosh (1995), Ghosh and Ostry (1995), and Obstfeld and Rogoff (1996), among others. The intertemporal approach to the current account draws on the work by Campbell (1987), who studied the present value model of saving-consumption decisions of a representative consumer in an intertemporal framework. Analogous to a consumer's decisions on intertemporal consumption and savings, a country's decision on current account, or decision on foreign borrowing and lending, may be modeled within a framework of a country's intertemporal welfare maximization problem.

4.1. The Intertemporal Approach to Current Account

We assume a small open economy with open capital market, in which the real interest parity condition holds. A representative consumer has infinite lives, and maximizes her lifetime utility by borrowing and lending in the capital market. Identical firms with constant return to scale technology operate in a perfectly competitive market, and rent capital from consumers. Each consumer also supplies one unit of labor for the firm. The firms will earn zero profit in the equilibrium. Given this setting, we render the general equilibrium model to a social planner's maximization problem.

A representative consumer maximizes the expected present value of her lifetime utility at any time t,

$$U_t = E_t\left\{\sum_{j=0}^{\infty} \beta^j u(c_{t+j})\right\} \tag{12}$$

where $u(\cdot)$ is time-separable, periodic utility function, c_{t+j} is per-person consumption, and β is subjective discount factor which takes a value between zero and one. The symbol E_t is the mathematical expectation operator, conditional on all available information held by the consumer at time t.

There is a world capital market in which a riskless bond is traded at constant, world real interest rate r. A consumer will maximize her lifetime utility by borrowing and lending in the world capital market. A consumer holds the bond b_t at the beginning of time t. b_t may be negative in which case she is a net borrower in the capital market.

The firm invests in capital stock at time t by the amount of i_t. By the assumption, this capital is in fact invested by the consumers. We assume there is no depreciation on the capital stock. There is the government sector whose spending g_t, in per-capita term, draws resources from the consumer.

We can derive the intertemporal budget constraint of a representative consumer as

$$E_t\left\{\sum_{j=0}^{\infty}\left(\frac{1}{1+r}\right)^j\cdot(c_{t+j}+i_{t+j})\right\}=E_t\left\{(1+r)\cdot b_t+\sum_{j=0}^{\infty}\left(\frac{1}{1+r}\right)^j\cdot(y_{t+j}-g_{t+j})\right\}, \tag{13}$$

where y_{t+j} is per-capita production, or per-capita GDP.

In deriving equation (13), we imposed the Non-Ponzi game condition, which precludes the case of growth rate of borrowing in excess of real interest rate.

The current account of this small open economy at time t, expressed in terms of per-capita term, is given by

$$ca_t=b_{t+1}-b_t=y_t+r\cdot b_t-c_t-g_t-i_t \tag{14}$$

where the current account ca_t equals capital account $(b_{t+1}-b_t)$.

Substituting consumption c_t from equation (14) into equation (12), we obtain the consumer's maximization problem in a reduced from,

$$\max_{\{b_{t+j},b_{t+j+1}\}} U_t=E_t\left\{\sum_{j=0}^{\infty}\beta^j u[(1+r)\cdot b_{t+j}-b_{t+j+1}+y_{t+j}-g_{t+j}-i_{t+j}]\right\}, \tag{15}$$

subject to the intertemporal budget constraint.

The first order condition with respect to b_{t+j+1} yields the Euler equation

$$E_t\{u'(c_{t+j})\}=(1+r)\cdot\beta\cdot E_t\{u'(c_{t+j+1})\}. \tag{16}$$

In particular, at period t where $j=0$,

$$u'(c_t)=(1+r)\cdot\beta\cdot E_t\{u'(c_{t+1})\}. \tag{17}$$

For analytical tractability, we use the linear-quadratic utility function,

$$u(c_t)=c_t-\frac{\alpha}{2}\cdot c_t^2. \tag{18}$$

Applied to the Euler equation, this liner-quadratic utility function yields random walk consumption path

$$E_t(c_{t+1})=c_t. \tag{19}$$

Conditional expectation of all future consumption path is equal to the current consumption. Using this property, the optimal consumption function is derived from the intertemporal budget constraint,

$$c_t=\frac{r}{1+r}\left\{(1+r)\cdot b_t+\sum_{j=0}^{\infty}\left(\frac{1}{1+r}\right)^j E_t[y_{t+j}-g_{t+j}-i_{t+j}]\right\}. \tag{20}$$

Given this consumption function, we can express the current account equation (14) in terms of the deviation of output, investment and government expenditure in current period, from the sum of discounted expected value of each item.

$$ca_t = b_{t+1} - b_t = y_t - \frac{r}{1+r}\sum_{j=0}^{\infty}\left(\frac{1}{1+r}\right)^j E_t(y_{t+j}) + i_t - \frac{r}{1+r}\sum_{j=0}^{\infty}\left(\frac{1}{1+r}\right)^j E_t(i_{t+j})$$

$$+ g_t - \frac{r}{1+r}\sum_{j=0}^{\infty}\left(\frac{1}{1+r}\right)^j E_t(g_{t+j}). \qquad (21)$$

Define net output z_t by

$$z_t \equiv y_t - g_t - i_t. \qquad (22)$$

Then, the current account can be written as

$$ca_t = z_t - E_t(\tilde{z}_t)$$

where

$$E_t(\tilde{z}_t) = \frac{r}{1+r}\sum_{j=0}^{\infty}\left(\frac{1}{1+r}\right)^j E_t(y_{t+j}) - \frac{r}{1+r}\sum_{j=0}^{\infty}\left(\frac{1}{1+r}\right)^j E_t(i_{t+j})$$

$$- \frac{r}{1+r}\sum_{j=0}^{\infty}\left(\frac{1}{1+r}\right)^j E_t(g_{t+j}). \qquad (23)$$

Adding and subtracting x_{t+k}, $k = 1, 2, ---$, where $x = y$, g, and i, from the right-hand side of the current account equation (21), we obtain Campbell's (1987) formula of present value equation, as applied to the intertemporal current account equation

$$ca_t = -\sum_{j=0}^{\infty}\left(\frac{1}{1+r}\right)^j \cdot E_t(\Delta z_{t+j}), \qquad (24)$$

where the symbol Δ is the first difference operator.

The current account in an intertemporal context is viewed as negative of sum of the present discounted value of expected change of future net output, where the world interest rate r is used as the discount rate. When future output, net of investment and government expenditure, is expected to be large, the country will run current account deficit now. In other words, the country will be a net borrower in the world capital market, if the economy is expected to grow faster on average in the future.

4.2. Estimation of the Intertemporal Current Account Model

We follow the methodologies used in Sheffrin and Woo (1990), Ghosh and Ostry (1995), and Obstfeld and Rogoff (1996) for setting up a bivariate first-order VAR model to estimate the current account equation (24). We assume that the consumer will forecast future change in net output Δz_s, $s > t$, based on one-period lagged realizations of net output and current account. Similarly, she is assumed to make a one-period ahead forecast of current account, using the information based on past realization of current account and change in net output.

The first-order VAR is specified as

$$X_t = \Phi \cdot X_{t-1} + e_t, \tag{25}$$

where

$$X_t = \begin{bmatrix} \Delta z_t \\ ca_t \end{bmatrix}, \quad e_t = \begin{bmatrix} e_{\Delta z_t} \\ e_{ca_t} \end{bmatrix}, \quad \text{and} \quad \Phi \equiv \begin{bmatrix} \phi_{11} & \phi_{12} \\ \phi_{21} & \phi_{22} \end{bmatrix}.$$

The variable e_t is assumed to be white noise process, but may be correlated with each other, $corr(e_{\Delta z_t}, e_{ca_t}) \neq 0$.

By successively iterating forward on equation (25), we obtain

$$E_t \begin{bmatrix} \Delta z_{t+j} \\ ca_{t+j} \end{bmatrix} = \Phi^j \cdot \begin{bmatrix} \Delta z_t \\ ca_t \end{bmatrix}. \tag{26}$$

Multiplication of both sides of equation (26) by a row vector [1 0] yields

$$E_t \Delta z_{t+j} = [1 \ \ 0] \cdot \Phi^j \cdot \begin{bmatrix} \Delta z_t \\ ca_t \end{bmatrix}. \tag{27}$$

Here we obtain the equation $E_t \Delta z_{t+j}$ that can be substituted into the current account equation. Substituting equation (27) into the current account equation (24), and using the formula for a geometric series $\left(\frac{\Phi}{1+r}\right) \cdot \sum_{j=0}^{\infty} \left(\frac{1}{1+r}\right)^j \cdot \Phi^j$, we obtain the predicted, by VAR, current account $\tilde{ca}_t$;

$$\tilde{ca}_t = -[1\ \ 0] \cdot \left(\frac{\Phi}{1+r}\right) \cdot \left(I - \frac{\Phi}{1+r}\right)^{-1} \cdot \begin{bmatrix} \Delta z_t \\ ca_t \end{bmatrix}, \tag{28}$$

where I is the 2×2 identity matrix.

Define a (1×2) row vector

$$[\phi_{\Delta z}\ \ \phi_{ca}] = -[1\ \ 0] \cdot \left(\frac{\Phi}{1+r}\right) \cdot \left(I - \frac{\Phi}{1+r}\right)^{-1}. \tag{29}$$

Then, the predicted current account at time t is expressed as

$$\tilde{ca}_t = [\phi_{\Delta z}\ \ \phi_{ca}] \cdot \begin{bmatrix} \Delta z_t \\ ca_t \end{bmatrix}. \tag{30}$$

The scalar $\phi_{\Delta z}$ and ϕ_{ca} can be computed from the estimated coefficients ϕ_{ij} of the first-order VAR (25), using the formula (29).

We first estimated the first-order VAR system (25). The data used are the current account in real terms ($CAREAL_t$), and the change in real net output ($\Delta NETOUT_t$). We deflated the nominal current account by the wholesale price index (1990 = 100). The components of net output, namely GDP, government expenditure, and investment, are also deflated by the wholesale price index (1990 = 100).[13]

Estimation of VAR yielded the following first-order coefficient matrix

$$\Phi = \begin{bmatrix} \phi_{11} & \phi_{12} \\ \phi_{21} & \phi_{22} \end{bmatrix} = \begin{bmatrix} -0.38615 & -0.12397 \\ (0.1324) & (0.0842) \\ 0.04824 & 0.90183 \\ (0.0906) & (0.0576) \end{bmatrix}, \tag{31}$$

where the figures in the parentheses are standard errors.

The computed row vector

$$[\Phi_{\Delta z}\ \ \phi_{ca}] = -[1\ \ 0] \cdot \left(\frac{\Phi}{1+r}\right) \cdot \left(I - \frac{\Phi}{1+r}\right)^{-1}$$

is found to be $[\phi_{\Delta z}\ \ \phi_{ca}] = [0.295874\ \ 0.681052]$, where we assumed the world real interest to be 3% per annum.

Given the computed vector $[\phi_{\Delta z}\ \ \phi_{ca}]$ we calculated the predicted, by VAR, current account ($CAPREDICT_t$) from the equation (28). The result is shown in Table 6. The actual current account is also included in the table for a

Table 6. Actual Current Account and Predicted (by VAR) Current Account.

Period (Quarter)	CAPREDICT	CAREAL	Period	CAPREDICT	CAREAL
1982.1	4.757	–2.725	1990.1	–19.499	–34.016
1982.2	–1.843	–3.796	1990.2	–22.644	–37.145
1982.3	–11.734	–11.504	1990.3	–36.371	–49.255
1982.4	–5.565	–10.378	1990.4	–48.094	–65.117
1983.1	–12.334	–15.401	1991.1	–37.511	–60.985
1983.2	–13.134	–17.515	1991.2	–37.911	–56.062
1983.3	–15.812	–22.033	1991.3	–26.812	–43.462
1983.4	–15.541	–23.992	1991.4	–4.052	–21.341
1984.1	–7.981	–17.451	1992.1	–19.769	–32.017
1984.2	–9.574	–15.631	1992.2	–28.244	–32.629
1984.3	–11.324	–15.894	1992.3	–26.966	–44.630
1984.4	–7.399	–12.085	1992.4	–23.036	–40.739
1985.1	–13.904	–17.270	1993.1	–39.411	–44.481
1985.2	–8.153	–17.073	1993.2	–13.629	–40.015
1985.3	–5.475	–11.908	1993.3	–9.362	–19.108
1985.4	–4.323	–5.591	1993.4	–29.504	–48.039
1986.1	2.318	–1.547	1994.1	–28.806	–35.506
1986.2	5.665	5.739	1994.2	–31.216	–46.925
1986.3	4.370	3.704	1994.3	–25.138	–47.866
1986.4	–1.675	0.477	1994.4	–39.770	–53.035
1987.1	0.657	–2.177	1995.1	–40.383	–61.785
1987.2	–0.068	0.863	1995.2	–41.348	–65.624
1987.3	–1.546	–3.675	1995.3	–45.304	–67.124
1987.4	0.621	–5.341	1995.4	–66.047	–86.305
1988.1	–9.897	–7.644	1996.1	–39.987	–79.297
1988.2	–8.463	–16.187	1996.2	–57.220	–80.398
1988.3	–5.941	–12.638	1996.3	–59.024	–78.635
1988.4	–5.562	–8.482	1996.4	–24.975	–58.167
1989.1	–29.940	–43.573			
1989.2	–6.699	–23.006			
1989.3	–30.739	–34.554			
1989.4	–31.058	–44.132			

Notes:
CAREAL = actual current account (Real seasonally adjusted: Billion Baht).
CAPREDICT = predicted current account by VAR (seasonally adjusted).

comparison. A comparison between the actual current account and predicted current account is shown in Fig. 3 too.

It appears that the first-order VAR can explain the actual movement of real current account fairly well. Contemporaneous correlation coefficient between

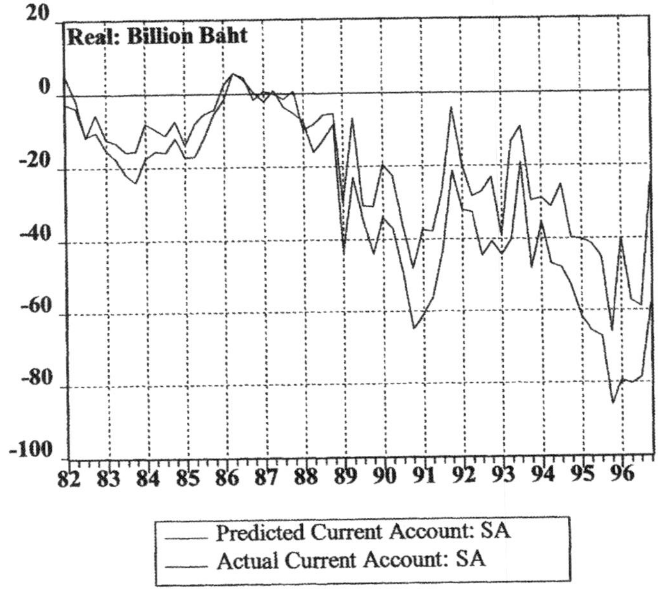

SA: seasonally adjusted series

Fig. 3. Predicted Current Account vs. Actual Current Account.

the actual and predicted current account is 0.9602, while one-period lagged and two-period lagged, on predicted current account, correlation coefficients are 0.8720, and 0.7458, respectively. The figure shows, however, that the deviation of actual current account from the predicted current account became more pronounced since the late 1980s, compared to the earlier period (Fig. 3 and Fig. 4). Standard deviation of prediction error, defined as the difference between the actual current account ($CAREAL_t$) and the predicted current account ($CAPREDICT_t$), is 3.8956 between 1981/1st quarter and 1986/4th quarter, 10.2558 between 1987/1st quarter and 1990/4th quarter, and 15.8372 between 1991/1st quarter and 1996/4th quarter (Fig. 4). We will examine next how the prediction error of the real current account is associated statistically with the measure of the real exchange rate misalignment, defined by the difference between trade-weighted REER and currency-weighted REER.

4.3. Real Exchange Rate Misalignment and Prediction Error of the Current Account

Table 7 presents cross correlation coefficients between seasonally adjusted prediction errors of the real current account $CAERR_{t-s}$ and the measure of the

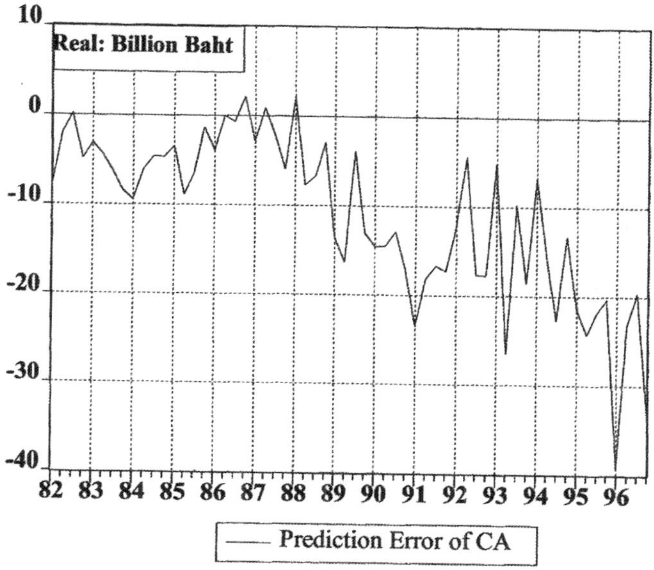

Mean and Standard Deviation of the Prediction Error

Period	Mean	Standard Deviation	Maximum	Minimum
1981/1st Quarter – 1986/4th Quarter	-1.8401	3.2654	3.8956	-8.2279
1987/1st Quarter – 1990/4th Quarter	-3.4098	6.0290	10.2558	-11.7574
1991/1st Quarter – 1996/4th Quarter	-7.8644	10.9075	15.8372	-21.8944

Note: Unit = Billion Baht

Fig. 4. Prediction Error of Current Account (Actual CA – Predicted CA: SA).

real effective exchange rate misalignment $RERDIF_{t-j}$, which is the difference between the trade-weighted real exchange rate and the currency-weighted real exchange rate, across different time period. Contemporaneous correlation coefficient is –0.5870, suggesting a negative correlation between the real exchange rate overvaluation and the magnitude of the under-predicted real current account deficit. The correlation coefficient of $CAERR_t$ and one-period lagged $RERDIF_{t-j}$ is –0.6133. The lagged (on $RERDIF_{t-j}$) cross correlation coefficients become larger in negative terms as the time difference becomes

Table 7. Cross Correlation between Prediction Error of Current Account and Real Exchange Rate Misalignment.

$Corr(CAERR, RERDIF(i)), i \in (-p, +p)$

No. of lag & lead (Quarter)	Lag $(-p)$	Lead $(+p)$
0	–0.5870	–0.5870
1	–0.6133	–0.5058
2	–0.6353	–0.4604
3	–0.6802	–0.4341
4	–0.6891	–0.3648
5	–0.6817	–0.3154
6	–0.6729	–0.3131
7	–0.6533	–0.3015
8	–0.6394	–0.2935
9	–0.6036	–0.2631
10	–0.5865	–0.2675
11	–0.5576	–0.2532
12	–0.4906	–0.2134
13	–0.4379	–0.1996
14	–0.4049	–0.1948
15	–0.3354	–0.1422
16	–0.2711	–0.1230
17	–0.2177	–0.1059
18	–0.1581	–0.0537
19	–0.1018	–0.0131
20	–0.0410	0.0229
21	0.0076	0.0611
22	0.0296	0.0970
23	0.0921	0.1558
24	0.0993	0.1775

Notes:
1. Sample period: 1982/1st Quarter–1996/4th Quarter.
2. *CAERR* = Actual CA – Predicted CA; Seasonally adjusted.
   RERDIF = Trade-weighted REER – Currency-weighted REER.

apart up to four lags (one year). After five lags, the (negative) cross correlation coefficients remain large, and die out quite slowly until 11 quarters (about three years).

When we use non seasonally-adjusted data of the prediction error, the picture looks similar, although cross correlations are a little smaller in magnitude. The contemporaneous correlation is –0.3269. The negative correlations stay at a high level around –0.31 ~ –0.35 up to 8th lags. After that, they die out slowly and become below –0.1 after 19th lags.

To study the dynamic relationship between $RERDIF_{t-j}$ with $CAERR_{t-s}$ for different lag lengths, we run a bivariate vector autoregression (VAR) as follows:

$$X_t = \alpha + \sum_{j=1}^{p} \Gamma_j X_{t-j} + e_t, \tag{32}$$

where

$$X_{t-j} = \begin{bmatrix} CAERR_{t-j} \\ RERDIF_{t-j} \end{bmatrix},$$

Γ_j is a (2×2) coefficients matrix,

e_t is a vector of serially uncorrelated random disturbances with zero mean and constant variance,

α is a coefficient vector on a constant term.

Since it is sometimes recommended that we run a standard VAR for stationary endogenous random variables, we first conducted the Augmented Dickey-Fuller unit root test on the variables (Table 8). The test results confirm that both variables are stationary, although the test statistic is sensitive to the number of lags included in the first difference term of the variable. The estimation result of the standard, reduced form VAR, the equation (32), is shown in Table 9. We selected a three-lag model as it has the smallest values of AIC (Akaike Information Criterion) and SBC (Schwarz Bayesian Criterion). Although the coefficients of a reduced form VAR do not have much meaningful interpretation, we nevertheless notice the negative statistical association of three quarters lagged value of $RERDIF_{t-j}$ $(j=3)$ and contemporaneous value of $CAERR_{t-s}$ $(s=0)$ in equation 1 with the prediction error of the current account as the dependent variable. The coefficient is –0.87841, and significant at 5% level. We also note that $RERDIF_t$ has strong, positive first-order autoregressive characteristics; the coefficient on one quarter lagged value $RERDIF_{t-1}$ (1.09529) is highly significant.

To gain more statistical support for the negative association of lagged $RERDIF_{t-j}$ with current value of $CAERR_t$, we examined dynamic interactions between the two variables. We first implemented the Granger-causality test, to test whether the past values of $RERDIF_{t-j}$ Granger-cause the current value of

Table 8. Augmented Dickey-Fuller Unit Root Test on $CAERR_t$ and $RERDIF_t$.

Dependent variable			
$\Delta CAERR_t$		$\Delta RERDIF_t$	
Explanatory variable	Coefficients	Explanatory variable	Coefficients
$CAERR_{t-1}$	–0.5532 (0.1636)	$RERDIF_{t-1}$	–0.4849 (0.1221)
$\Delta CAERR_{t-1}$	–0.2853 (0.1323)	$\Delta RERDIF_{t-1}$	0.3716 (0.1428)
Constant	1.3563 (1.7448)	$\Delta RERDIF_{t-2}$	0.1904 (0.1468)
Trend	–0.2644 (0.0837)	$\Delta RERDIF_{t-3}$	0.3390 (0.1388)
		$\Delta RERDIF_{t-4}$	0.3925 (0.1360)
		$\Delta RERDIF_{t-5}$	0.1817 (0.1438)
		$\Delta RERDIF_{t-6}$	0.2050 (0.1455)
		$\Delta RERDIF_{t-7}$	0.3390 (0.1416)
		$\Delta RERDIF_{t-8}$	0.2979 (0.1588)
		Constant	–3.7170 (1.2021)
		Trend	0.1252 (0.0378)
Adjusted R^2	0.3915	Adjusted R^2	0.2217
Durbin-Watson	2.1875	Durbin-Watson	1.9257
F-statistic	13.227	F-statistic	2.539
ADF test statistic	–3.3807*	**ADF test statistic**	–3.9724**

Notes:

1. The equation is, $\Delta y_t \equiv y_t - y_{t-1} = c + \alpha \cdot t + (\beta - 1) \cdot y_{t-1} + \sum_{j=1}^{p} \gamma_j \Delta y_{t-j} + \varepsilon_t$. t is a linear trend term. Null hypothesis is a unit root; $\delta \equiv \beta - 1 = 0$.
2. The figure in parenthesis is standard errors.
3. For $RERDIF_t$, a total of eight lags on $\sum_{j=1}^{p} \gamma_j \Delta y_{t-j}$ generate the smallest value of Akaike Information Criterion (AIC). For $CAERR_t$, the smallest value of AIC obtains with nine lags on $\sum_{j=1}^{p} \gamma_j \Delta y_{t-j}$ (AIC = 6.3808 as opposed to 6.4802 for one lag model). Schwartz Bayesian Criterion

Table 9. Prediction Error of Current Account and Real Exchange Rates Misalignment – VAR Result.

Dependent variable	CAERR	RERDIF
CAERR(–1)	–0.01547	0.06530
	(0.11965)	(0.04236)
CAERROR(–2)	0.21915*	0.00300
	(0.11409)	(0.04039)
CAERR(–3)	0.46141***	–0.03750
	(0.11681)	(0.04135)
RERDIF(–1)	–0.12608	1.09529***
	(0.39294)	(0.13910)
RERDIF(–2)	0.52749	–0.23179
	(0.57712)	(0.20430)
RERDIF(–3)	–0.87841**	0.12517
	(0.41475)	(0.14682)
Constant	–1.77373	0.57777
	(1.21452)	(0.42993)
Adjusted R^2	0.6784	0.9175
F-statistics	20.691	104.836
Akaike criterion	6.2199	4.1429
Schwarz criterion	6.4708	4.3938
(VAR system)		
Log likelihood	–281.105	
Akaike criterion	10.3545	
Schwarz criterion	10.8564	

Notes:
1. Sample period: 1982/1st Quarter–1996/4th Quarter.
2. CAERR = Actual current account – Predicted current account; Seasonally adjusted series, by Census 11 method.
RERDIF = Trade-weighted REER – Currency-weighted REER.
3. Asterisks ***, ** and *show the significance level at 1%, 5% and 10%, respectively.

$CAERR_t$ (Table 10). We tried several lag lengths, starting from a large number of lags such as $j = 18$ in $RERDIF_{t-j}$, i.e. four and half years. From eighteen lags to fifteen lags, we cannot reject the null hypothesis that $RERDIF_{t-j}$ does not Granger-cause $CAERR_t$. We reject, however, the null hypothesis from fourteen lags to smaller number of lags with one exception; six lags which fails to reject the null at 10% significance level. In most of the number of lags, we reject the null at 5% significance level. In the case of three lags, our VAR model's specification, the F-statistic is 3.61711, which is significant at 5% level. Thus, the past values of the measure of real exchange rate misalignment could improve on the explanatory power of the past values of the prediction error of the current account in predicting the latter's future value. On the other hand, the reverse causality test, i.e. the null hypothesis that $CAERR_{t-s}$ does not Granger-cause $RERDIF_t$, fails to reject the null in most of the lags.

Next, based on the estimation result of the three-lag VAR system, we computed two dynamic statistics: impulse response functions and forecast error variance decomposition. In computing the impulse response functions, we put $CAERR_t$ in the first position, followed by $RERDIF_t$, for the Choleski decomposition. That is, we assumed that $CAERR_t$ would not have a contemporaneous effect on $RERDIF_t$. The impulse response functions of $CAERR_t$, in response to one standard deviation shock to the innovation in $RERDIF_{t-j}$ of the underlying structural model, are plotted in Fig. 5. The

Table 10. Granger-causality Test.

Lag length p in $\sum_{j=1}^{p} \alpha_j \cdot RERDIF_{t-j}$	F statistic	Probability	*Reject* or *Not reject* the null hypothesis
1	9.36594	0.00339	reject at 1% level
2	3.22394	0.04772	reject at 5%
3	3.61711	0.01929	reject at 5%
4	2.81757	0.03550	reject at 5%
5	2.05337	0.08960	reject at 10%
6	1.70446	0.14442	not reject at 10%
7	1.96652	0.08567	reject at 10%
8	2.82747	0.01569	reject at 5%
9	2.52822	0.02575	reject at 5%
10	2.79553	0.01490	reject at 5%
11	2.30036	0.03956	reject at 5%
12	2.63295	0.02224	reject at 5%
13	2.31437	0.04444	reject at 5%
14	2.18415	0.06390	reject at 5%
15	1.91870	0.11546	not reject at 10%
16	1.68916	0.19080	not reject at 10%
17	1.46111	0.30022	not reject at 10%
18	0.76691	0.69395	not reject at 10%

Notes:
1. Null hypothesis: $RERDIF_{t-j}$ *does not Granger-cause* $CAPRDCTERR_t$.
2. Sample period: 1982/1st Quarter – 1996/4th Quarter.
3. $CAERR_t$: Prediction error of real current account = Actual real current account – Predicted real current account, seasonally adjusted
   $RERDIF_{t-j}$: Difference in real effective exchange rates = Trade-weighted REER – Currency-weighted REER.

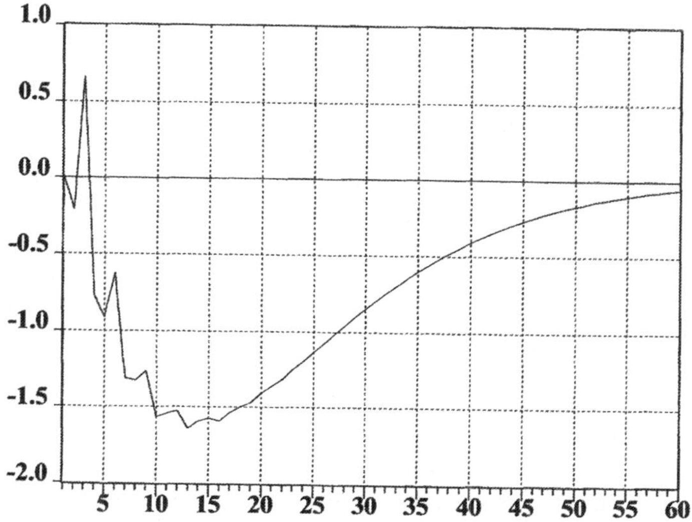

The VAR model

$$X_t = \alpha \cdot C + \sum_{j=1}^{3} \Gamma_j X_{t-j} + \varepsilon_t$$

$$\text{where } X_t = \begin{bmatrix} CAERR_t \\ RERDIF_t \end{bmatrix}$$

Fig. 5. Impulse Response of CA Prediction Error (CAERR) to One S.D. Innovation of REER Differential (RERDIF).

prediction error of the real current account shows negative responses to the shock to the innovation in $RERDIF_{t-j}$. One exception is three-quarters ahead prediction: the impulse response function of $CAERR_t$ on three-quarters lagged $RERDIF_{t-j}$ is positive. This positive impulse response may reflect a kind of J-curve effects.

As the variation of the real exchange rate misalignment increases, the prediction error of the current account becomes negative. Furthermore, the negative response of the prediction error to the impulse in $RERDIF_{t-j}$ increases as time goes by, until thirteen quarters. Although after fourteen quarters it gradually dies out, the speed of convergence to zero impulse response looks quite slow: it takes more than ten years to approach zero. In summary, the shape of the impulse response functions suggests that an increase in the overvaluation of real exchange rate would lead to an increase in the current account deficit in real terms. Moreover, it would have a long lasting impact on the current account deficit.

The variance of the error of s-period ahead forecast $CAERR_{t+s}$ is represented by the forecast error variance decomposition (Fig. 6). The portion of the forecast error variance of $CAERR_{t+s}$, explained by the variance of the innovation in $RERDIF_t$ of the underlying structural model, reaches about 20% two years later, and becomes 50% four-and-half years later. The variation in the measure of the real exchange rate misalignment at present would effect 20%, and 50% of the future path of the variation in the predicted error of the current account, two years and four-and-half years later, respectively. It appears that the percentage hits the peak at around 65% eight years later.

From the various statistical analyses in this section, we conclude that the statistical evidences suggest that the measure of the real exchange rate misalignment $RERDIF_t$ is negatively associated with the prediction errors of the actual current account in real terms. As the magnitude of overvaluation of the real effective exchange rate and its variation increase, the actual current

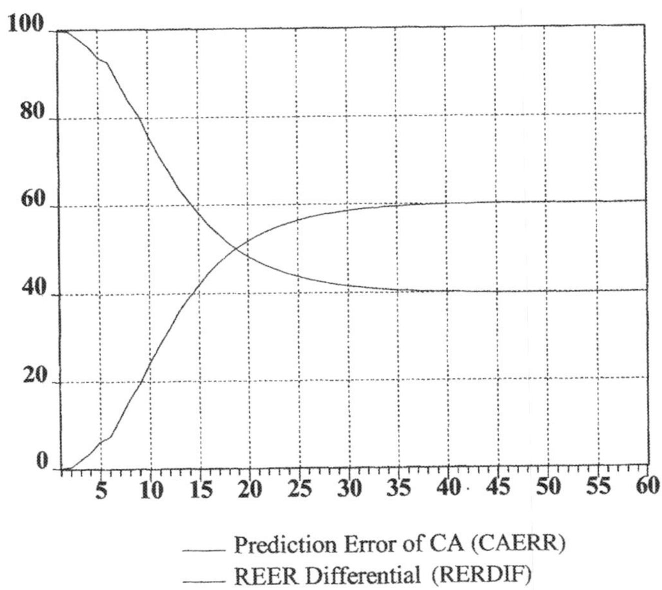

Fig. 6. Forecast Error Variance Decomposition of CA Prediction Error.

account deficit tends to increase more than the predicted value that is derived from the intertemporal optimization model. This statistically significant negative association may be due to the intertemporal model's appropriateness in predicting the actual current account. Taking into account the persistent and negative correlation between the real exchange rates differentials $RERDIF_t$ and the actual current account (about –0.7 for several lags), however, we think that the overvaluation of the real effective exchange rates is responsible, as one of the causes, for the widening current account deficit, particularly since the late 1980s. This overvaluation of the real effective exchange rates is regarded as being associated with the Thai authority's nominal anchor foreign exchange rate policy under the basket method.

5. CONCLUSION

In this paper, we studied potential consequences of the foreign exchange rate policy adopted by the Thai authority during the 1980s and 1909s, on the real effective exchange rate and the current account. First, we postulated that there would exist the equilibrium real exchange rate determined by fundamental, real economic factors, and estimated realized real exchange rate function by vector error correction model (VEC). A cointegrating equation, which corresponds to the equilibrium real exchange function, was identified. Although being not perfectly consistent with the equilibrium exchange rate model, many important real factors such as terms of trade, technology progress, and capital accumulation prove to be the determinant of the equilibrium real effective exchange rate. The estimation result of VEC also indicates that when realized real effective exchange rate deviates from the equilibrium rate, an autonomous force will work to correct for the equilibrium error in the subsequent period.

One innovation in this paper is we computed a hypothetical measure of real effective exchange rate misalignment, using currency-weighted real effective exchange rate. There is a substantial, and widening discrepancy between the currency-weighted REER and usual trade-weighted REER, particularly after the mid-1980s when the Thai authority switched exchange rate policy to the currency basket method. Given this, we investigated the potential consequences of this measure of real exchange rate misalignment on realized trade-weighted real effective exchange rate and on the real current account.

Although the measure of real exchange rate misalignment turns out to have a negative lagged correlation with realized trade-weighted real effective exchange rate, it does not imply that this policy disturbance has a self-correcting mechanism on the realized real rate. Rather, investigation of time series properties of the data suggests that potential overvaluation of REER due to currency basket method appears to have long-lasting impacts on subsequent currency overvaluation.

Next, we studied the consequence of the real exchange rate misalignment on the evolution of current account deficit. We first estimated a theoretical intertemporal current account model by vector autoregression (VAR). The theoretical, predicted current account fits the actual current account fairly well. The prediction error, however, becomes increasingly significant after the late 1980s, which corresponds to the period of increasing real exchange rate misalignment. Therefore, we estimated a bivariate VAR with the variables of the prediction error and the measure of real exchange rate misalignment, to examine their dynamic interactions. The results of impulse response functions, variance decomposition, and Granger-causality test suggest that the real exchange rate misalignment has a negative association with the prediction error of the real current account. This implies that the foreign exchange rate policy by the Thai authority might be responsible, as one of the causes, for the widening current account deficit since the mid 1980s.

NOTES

1. This paper was presented at the international conference "A Macroeconomic Core of an Open Economy for Progressive Industrialization and Development in Asia in the New Millenium", organized by Chularongkorn University (Thailand) and the American Committee on Asian Economic Studies, on December 16, 1998. We thank helpful comments from the participants.

2. We can be contacted via email at kasajimaeecon.hokkai-s-u.ae.jp and sukanda@econ.tu.ac.th

3. See Tasaka 1996.

4. The measured real effective exchange rate is defined as $RER = (NER \cdot P_f/P_{TH})$, where NER is the weighted nominal effective exchange rate index, P_f is the weighted foreign price index, and P_{TH} is the price index of Thailand.

5. The real exchange rate misalignment is usually defined as the discrepancy between the realized real exchange rate and the equilibrium real exchange rate. The trade weighted real exchange rate in this paper is not interpreted as the equilibrium exchange rate. It may also deviate from the equilibrium rate by macroeconomic policy disturbances. In section 3, we examine the dynamic adjustment of actual trade weighted real exchange rate toward its equilibrium rate, and the impact of currency basket method on the deviation of actual real rate from its equilibrium rate.

6. Suwanmana (1993).

7. Tasaka (1996), Chapter 1.

8. We computed currency-weighted real effective exchange rate, using 75% weight assigned to U.S. dollars on average during the sample period 1981/4th quarter and 1996/4th quarter. Remaining weight 25% is assigned to the currencies of major trading partner countries, with each trade weight, net of U.S. trade, as the currency weight.

9. The contemporaneous correlation coefficient between $\Delta RERDIF_t$ and q_t is as low as 0.21.

10. Edwards (1989) estimated equation (5) without the variable $RERDIF_{t-1}$ by OLS for a panel of large number of countries, without considering non-stationarity of each variable. We suspect that his estimation results may be biased because of non-stationarity of the data.

11. It is often argued that when a series is near-random walk, the power of ADF test statistics is weak.

12. The correlation $Corr(q_t, RERDIF_{t-j})$ is 0.793 for j = 0, 0.713 for j = 1, 0.634 for j = 2, and 0.558 for j = 3. The correlation falls below 0.10 after 8th lags too.

13. Thailand does not have quarterly GDP data. We used interpolated quarterly GDP series, estimated by Inthisuang (1998).

14. Note that the production error is defined as the actual real current account minus the predicted real current account. Therefore, negative prediction errors imply under-prediction of the predicted real current account.

REFERENCES

Balassa, B. (1964). The purchasing power parity doctrine: A reappraisal. *Journal of Political Economy, 72*, 584–596.

Campbell, J. (1987). Does saving anticipate declining labor income? An alternative test of the permanent income hypothesis. *Econometrica, 55*, 1249–1273.

Corden, W. M. (1981). The exchange rate, monetary policy and North Sea oil. *Oxford Economic Papers, 33*, 23–46.

Davidson, R., & MacKinnon. J. G. (1993). *Estimation and Inference in Econometrics*. New York, New York: Oxford University Press.

Edwards, S. (1989). *Real Exchange Rates, Devaluation, and Adjustment*. Cambridge, Massachusetts: The MIT Press.

Edwards, S. (1994). Real and monetary determinants of real exchange rate behavior. In: J. Williamson (Ed.), *Estimating Equilibrium Exchange Rates*. Washington, D.C.: Institute for International Economics.

Engel, R. E., & Granger, C. W. J. (1987). Cointegration and error correction: Representation, estimation, and testing. *Econometrica, 55*, 251–276.

Ghosh, A. R. (1995). International capital mobility amongst the major industrialized countries: Too little or too much? *Economic Journal, 105*, 107–128.

Ghosh, A. R., & Ostry, J. D. (1995). The current account in developing countries: A perspective from the consumption-smoothing approach. *The World Bank Economic Review, 9*(2), 305–333.

Inthisuang, J. (1998). *Economic fluctuation and its origin in Thailand. The Master's Thesis.* Faculty of Economics, Thammasat University.

Johansen, S. (1991). Estimation and hypothesis testing of cointegration vectors in Gaussian vector autoregressive models. *Econometrica, 59*, 1551–1580.

Obstfeld, M., & Rogoff, K. (1996), *Foundations of International Macroeconomics*. Cambridge, Massachusetts: The MIT Press.

Osterward-Lenum, M. (1992). A note with quantiles of the asymptotic distribution the maximum likelihood cointegration rank test statistics. *Oxford Bulletin of Economics and Statistics, 54*, 461–471.

Samuelson, P. (1964). Theoretical notes on trade problems. *Review of Economics and Statistics, 46*, 145–154.

Sheffrin, S. M., & Woo, W. T. (1990). Present value tests of an intertemporal model of the current account. *Journal of International Economics, 29*, 237–253.

Suwanmana, P. (1993). *The Evaluation of Exchange Rate Policy in Thailand during 1970–1990, The Master's Thesis*. Faculty of Economics, Thammasat University.

Tasaka, T. (1996), *Baht Economy, and Financial Liberalization (in Japanese)*. Ochanomizu Shobou Publishing Co., Tokyo: Japan.

Tinakorn, P., & Sussangkarn, C. (1996). *Productivity Growth in Thailand, Research Monograph No. 15, Thailand Development Research Institute*. Bangkok: Thailand.

17. APPROPRIATENESS OF FLEXIBLE EXCHANGE RATE RÉGIME IN ECONOMIC ADJUSTMENTS: AN ASIAN EXPERIENCE

Chartchai Parasuk and Wimonrat Sukcharoen[1]

A choice between the fixed exchange rate régime and the flexible exchange rate régime is an age-old argument among economists. Theoretically, the two régimes are equally effective in bringing the economy to the equilibrium. The difference is the adjustment mechanisms whereby the fixed exchange rate régime works through the income effect while the flexible exchange rate régime operates through the price effect. In practice, the price effect is favorable as the domestic economy does not fluctuate in response to the external imbalances. Thus, the flexible exchange rate system becomes the preferred choice of the international community, including the International Monetary Fund.

Asia, although most countries in this region has long adopted flexible exchange rate system, was able to maintain the exchange rate stability until the recent crisis in mid-1997. The stable exchange rates undeniably helped facilitate international trade and brought foreign investments to this part of the world. However, the excessive inflows of capital caused the Asians to over-consume and over-invest. As a result, the current account deficits widened from the average of 1 to 6% of GDP around early 1990s to 5 to 8% of GDP by mid-1990s. In the mean time, foreign debts stated to balloon to an alarming level.

Restructuring Asian Economics for the New Millennium, Volume 9A, pages 323–339.
2001 by Elsevier Science B.V.
ISBN: 0-7623-0709-9

In early 1997, it was clear to foreign investors that "Asian (Growth) Miracle" was not sustainable, and they no longer wanted to finance these economies. On contrary, giving the high level of current account deficits and mountaining foreign debts, foreign investors and creditors seriously questioned the ability to service debt of the countries. Capital then began to leave these countries.

Even with capital flights, Asian countries tried to maintain the stable exchange rates by heavily intervening in the exchange markets. The heavy interventions attracted speculators who knew for sure that, at this rate, the currencies could not hold. Large deprecations were necessary for these economies to clear the current account gaps.

No force can be against the market force. Thailand, the first causality of the crisis, has to abandon its Basket of Currencies system, which is a form of the fixed exchange rate system, for a fully floating exchange rate system on July 2nd, 1997. After the Baht, the attacks moved to the Ruphia and the Won, and, to certain extents, all Asian currencies, including the Hong Kong dollars.

At the end, apart from the Hong Kong dollar and the Chinese Renminbi, other currencies depreciated sharply. Theoretically, this is what Asia needs. Currency depreciation is necessary for any country facing the balance of payments crisis. Without new financing, there can be no current account deficit, and, in fact, a positive current account is preferable. Lower currency values, according to the text book, will boast exports while discourage imports, which should make the adjustment process easier.

Unfortunately, that text-book phenomenal did not happen in Asia. It is true that import demands in all countries in crisis have contracted substantially, but exports hardly move. In Thailand, exports (in dollar terms) in 1998 is about the same level as in 1997, and even lower than 1996 level. Similar conclusion could be reached for South Korea and Indonesia. What is wrong with the glory of the flexible exchange rate régime which supposes to automatically restore the balance of payments stability? Why export growth in these countries remains flat despite severe devaluation? These are the points of investigation of this paper.

This paper has three objectives. The first objective is to investigate the causes of failing price mechanism. Theoretical and empirical studies are utilized to find the reasons behind the failure of export stimulation through depreciation. Based on the findings, a conclusion will be drawn on the effectiveness and appropriateness of the flexible exchange rate régime in the time of financial crisis. The final objective is to provide a policy suggestion on the choice of the exchange rate régime which is better suited to the management of economy in Asia.

There are four sections to this paper. The first section explains the exchange rate developments in comparison to the external movements of selected Asian economies. The second section investigates the theoretical and empirical evidences of the effectiveness of depreciation on export demand stimulation. This section is divided into two parts. The first part studies the price transfer mechanism from the depreciation of the exporter's currency to the import prices at the consumer's country; the second explore the relative importance between the price effects derived from the depreciation of the exporter's countries and the income effects of the consumers. Section 3 concludes the effectiveness of the flexible exchange rate régime during the time of financial crisis. The last section provides policy recommendations regarding the choice of exchange rate régime.

I. EXCHANGE RATE MOVEMENTS AND EXTERNAL DEVELOPMENTS

Despite the booming economies of Asia since the beginning of this decade, the currencies in this region were rather stable (Table 1). Of course, this was achieved by the intervention of the central banks in the foreign exchange markets. For Thailand, the central bank's role was required as the country adopted a fixed exchange rate system called the Basket of Currencies.[2] Other central banks, even with the adoption of the flexible exchange rate régime, bought dollars and other hard currencies into the foreign reserves to keep the value of their currencies artificially low to support export. It is evident from the Table below that, even with heavy inflows of foreign capital, the real effective exchange rate indices of South Korea, Malaysia, and Thailand consistently declined from the start of the boom period until the time of crisis after mid-1997.

After the crisis, these currencies tremendously lost their values (Table 2). The Thai Baht lost 112% of its value compared to the pre-crisis value in

Table 1. Real Exchange Rate Index 1992–1998 Q2.

Currency	1992	1993	1994	1995	1996	97-Q1	97-Q2	97-Q3	97-Q4	98-Q1	98-Q
Won	103.29	104.34	100.87	95.28	97.44	104.47	106.30	106.55	134.19	179.42	156.22
Ringgit	91.19	91.63	92.35	86.07	85.93	84.67	85.65	94.78	118.15	132.55	126.66
Peso	87.85	89.39	81.86	75.79	73.35	72.87	72.41	80.84	94.97	106.86	102.27
Baht	98.59	97.91	94.88	91.37	90.36	91.43	91.01	113.29	136.57	154.96	131.35

Note: Using Purchasing Power Parity Method (1991 = 100).

Table 2. Comparison of Selected Asian Currencies.

Currencies	Pre-Crisis Value* Per U.S. Dollar	Worst Value Per U.S. Dollar	Present Value** Per U.S. Dollar
Indonesian Ruphia	2,383	10,375	7,400
% Change	(–3.2)	(–335.4)	(–210.5)
Korean Won	804.5	1,1701.5	1,245.0
% Change	(–4.3)	(–111.5)	(–54.7)
Malaysian Ringgit	2.5159	4.4141	3.7997
% Change	(–0.5)	(–75.5)	(–51.0)
Philippine Peso	26.216	42.661	39.200
% Change	(–2.0)	(–62.7)	(–49.5)
Singaporean Dollar	1.6449	1.7545	1.6475
% Change	(–0.9)	(–6.7)	(–0.2)
Thai Baht	25.342	53.810	36.100
% Change	(–1.7)	(–112.3)	(–42.5)

Note: * At End of 1996
** At End November 1998

January 1998. At the same time, the Korean Won suffered similar level of depreciation. The biggest loser is the Indonesian Ruphia which was forced to depreciated over 300%. The currency turmoil does not only limit to those countries resorted to the help of the International Monetary Fund. Other currencies – the Philippine Peso, the Malaysian Ringgit, the Singaporean dollar, and the New Taiwan dollar – all lost 7 to 75% during this period. Fortunately, Asian currencies have been stabilized since the second half of this year. Sill, these currencies lost upto 50% of the value compared to the pre-crisis era.

While the currencies fluctuate widely after the crisis, export figures did not seem to response to the change in the value of the currencies. Somehow there seem to be other factors which are much more important than the exchange rate factors.

II. FROM CURRENCIES TO IMPORT DEMAND

This Section wants to demonstrate that the transfer mechanism from devaluation to a rise in import demand in the consumer's countries is far from being perfect. Firstly, an *x*% devaluation often results in much less than *x*% decline in prices in the destination country. Secondly, even with a fall in import prices, a rise in import demand is normally small. This is the effects of the inelasiticity of import demand.

A. Depreciation to Import Prices

The first place to start investigate is the price transfer mechanism, i.e. the translation of depreciation to cheaper import prices at the consumer's countries. Theoretically, the import prices at the consumer's country should fall in proportion to the depreciation of the exporter's currency. For instance, when the Thai Baht depreciates 50% against the U.S. dollars, Thai products should be 50% cheaper to the U.S. consumers. This does not seem to be the case. While the Asian currencies depreciated from 100 to 300% in the last quarter of 1997, the U.S. import prices fell merely 4.3% (Table 4). Japan did not feel the effects

Table 3. Percentage Changes of Exports in U.S. Dollar Terms and Currency Movements (Compared to the Same Quarter the Year Before.

(% Change)

	Q1'97	Q2'97	Q3'97	Q4'97	Q1'98	Q2'98
Korea						
Exports	–5.6	7.1	16.1	4.4	8.7	–2.9
Currencies	(–11.2)	(–13.4)	(–10.0)	(–37.6)	(–84.4)	(–56.4)
Malaysia						
Exports	(5.84)	(–0.1)	(0.5)	(5.1)	(11.4)	n.a.
Currencies	(2.4)	(–0.3)	(–11.3)	(–38.2)	(–60.9)	(–53.4)
Philippines						
Exports	56.6	27.9	35.3	19.6	23.3	n.a.
Currencies	(–0.5)	(–0.7)	(–13.7)	(–34.7)	(–54.5)	(–49.3)
Thailand						
Exports	0.5	2.3	6.5	4.8	(–6.2)	(–7.7)
Currencies	(–2.4)	(2.4)	(–30.4)	(–59.5)	(–82.1)	(–55.7)

Table 4. Movement of Import Prices in Major Markets.

(% Change)

	1996	Q1'97	Q2'97	Q3'97	Q4'97	Q1'98
USA	4.5	—	–2.8	–2.6	–4.3	–6.2
Japan	2.8	17.0	14.5	14.0	8.0	–2.8
Germany	2.0	2.8	2.7	4.6	n.a.	n.a.

of cheap Asian imports until the first quarter of the following year when the Yen strengthened in the early part of the year.

One of the key reasons to this phenomenal is that exports from Asia generally have high level of import contents. A 100% devaluation, thus, will translate to much less reduction in product prices. A study carried out by the SCB Research Institute in Thailand found that, on average, Thai exports have import contents around 42%. The import contents are as high as over 90% for computer and computer parts; garments and textiles have around 60% of import contents. Only agricultural products has low level of import contents of 10% or less. Based on this data, a 100% devaluation of Thai Baht results in only 60% reductions in export prices. Similar conclusions can be reached from other Asian economies, it is estimated that Korean exports have import contents around 35%. Moreover, Asia has relatively small share in total imports to those countries. A drop in prices of Asian imports will have little effects on the overall consumer prices, and, thus does not significantly alter their consumption patterns.

Another explanatory factor for the low level of import price changes in the consumer's country is the imperfection competition and dynamics of the product markets – so called the Price to Market theory. Because of imperfect competition, U.S. importers do not necessary lower their prices even the cost of imports has fallen. Unfortunately, there is no study on the actual changes of import prices from Asia in the U.S. A study by Krungman (1987) founds that a 30% depreciation of the real exchange rate of German Mark resulted in only 1% decline in manufactured imports from Germany. Similar conclusions are reached for imports from other countries. Krugman's study further concludes that the product cost is only a portion of the final prices. The marketing and distribution costs are also significant, and, these costs are not affected by the devaluation. Thus, the final changes in cost to the consumers can be far less than the depreciation. Another reason could be the dynamics of demand whereby product importers tend to look at the long-run cost rather than the short-run price fluctuations caused by devaluation. In light of this, importers is unlikely to immediately reduce their prices to fully reflect the cost savings arising from depreciation.

B. Price and Income Effects

Even with perfect price transfer mechanism, the depreciation might not produce the needed effect of export stimulation as price elasticities are generally less than one. Numerous works have been done on the estimations of income and price elasticities of international trade. These works include Cline

Table 5. Short-run and Long-run Import Elasticities of G-7 Countries.

	Income		Price	
	Short-run	Long-run	Short-run	Long-run
Canada	1.3	1.4	–0.1	–0.9
France	1.6	1.6	–0.1	–0.4
Germany	1.0	1.5	–0.2	–0.06
Italy	1.0	1.4	–0.0	–0.4
Japan	1.0	0.9	–0.1	–0.3
United Kingdom	1.0	2.2	–0.0	–0.6
United States	1.0	1.8	–0.1	–0.3

(1989), Goldstien and Kahn (1985), Hooper and Marquez (1995), Houthankker and Magee (1969), and Marquez (1990, 1998). All estimations find that price elasticities are smaller than those of income, and are less than unity.

The most recent study on price elasticities of developed economies was done by Hooper, Johnson, and Marquez (1998). They estimated, both short-run and long-run trade elasticities of the G-7 countries. Their findings on the import elasticities are summarized in Table 5 above.

The estimation results are quite interesting. It is clear that changes in price have no significant effects on import demand in the short-run. Even over the longer-run, the import demand is primarily determined by the income factors, except in Canada where the price elasticity is close to unity.

III. APPROPRIATENESS OF THE FLEXIBLE EXCHANGE REGIME

As seen in Section II, the depreciation generate little actual effects on export growth. This does not mean that the flexible exchange rate concept is not valid beyond the realm of the text books. The flexible exchange rate system does work, but it is not supposed to work in an abnormal situation where exchange rates can fluctuate widely in either direction. And there are two reasons why the exchange rates do fluctuate frequently and widely: international capital movements and currency speculations.

Let's look at the case of the Yen/U.S. Dollar as an example. Figure 1 shows the movements of the Yen/U.S.$ exchange rates during 1997–98. It can be seen that the Yen lost about 20% of its value from mid-1997 to mid-1998, and then regained that 20% back within a few short months. Assuming that the relative inflation rates between these two countries do not change, the 20% depreciation

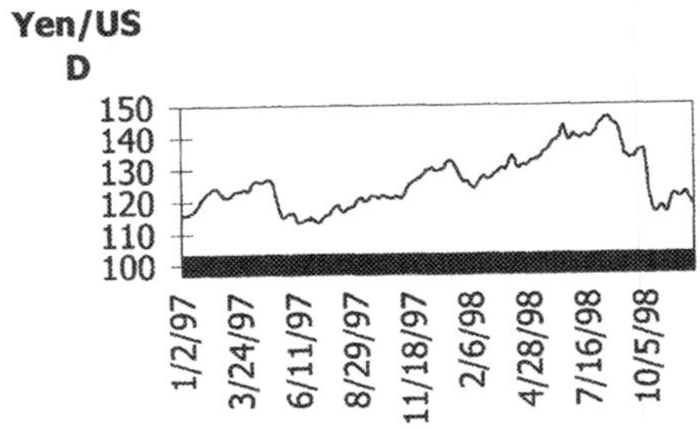

Fig. 1. The Japanese Yen/U.S. Dollar Movements from 1997–1998.

means that Japan lost 20% of its competitiveness over the year, and it needs the depreciation to compensate for that. And in the past few months, that competitive loss has miraculously been regained, and so the currency rebounds. This is because, theoretically, the movements between the two currencies reflects the relative change in productivity growth of the two economies and the relative change in inflation. Of course, that could not actually be the case.[3] It is likely that the Yen/U.S.$ rate is primarily determined by the currency traders, not by economic fundamentals. If the economic fundamentals rule, the Yen must sharply appreciate against the U.S.$ in order to clear the current account surplus which increases every single year. However, the *under-valued* Yen does both parties no harm. Japanese manufacturers benefit from strong exports while U.S. consumers can enjoy cheap imports. So both countries never try hard to correct this imbalance.

Theoretically, exchange rate adjustments can be effective in balancing the trade accounts when the Marshall-Lerner condition is met, i.e. the absolute values of the import and export elasticity exceed unity: $|d_x + d_m| > 1$. Under such condition, the market will force the exchange rates to gradually move to restore the trade balances. Here, the flexible exchange rate régime works wonder as the economy will never have a prolonged external imbalances. Moreover the adjustments will be quite gradual and orderly in such a way that they never disrupt the domestic economy.

That is not the case in the modern world as proved in the Yen/U.S. Dollar example above. With the advancement of the world financial systems, capital markets are inter-connected and money can move swiftly from one country to another. The important point is that this capital movements can be in large magnitude and can easily outpace trade movements. A case in point is Thailand, its trade balances have turned positive right after the crisis. But that trade gains are overshadowed by capital outflows (Table 6). Thus the Baht

Table 6. Thailand: Trade Balances and Capital Movements 1997–1998 (in millions of U.S. dollar).

	Q1'97	Q2'97	Q3'97	Q4'97	Q1'98	Q2'98	Q3'98
Trade Balances	–1,493	–1,431	710	3,778	3,128	2,599	3,173
Capital Movements	2,556	–4,160	–6,178	–7,659	–3,858	–2,932	2,707

depreciated very rapidly during such period, particularly in the second half of 1997. At the end of the year, the Baht lost over half its value. It is not until the trade surpluses and the level of capital flows become close that the Baht stabilizes. Therefore the exchange rate movements, largely in response to the capital flows, is certainly not what the economy needs to adjust to equilibrium. It is quite possible and, often times, it is the case that the flexible exchange rate régime actually moves the economy away from equilibrium.

For an economy to be at equilibrium, there will be no internal and external imbalances. Internal balance means low inflation,[4] and external balance means no current account deficit or surplus. Since the crisis, current account balances of many Asian economies have turned very positive, and inflation has been lower than expected. These are actually signs of disequilibrium. Thailand is estimated to record an 8% of GDP current account surplus, and inflation around 8%. Taking supply factors such as currency devaluation out, the demand-pull inflation is likely to be much less than 8%, and could possibly be negative. The large current account surplus and unnaturally low level of inflation means that Thailand is at a serious stage of disequilibrium, where the economy is performing way below its potential. Prior to the crisis, the Thai economy – with large current account deficits and rising demand-pull inflation – was also at disequilibrium. But, at that time, the economy was performing way above its potential.

According to the theory, the exchange rate depreciates when the country has current account deficits and appreciates when the current account turns around. The *fundamental equilibrium exchange rates* are the exchange rate that create the long-run current account equilibrium[5] (Williamson 1983). This has not been the experience of Asia. All the Asian countries in crisis have quickly turned their current accounts into surpluses, large surpluses in fact. But the exchange rates fall, rise, or stabilize in no relation to the trade movements. Rather, the exchange rates response to capital flows and speculation. In Thailand, there is an evident that the Baht/U.S. Dollar movements track the Yen/Dollar movements closely, except when the capital flows are strong. Figure 2 depicts

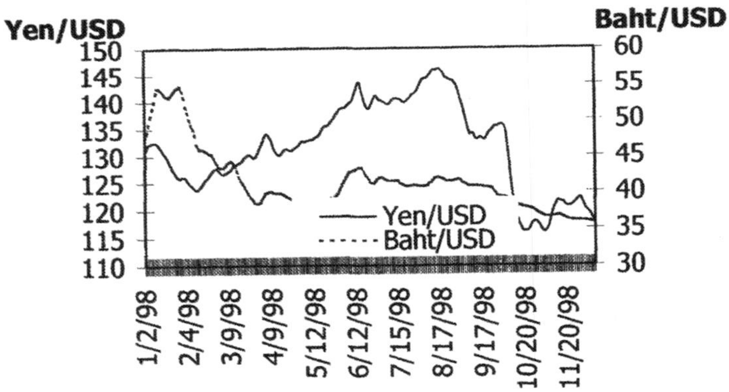

Fig. 2. Baht/U.S. Dollar and Yen/U.S. Dollar Movements.

the co-movements of the Baht-Yen against the U.S. Dollar. It is interesting to see that the two currencies move in tandem although the economic fundamentals of the two countries and not at all alike. The period that the co-movements deviate are the period that the capital flows are unusual in Thailand, normally higher or lower than U.S.$500 million per month. For instance, the Baht moves in the opposite direction against the Yen in March and April 1997 because of the U.S.$1.8 billion inflows from the capital raising exercises of Bangkok Bank and Thai Farmers Bank.

The influence of capital flows and currency speculation defeats the usefulness of the flexible exchange rate system. On contrary, the adoption of the fully flexible exchange rate system destabilizes the economy and damages the long-run competitiveness. The over-depreciation caused by capital outflows creates an undue debt burden to the domestic businesses. The evident is very clear in Asia. Because of high level of private foreign debt borrowing, the depreciation raises the debt obligations proportionally. For the case of Thailand, every one single Baht of depreciation means an additional 80 billion Baht of debt burden to local businesses. It is undeniable that the undue depreciation of Asian currencies is one of the key factor to the mountaining non-performing loan problems in the region. The caveats of the non-performing loan problems to the financial sectors and to the real economies are already well publicized, and is not the scope of this paper.

The more important issue of the over-depreciation is the long-term damages to the competitiveness of the economy. Because of rapid capital outflows, Asian currencies have over-depreciated beyond their equilibrium levels. This *disequilibrium* exchange rate in effect subsidized exporters. Such price distortion creates an artificial competitive edge to Asian exporters, and provides disincentive to improve their production efficiency. Once, the capital outflow episode is over, Asian exporters will be at a weak position to compete effectively in the international markets. Inefficient export sector can damage the long-run growth potential of these economies.

IV. POLICY CHOICE ON EXCHANGE RATE RÉGIME

There are two reasons why Asia should adopt the flexible exchange rate régime. One is improve the current account balance; the other is to restore the confidence of investors and creditors. The flexible exchange rate obviously does not fulfill its first duty. The paper has proven that the depreciation has done very little to stimulate exports. The market imperfections prevent the devaluation to fully translate into cheaper import prices at the consumer's countries. Furthermore, cheaper import prices, if achieved, do not mean more consumption as the income factor is much stronger than the price factor. Import reductions, handsomely achieved in all Asian economies, are not a result of depreciation either. The reduction in imports is a result of the sharp contraction of income and consumer's confidence loss. Again, the income factor is much stronger than the price factor.

The flexible exchange rate régime also fails to perform its second duty – restoring investor's and creditor's confidence – successfully. It is correct to say that before the crisis Asian currencies were over-valued, and investors and creditors were seriously concerned. They were concerned that the currency support provided by the central banks was not sustainable. The adoption of fully flexible exchange rates after the crisis alleys this fear, and, more importantly, narrows the room for speculation. However, the move does not restore the investor's confidence. On contrary, the confidence is even less as the depreciation creates huge losses to local businesses. This is evident by a continued large outflows of capital despite the adoption of a fully flexible exchange rates.

Based on the findings of this paper, Asian countries facing the financial crisis is likely to be better off with the fixed exchange rate system because the flexible exchange rate system are not effective in doing its job, i.e. removing the external imbalances. The fixed exchange rate system, on the other hand, is more effective in restoring the equilibrium, and caused less economic contractions. First, the fixed exchange rates certainly will produce less inflationary effects.[6] This is important for countries depending on food imports like Indonesia. Lower inflation means less restricted fiscal and monetary policies are needed, and, thus, less contraction of the economy. Second, stronger currencies under the fixed exchange rate system will result in lesser debt burdens to Asian businesses which heavily borrowed from abroad. Without additional debt burdens arising from undue depreciation, the stability of the financial sectors will be easier to restore. The stability of the financial sectors of Asian countries is a key, if not the most important, factor to the solution of the crisis. Third, the wild exchange rate fluctuation under the flexible régime negatively affected the stability of the export sector – another key factor to the solution of the crisis. Many exporters have incurred big financial losses after the Asian currencies quickly turn around since the second half of this year. The losses damage the already thin working capital which is needed to keep the operation running. Moreover, the *unstable* exchange rates make it difficult for exporters to calculate the right cost. This problem is rather pronounced in Asia where the exchange markets are underdeveloped, and exporters have no knowledge or find it difficult to hedge their foreign exchange positions.

A better strategy for Asia is not to adopt a fully flexible exchange rate system, but to fixed their currencies at an *appropriate* value. The key question is what is the appropriate value for each economies. Theoretically, the currency can be fixed at any value, and the domestic economy will have to adjust accordingly to regain the competitiveness.[7] In practice, these appropriate rates should be the rates that foreign investors and creditors can be confidence that the economy is able to support. This a more psychological factor than economic reasoning as confidence is crucial to the management of the crisis.

Some might worry about the adoption of the fixed exchange rate régime if other Asian countries do not, and exporters of that country could be put at an disadvantage position. This could be the case if two countries – one fixes its exchange rate and the other over-depreciates its exchange rate – have same products to export. This is unlikely the case. If such argument is correct, country like Indonesia, whose currency depreciates 200%, would have dominated the world export markets by now. On contrary, the export patterns of Asian countries hardly change after the crisis, despite the different magnitude of depreciation of each currencies (Table 7). There is also no need to worry about the impacts on import reductions. The import reductions achieved in Asia is a result of the collapse of the domestic demand more than currency depreciation.

The key issue of Asian crisis is not the management of exchange rates, but it is the management of capital flows and domestic demand control. However, an inappropriate choice of exchange rate régime makes the solutions to the

Table 7. Exports and Imports of Selected Asian Countries.

(Unit: %)

	1995	1996	1997	Q1'97	Q2'97	Q3'97	Q4'97	Q1'98	Q2'98
Korea									
Exports	30.7	3.7	5.3	(5.6)	7.1	16.1	4.4	8.7	(2.9)
Imports	32.0	11.3	(3.8)	3.9	0.8	(3.8)	(15.0)	(35.4)	(79.1)
Malaysia									
Exports	26.0	6.0	0.6	6.1	0.3	2.7	(5.1)	(11.4)	n.a.
Imports	30.8	1.0	0.2	(1.1)	11.2	1.2	(8.3)	(19.3)	n.a.
Philippines									
Exports	32.2	16.5	23.8	18.0	28.7	26.1	19.6	23.3	n.a.
Imports	25.7	20.2	13.3	11.3	8.2	14.5	12.4	(3.8)	n.a.
Thailand									
Exports	24.8	(1.3)	3.4	2.0	2.8	5.5	4.1	(2.7)	(6.8)
Imports	30.1	2.2	(15.2)	(10.3)	(7.3)	(15.2)	(25.2)	(37.3)	(38.2)

crisis much more complicated. In some cases, the hyperinflation resulting from sharp depreciation can destabilize the economy, and greatly lessen the chance to restore the equilibrium to the country. Indonesia will be an interesting case to watch.

V. CONCLUSIONS

The paper has demonstrated that the fully flexible exchange rate is not an appropriate choice for Asia in the midst of the financial crisis. The benefits of the flexible system, which allow the external imbalances to adjust automatically, are outweigh by the caveats of erratic capital flows. International trade theories that normally advocate the beauty of the flexible exchange rate system never incorporate the effects of capital flows. The problem with the flexible exchange rate régime is certainly not its theoretical deficiencies. It is the practical problem of the real world where capital movements can be swift and strong.

In the real world, the effects of the exchange rate adjustments are also overshadowed by the imperfections of the product markets and the low response to the price effects. Furthermore, even in developed economies where the market supposes to be near perfect, the exchange rates are primarily determined by currency traders, speculators, and capital movements. Economic fundamentals are hardly the determining factors of the exchange rate movements.

It is not recommended to replace the existing flexible exchange rate system with a fixed one as the capital outflows from Asia is already stable. The change in the exchange rate system, at this point in time, will create confusion. However, this has been an expensive lesson to be learnt that an adoption of the flexible exchange rate system is not a solution to the external balance problems as the text books might recommend. Rather, when the capital flows are erratic, the flexible exchange rate will do more harm than good.

NOTES

1. Mr. Parasuk is presently the First Vice President, Corporate Finance Department at National Finance Public Company, Bangkok, Thailand. He was previously an economist in the Southeast Asia and Pacific Department at the International Monetary Fund, Washington D. C. Ms. Sukchareoan is an economist at National Finance Pubic Company.

2. Under this system, the value of the currency is fixed to the valued of the foreign currencies in the basket. In case of Thailand, it was estimated that the basket mainly comprised of U.S. dollars, the German Marks, and the Japanese Yen in the ratio of 80–15–5.

3. It has been estimated that the productivity growth in Japan is about 5% per annum and the productivity growth in the U.S. is about 3% per annum. If the flexible exchange rate system works, Yen will consistently appreciate 2% a year against the dollar.

4. Internal balance does not require zero inflation. Even at the equilibrium stage, inflation will be equal to the quality improvement of domestic products. For advanced economy, this product improvement is estimated to be around 2%. Thus, a 2% inflation is a sign of an economy in equilibrium.

5. There have been arguments that, for developing countries, the equilibrium exchange rates are the rates that generate *sustainable* current account deficits. This is because developing countries always need foreign capital to finance their economic developments. Therefore their current accounts do not need to be in balance as long as the deficits can be financed through long-term investments.

6. This follows the conclusion of the paper that the *equilibrium* exchange rates for each countries are theoretically lower than the *market determined* rates under the freely floated currency system.

7. For instance, if the currency is fixed at a too high level, real wage (in local currency) will need to be deflated in order to maintain the international competitiveness. This is not an unusual phenomena. Hong Kong, a country with fixed exchange rate system, is a classic example. To maintain its competitiveness when other Asian currencies depreciate, the cost of production in Hong Kong has to be cut. And, for Hong Kong, the major production factors are real estate rents and wages. Therefore, it is not

surprising to see real estate prices in Hong Kong swing sharply. The Hong Kong's economy needs that swing to counter the gain or loss in competitiveness.

REFERENCES

Adams, F. et al. (1969). *An Econometric of International Trade*. Paris: OECD.

Adams, F., Gangnes, B., & Shishido S. (1993). *Economic Activity, Trade, and Industry in the U.S.-Japan-World Economy*. Westport, CT: Praeger.

Campbell, B. (1981). Flexible and Fixed Target Stabilisation in an Open Economy. *Journal of Economic Dynamics and Control, 3*, 235–263.

Cline, W. (1989). *United States External Adjustment and the World Economy*. Washington D.C.: Institute for International Economics.

Dornbusch, R. (1986). Special Exchange Rates for Capital Account Transactions. *The World Bank Economic Review*, 3–33.

Goldstein, M. (1980). Have Flexible Exchange Rates Handicapped Macroeconomic Policy? *Special Papers in International Economics, 14*, Princeton University.

Goldstein, M., & Khan M. S. (1985). Income and Price Effects in Foreign Trade. In: R. Jones & P. Kenen (Eds), *Handbook of International Economics* (Vol. 2, pp. 1041–1105). Amsterdam: North Holland.

Hoontrakul, P. (1997). *Exchange Rate Theory: A Review*. Discussion Paper, SASIN Graduate School.

Hooper, P., & Marquez, J. (1995). Exchange Rates, Prices, and External Adjustment in the United States and Japan. In: P. Kenen (Ed.), *Understanding Independence* (pp. 107–168). Princeton: Princeton University Press.

Hooper, P., Johnson, K., & Marquez, J. (1998). Trade Elasticities for G-7 Countries. *International Finance Discussion Papers, 609*, Board of Governors of the Federal Reserve System.

Houthakker, H., & Magee, S. (1969). Income and Price Elasticities in World Trade. *Review of Economics and Statistics, 51*, 111–125.

Khan, M., & Lizondo J. (1987). Devaluation, Fiscal Deficits, and the Real Exchange Rate. *The World Bank Economic Review*, 357–374.

Kindleberger, C., & Lindert, P. (1978). *International Economics, 6*, Illinois: Richard D. Irwin, Inc.

Krugman, P. (1987). Pricing to Market When the Exchange Rate Changes. In: S. Arndt & J. Richardson (Eds), *Real-Financial Linkages among Open Economies*. Cambridge: The MIT Press.

Marquez, J. (1990). Bilateral Trade Elasticities. *Review of Economics and Statistics, 72*, 70–77.

Marston, R. C. (Ed.) (1988). *Misalignment of Exchange Rates: Effects on Trade and Industry*. National Bureau of Economic Research, The University of Chicago Press.

McKinnon, R. I. (1988). Monetary and Exchange Rate Policies for International Financial Stability: A Proposal. *Journal of Economic Perspectives, 2*(1), 83–103.

Parasuk, C. (1989). *Application of Optimal Control Techniques in Calculating Equilibrium Exchange Rates*. Ph.D. Dissertation, University of Texas at Austin.

Siam Commercial Bank Research Institute (1996). *Analysis of Import contents of Thai Exports*, Research Paper.

Stern, R., Francis, J., & Schumacher, B. (1976). *Price Elasticities in International Trade: An Annotated Bibliography*. Macmillan of Canada.

Williamson, J. (1983). *The Exchange Rate System*. Washington D.C.: Institute for International Economics.

APPENDIX

Table A1. GDP Growth in Selected Asian Countries.

(Unit: %)

Countries	1992	1993	1994	1995	1996	1997
Korea	5.1	5.8	8.9	8.9	7.1	5.5
Malaysia	7.8	8.3	9.2	9.5	8.6	7.8
Philippines	0.3	2.1	4.4	4.8	5.8	9.0
Thailand	8.1	8.4	8.9	8.8	5.5	(0.4)

Table A2. Current Account Deficit in Selected Asian Countries.

(in % of GDP)

Countries	1991	1992	1993	1994	1995	1996	1997
Korea	(2.9)	(1.3)	0.3	(1.0)	(1.9)	(5.0)	(3.3)
Malaysia	(8.6)	(3.8)	(4.9)	(6.1)	(8.6)	n.a.	n.a.
Philippines	(2.2)	(1.9)	(5.7)	(4.3)	(2.7)	(4.9)	(7.1)
Thailand	(8.0)	(6.0)	(5.4)	(5.6)	(8.1)	(7.9)	(2.9)

Table A3. Capital Flows in Selected Asian Countries.

(in Millions of U.S. Dollar)

Countries	1996	1997	97Q1	97Q2	97Q3	97Q4	98Q1	98Q2
Korea	23,335	(9,803)	4,038	6,572	618	(21,030)	(967)	2,393
Malaysia	n.a.	n.a.	n.a.	n.a.	n.a.	n.a.	n.a.	n.a.
Philippines	11,277	6,396	2,521	3,244	1,865	(1,234)	n.a.	n.a.
Thailand	19,487	(15,441)	2,556	(4,160)	(6,178)	(7,659)	(3,858)	(2,932)

Table A4. Exchange Rate Movement.

Exchange Rate in % Change	1995	1996	1997	Q1'97	Q2'97	Q3'97	Q4'97	Q1'98	Q2'98
Won: U.S.$	0.5	(9.0)	(100.8)	(14.6)	(9.5)	(11.4)	(100.8)	(54.2)	(54.6)
Ringgit: U.S.$	0.7	0.5	(53.9)	2.3	(1.2)	(27.3)	(53.9)	(47.2)	(65.4)
Peso: U.S.$	(7.4)	(0.3)	(52.1)	(0.6)	(0.7)	(29.0)	(52.1)	(40.6)	(59.5)
Baht: U.S.$	(0.4)	(1.7)	(75.2)	(8.4)	(7.1)	(43.7)	(84.5)	(41.9)	(55.8)
Yen: U.S.$	10.8	(3.1)	(12.8)	(5.7)	(16.7)	(4.6)	(9.0)	5.4	(6.4)
DM: U.S.$	10.3	7.4	(8.5)	13.3	(13.7)	(14.6)	(15.6)	(15.3)	(10.1)

Table A5. Import Price Index of Major Import Countries.

% Change of Input Price	1995	1996	1997	Q1'97	Q2'97	Q3'97	Q4'97	Q1'98
USA	4.5	0.9	–2.4	0.0	–2.8	–2.6	–4.3	–6.2
Japan	2.8	–4.9	–4.5	17.0	14.5	14.0	8.0	–2.8
Germany	2.0	0.5	3.2	2.8	2.7	4.6	n.a.	n.a.
Korea	5.8	0.6	9.6	7.7	8.6	5.0	16.4	53.6

Table A6. Exporters and Importers of Industrial Countries in 1996.

% Share	USA		Japan		Germany	
	Export	Import	Export	Import	Export	Import
Industrial Countries	56.4	54.2	47.1	46.1	74.2	74.9
Developing Countries	43.6	45.8	52.9	53.9	25.8	25.1
Asia (excluding China)	17.3	17.8	38.8	26.1	6.7	6.5

Table A7. Major Exporters and Importers of Asian Countries in 1996.

% Share	Korea		Malaysia		Philippines		Thailand	
	Export	Import	Export	Import	Export	Import	Export	Import
USA	16.7	22.1	18.2	15.2	33.9	19.7	18.0	12.6
Japan	12.3	20.9	13.4	24.1	17.9	21.8	16.8	27.9
Germany	3.6	4.8	3.0	4.2	4.1	3.8	2.9	5.0
Total Share	32.6	47.8	34.6	43.5	55.9	45.3	37.7	45.5

18. MONETARY CONDITIONS AND MONETARY POLICY IN SMALL OPEN ECONOMIES: EMPIRICAL RESULTS FOR THAILAND UNDER FLOATING RATE RÉGIME

Rungsun Hataiseree

1. INTRODUCTION

Over the past decade or so, the framework and strategy of central banks in implementing monetary policy has continually evolved along with a rapidly changing economic and financial environments at home and abroad. An increase in the volume and volatility of the international capital flows coupled with an intensified financial innovation have made financial markets and economic systems more and more interdependent. As the domestic financial market becomes more closely linked to the global financial system, the exchange rate becomes an increasingly important factor as a channel through which monetary policy may have potential impacts on the real sector.

During the 1950s and 1960s, most central banks tended to place more emphasis on using the *fixed exchange rates* system as the main framework in the conduct and implementation of monetary policy. The framework of this kind is normally characterized by pegging the domestic currency to major international currencies. Monetary policy management through the pegged exchange rate serving as a nominal anchor has been, to a certain extent,

Restructuring Asian Economics for the New Millennium, Volume 9A, pages 341–362.
2001 by Elsevier Science B.V.
ISBN: 0-7623-0709-9

claimed to have exerted a positive impact on economic performance of a number of counties. Despite such perceived benefits, however, the risks of currency misalignment have still remained a serious concern.

In the 1970s, central banks of leading industrialised nations – such as Germany, the United States and Japan – have promoted *monetary targeting* as the primary strategy for monetary policy.[1] This has primarily been in response to the escalating inflation during 1973–74 and 1979–80 that resulted from the oil crisis. However, since the early 1980s, the central banks of many industrialised nations have gradually shifted away from monetary targeting. This was because the targeted monetary levels have been made more volatile by the increased pace of financial innovations and the intensified movements of international capital transactions.

Towards the late 1980s, central banks of many industrialised nations have turned their attention towards an *inflation targeting* régime as the focal point of their monetary policy. The list includes the central banks of New Zealand, Canada, England, Sweden, Finland, Australia, Spain and Israel (Green, 1996; Svensson, 1997).

The framework and strategy in implementing an inflation targeting-oriented monetary policy rests upon the basic assumption that: ". . . .*Monetary policies affect the economic system and in particular, the inflation rate, through two main transmission mechanisms: (i) the interest rate, which influences the level of expenditure and investment, and (ii) the exchange rate, which influences the price of imports, and ultimately the inflation level* . . ." (Fig. 1). In view of this, an continued depreciation (appreciation) in the exchange rate would require an increase (decrease) in the interest rate in order to sustain the target rate of inflation.

As interest rates and exchange rates are both important channels through which monetary policy affects economic activity and inflation, it has been argued that, combining both interest and exchange rates in a single policy indicator, a Monetary Conditions Index (MCI), may serve as a better indication of the overall policy stance (Freedman, 1995; Nadal-De Simone et.al., 1996). For this reason, central banks of many industrialised countries, e.g. The Reserve Bank of New Zealand, Bank of Canada, place importance on the construction and implementation of the MCI. The MCI is designed to indicate the stance (the degree of tightening or loosening) of monetary policy during a given period.

Although an inflation-targeting régime has not yet been adopted officially, the Bank of Thailand has for some time paid a close watch to the developments and behaviour of interest rate and the exchange rates in its periodical assessment of Thailand's economic and monetary conditions (Hataiseree, 1999).

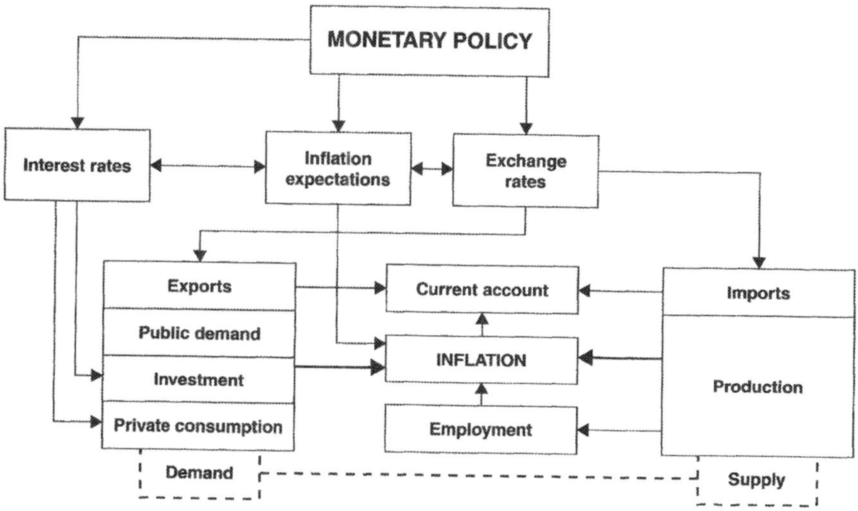

Fig. 1. Channels of Monetary Transmission Mechanism.

More importantly, the floatation of the Thai baht since July 2, 1997 has been claimed to be the major development which has heightened the role of the exchange rate as the main channel through which monetary policy gets transmitted to economic activity and the general price level. Incorporating the effect of both interest and exchange rates seems to be particularly relevant in small open economies like Thailand, where interest and exchange rates tend to be highly linked with the domestic/foreign interest rate deferential off set by expected exchange rate changes (Hataiseree & Phipps, 1996a). This close relationship implies that shocks are rapidly transmitted between money markets and foreign exchange markets.

Using the MCI which encompasses movements in both interest rates and exchange rates may help the Thai authorities gain a better assessment of the overall monetary conditions. This is because the information content contained in the MCI would characterize the degree of pressure that monetary policy is placing on the economy and, therefore, on inflation rate. Nonetheless, the potential adopting of an MCI-oriented monetary policy, and therefore an inflation-targeting régime, calls for a more detailed analysis of whether interest rate and exchange rate variables are particularly important factors determining future inflationary pressures in the context of Thailand.

This paper aims to introduce the concept and framework of implementing an MCI-oriented monetary policy by considering the benefits of and issues involved in putting this concept into practice in the case of a small open economy like Thailand. It also discusses how the implementation of the MCI would help enhance the clarity and accuracy in determining the policy stance of the central bank when compared to a policy assessment based on either interest rate or exchange rate alone. In this paper, we also attempt to build an MCI for Thailand.

The remainder of this paper is divided into four sections. It begins in Section 2 by describing the main features of using the MCI in monetary policy making. Section 3 discusses the framework and strategy for the conduct of monetary policy and implementing the MCI for Thailand. Section 4 provides estimates with respect to the relative weights of interest rate and the exchange rate in an MCI for the overall price level for Thailand. This section also presents research findings on the relationship between the MCI and the rate of inflation. Section 5 concludes the paper and considers the implications of using the MCI in the administration of monetary policy in the future.

2. THE THINKING UNDERLYING NEW MONETARY POLICY ARRANGEMENTS

In the formulation and implementation of monetary policy, the central bank needs to select an appropriate set of policy tools to implement its monetary policy. This is deemed necessary for the smooth attainment of ultimate targets of monetary policy. Over the past decades, central banks, especially those in developing countries, have not been clear in establishing their ultimate targets and have not yet prescribed explicitly what to do in order to attain such ultimate targets. Apart from this, there has been no in-depth analysis of the relationship between monetary policy instruments and the ultimate targets of the monetary policy instruments (Duguay, 1994).

Nonetheless, in recent years, central banks of many industrialized countries have begun to more seriously consider the importance of clearly stating their ultimate targets of monetary policy and announce those targets to the public, as it has been argued that this kind of information seems to serve as a guide for private companies in conducting business transactions. In doing so, the conduct and implementation of monetary policy tends to become clearer and more transparent. As a result, the credibility of the central banks' actions tends to have been enhanced. For instance, in the case of the Reserve Bank of New Zealand, the term of the governor has been specified as being contingent upon his ability to maintain the inflation level at the pre-established target level that has been announced by the authorities. Most significantly, the attainment of a low inflation rate in accordance with the target agreed with Government (initially 0 to 2%, now 0 to 3%) has been clearly codified by the 1989 Reserve Bank Act (Brash, 1998).

The central banks of many industrialised countries, such as the central banks of New Zealand, Canada, Sweden, and Norway, are aware of the necessity of creating a new framework for conducting monetary policy so as to be a clear indicator of the central banks' policy stance and allow for a better communication with participants in the money markets. At the same time, the adoption of an MCI-oriented monetary policy is seen as an essential framework for use to follow and gauge the likely effects that monetary policy actions may have on the economy – especially in terms of the level of inflation – which is a direct responsibility of the central bank. As can be seen from Fig. 2, an inflation targeting framework for monetary policy was first adopted by New Zealand's central bank in March 1990, followed by the central bank of Canada in February 1991.

The framework and strategy of implementing inflation targeting-oriented monetary policy fundamentally stresses on the condition that "Inflation Targets" must clearly be the ultimate target of monetary policy. Under an inflation-targeting régime, the monetary authorities normally have to make announcement of the target or a range of inflation target for the future. If the inflation projection for the next 1–2 years is believed to fall outside the range of the official target, a series of policy actions needs to be carried out in order to bring the inflation level back into the targeted range. The monetary authorities may have to send a signal reflecting a change in the policy stance by adjusting short-term interest rates or intervening in the foreign exchange market.

It is interesting to note, however, that the management of monetary policy under the framework of monetary targeting primarily targets the inflation level – just as the management of monetary policy under the inflation targeting régime. For monetary targeting, monetary aggregates are used as intermediate targets for the conduct and implementation of monetary policy. In the case of inflation targeting, however, inflation forecasts are used as intermediate targets for the conduct and implementation of monetary policy (Fig. 3). Under an inflation targeting régime, short-term interest rates tends to be pushed upwards in the event that forecasted inflation shows a tendency of stabilizing at a higher level than the "established targeted inflation." Further details on this matter may be found in Svensson (1997), among others.

Under an inflation targeting framework for monetary policy, there will be monetary indicators that uses the Monetary Conditions Index, which is a kind of monetary indicator that shows whether a central bank's monetary policy at any one point in time is relatively loose or tight, and to what degree. This

Country	New Zealand	Canada	United Kingdom	Sweden	Finland	Australia	Spain
Date first Instituted	March 1990	February 1991	October 1992	January 1993	February 1993	Approximate April 1993	Summer 1994
Current target	0-3%	1-3%	1-4%	2%	2%	2-3%	less than 3%
Time-frame	5 years (to 1998)	through end - 1998	by spring 1997, 2.5% or less thereafter	1996 onwards	1996 onwards	On average over the cycle	by late 1997, less than 2% thereafter
Inflation measure	Underlying CPI	CPI	Retail Price Index excl. Mortgage interest payment (RPIX)	CPI	Underlying CPI	Underlying CPI	CPI
Factors excluded from CPI	Interest cost component, indirect taxes, government charges and significant change in terms of trade	None. Underlying inflation rate used operationally	Mortgage interest payment	None	Mortgage interest payments, indirect government taxes, govt. subsidies, house prices	Mortgage interest payments, indirect taxes, other volatile items	None
Target announcement	Defined in policy target agreement (PTA) between the Minister of France and the Governor of the central bank	Joint agreement between the Minister of France and the Governor of the central bank	Chancellor of the Exchequer	Bank of Sweden (Riksbank)	Bank of Finland	Reserve Bank of Australia	Bank of Spain
Inflation report	Quarterly since March 1990	Half-yearly since May 1995	Quarterly since February 1993	Quarterly since February 1993	No	No	Semi-annual
Inflation forecasts published?	Yes	No	Yes	No	No	No	No

Note: *The New Zealand inflation target band width was widened from 0-2% in December 1996.*
Source: *Debelle (1997)*

Fig. 2. Salient Features of Inflation Targeting.

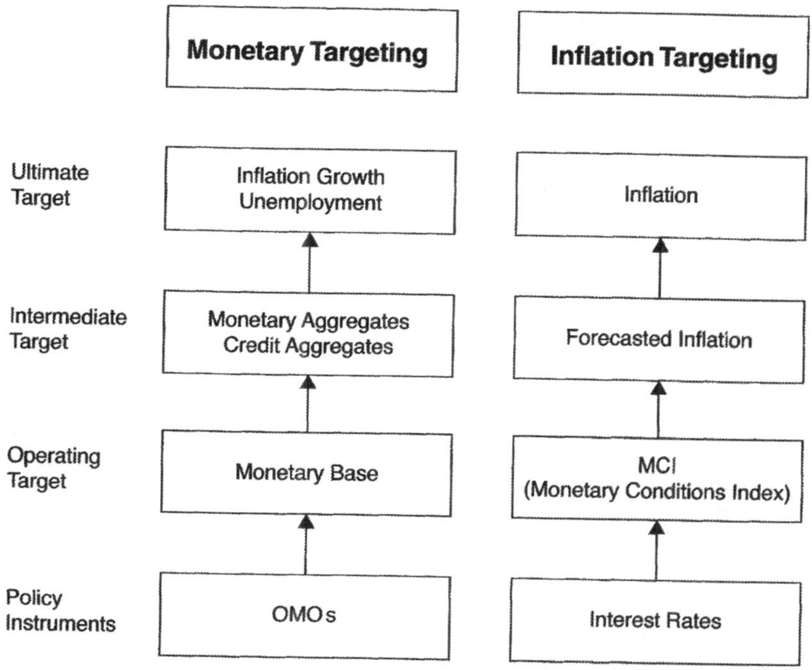

Fig. 3. Major Frameworks for the Conduct of Monetary Policy.

indicator therefore acts as an indicator of operating target within an administration of monetary policy.

Indicator of this kind reflects the degree of influence that the monetary policy has on the overall economy – especially on the level of inflation. Overall, an MCI index has a base-year equivalent to 100, as is the case of New Zealand, and serves as a benchmark indicating the direction and outlook of the future inflation. Formulating and conducting monetary policy under the MCI framework (in addition to other economic and monetary indicators) is therefore considered a policy strategy that is forward looking.

The use of the MCI as part of the central bank's monetary policy administration is based on the premise that: "*Both the interest rate and the exchange rate are important and influential factors of the overall economic condition – especially to the inflation rate.*" When the interest rate rises or the exchange rate strengthens, the effect is for the economy to decelerate in the future and eventually lead to a weakening of the pressure on price levels. In contrast, when the interest rate falls or the exchange rate weakens, the effect is for expenditures, consumption and investments to rise in the future, which may eventually lead to a higher level of inflation (Fig. 1).

Because the interest rates and the exchange rate are both important and influential channels that link the monetary policy to the real sector, the central banks of many countries tend to be faced with an increasing difficulty in sending a clear signal to the market about the direction and tendency of the monetary policy. As such, on some occasions, the interest rates and the exchange rate may have influenced the economy and thus the level of inflation in *the same directions* as those characterized in Row I and II of Fig. 4. Nonetheless, in many cases, it is also possible that the interest rates and the exchange rate may have influenced the economy and, thus, the inflation in *the opposite direction.*

This is the case in the periods when the exchange rate is strengthening, but interest rates are falling (and vice versa) with the MCI staying fairly steady as those characterized in Row III of Fig. 4. It is at those periods when the MCI is most useful. If the authorities concerned were watching just interest rates, or just the exchange rate, as a separate indicator of monetary policy, it would be misleading because both parts are important in determining future inflationary pressures.

From the stated situation, coupled with the fact that both the interest rate and the exchange rate are continuously changing makes it very difficult for the central bank of many nations to estimate whether the monetary conditions at a certain point in time are relatively tight or relaxing and, thus, may cause the

	Δr	Δe	**Policy Stance (Monetary Conditions)**
I.	**r ↓ (i.e. 15% → 10%)**	**e ↑ (i.e. 35 → 40)**	**Easy Monetary Policy**
II.	**r ↑ (i.e. 10% → 15%)**	**e ↓ (i.e. 40 → 35)**	**Tight Monetary Policy**
III.	**r ↓ (i.e. 15% → 10%)**	**e ↓ (i.e. 40 → 35)**	**Inconclusive (Depending on the Relative Importance of r and e)**
IV.	**r ↑ (i.e. 10% → 15%)**	**e ↑ (i.e. 35 → 40)**	**Inconclusive (Depending on the Relative Importance of r and e)**

Notes : **r = short-term interest rates**
e = the exchange rate (e.g. baht / US dollar)

Fig. 4. Interest Rate and the Exchange Rate: Their Role as Indicators of Monetary Policy Stance.

inflation rate to fall or to rise. This is especially the case where the interest rate is adjusted upwards (downwards) while the exchange rate weakens (strengthens).

The assessment of liquidity conditions in the financial system and the monetary policy stance of the central banks requires a careful consideration of the behaviour of the interest rates and the exchange rate. Therefore, an MCI index can serve as an informative indicator for liquidity conditions in the financial system. It also provides useful information regarding the central bank's monetary policy stance by comparing the effects of interest rate and exchange rate on the inflation rate (Hanson & Lindberg, 1994).

Figure 2 characterises some principal features associated with an inflation targeting-oriented monetary policy. As can be seen from the figure, the MCI is an important tool in the administration of monetary policy for a number of central banks. This is particularly the case for the central banks of New Zealand and Canada, which have used the MCI as an operating target variable for the administration of monetary policy.

3. POLICY FRAMEWORK AND THE APPLICATION OF AN MCI INDEX IN THE THAI CONTEXT

3.1. Thailand's Monetary Policy Framework in the New Environment

Since the early 1990s, the Thai economy has been claimed to be more open.[2] The monetary environment has been relaxed and the financial reform has been progressively implemented (Hataiseree, 1995d, 1995f, 1996a). Example in this regard included the official acceptance of IMF's Article VIII in 1990 and the elimination of all interest rate ceilings for financial institutions in 1992. Such a relaxation of financial policies have allowed the Thai economy and financial system to become increasingly integrated with the economy and financial system of other nations. This can be observed from the increased flow of international capital that came into the country since the beginning of the 1990s until the economic crisis in July 1997. In such a rapidly changing economic and financial environment, the movements of the exchange rate have become increasingly important in the Thai economy. Further details may be found in the recent study of Hataiseree (1997a).

As Thailand is regarded as a small open economy, the exchange rate, as it has been claimed, has gained a relatively important role in the Thai economy and financial system through international trade and investments. With such rapidly changing economic and financial environments together with the fact that the exchange rate system of Thailand was changed from a basket-pegging régime to a managed floating régime on July 2, 1997, the role of the exchange rate has become increasingly important as a channel through which monetary policy may have affected the real sector and, thus, the inflation rate (Hataiseree, 1997a).

It has been argued that this new exchange rate régime has given the Thai monetary authorities more room to manoeuvre its exchange rate policy in the sense that an adjustment in the nominal exchange rate under the new régime can be implemented continuously in response to developments at home and abroad and without the risk of being politically unacceptable, which was the case under the previous exchange rate régime.

Under the present circumstance, where the understanding of the monetary transmission mechanism in Thailand is still far from complete, an evaluation of the stance of monetary policy needs to rely on various sets of monetary indicators. It has been claimed that in evaluating the economic conditions and inflation levels in Thailand, the Thai authorities rely on a various sets of economic information and monetary indicators, including in particular the manufacturing production index, import and export levels, the behaviour of interest rate and exchange rates, monetary aggregates, credit aggregates, to name but a few. The various economic information and monetary indicators mentioned above have, to a certain extent, assisted the authorities in forecasting the direction and likely outlook of the economy and the inflation levels.

In the near future, however, the authorities tend to have encountered an increasing difficulty in evaluating and estimating the effects that the interest rate and the exchange rate may have on inflation. This is because the effects of the changes in the interest rate and exchange rate on the direction and level of inflation cannot be clearly observed. The variables influencing the future direction of the inflation tend to change rather easily, as a result of shocks from both domestic and internal factors.

The application of an MCI index in the case of Thailand is, therefore, an important issue which deserves further study. The exchange rate has played an increasingly important role as a channel through which the monetary policy may have affected the real economy, particularly the future direction of the inflation. Thus, an evaluation of the overall economic picture, monetary conditions and the level of inflation in the case of Thailand requires a careful consideration of the movements of an MCI index, together with some other economic and monetary indicators mentioned above.

3.2. The Application of an MCI Index in the Case of Thailand

As mentioned in Sections 2 and 3.1, it is important to evaluate monetary conditions in order to show how tight (easy) monetary policy is and thus its likelihood to lead to a lower (higher) inflation level. In order to do this effectively, it is crucial for the monetary authorities to simultaneously consider the behaviour of the movements of both the interest rates and the exchange rate.

Such an interactive movement may be expressed in equation (1) as follows

$$MCI = w_r^*(r_t - r_b) + w_e^*(e_t - e_b);\ w_r + w_e = \qquad (1)$$

Where r_t and e_t are interest rates and exchange rates during time t, respectively; r_b and e_b are interest rates and exchange rates during a given base year. The exchange rate variables in equation (1) are expressed in terms of logarithms.

Within an analytical framework of the CPI, the base indicator of economic activity and inflation is a variable that appears in equation (1) which in turn is the interest rate r, and the exchange rate e. The most important factor is weight w, derived from the subsequent empirical analysis. The value of this weight provides a useful information regarding the relative importance of the weight given to the interest rates (w_r) compared to the weight given to the exchange rates (w_e), which stipulates the direction of demand (economic activity) or inflation level.

The model explaining the behaviour of the inflation in Thailand can be expressed in a reduced form as follows:

$$\pi_t(\Delta P_t) = \alpha + \beta\Delta r_t + \gamma\Delta e_t + \delta\Delta pm_t + \phi\Delta pag_t + \upsilon\Delta g_t + \varepsilon_t. \qquad (2)$$

Where, $\beta < 0, \gamma > 0, \delta > 0, \phi > 0, \upsilon > 0$

Δ = Difference operator
P = Consumer Price Index or CPI (base year 1990 = 100)
π = Inflation (rate of change in CPI)
r = Interest rate (an average weight between 1-day interest rate in the repurchase market and the minimum lending rate or MLR)
e = Nominal effective exchange rate (NEER)
pm = Import Price Index (Unit value index, base year 1985 = 100)
pag = Agricultural Price Index (weighted average of major crop and livestock)
g = Government Fiscal Indicator (ratio of expenditure to income of the government)
ε = Error variable

It would be worth noting, however, that some important factors that might have potential influence on behaviour of the inflation rate have not yet been included in equation (2). It is planned to incorporate some additional factors such as world income and potential output in the subsequent study.

It is important to note also that the weight of interest rate (w_r) together with the weight of the exchange rate (w_e) in equation (1) can be calculated from the coefficients from equation (2) which equal:

$$\frac{|\beta|}{|\beta|+\gamma} \text{ and } \frac{|\gamma|}{|\beta|+\gamma} \text{ where } \beta<0 \text{ and } \gamma<0$$

Equation (2) is a reduced form of the structural equation system which is composed of the following equations:

$$y_t^s = \gamma(p_t - E_{t-1}p_t) + \infty_t^s \tag{3}$$

$$y_t^d = -\alpha r_t - \beta(e_t + p_t - p_t^f) + \varepsilon_t^d \tag{4}$$

$$R_t = R_t^f - E_t e_{t+1} + e_t + \varepsilon_t^x \tag{5}$$

$$r_t = R_t - E_t(p_{t+1} - p_t) \tag{6}$$

$$p_t = \phi p_t^d + (1-\phi)(e_t + p_t^f) \tag{7}$$

Equation (3) is an aggregate supply equation which assumes the production level to rely on expected price and supply shocks (ε_t^s). Equation (4) represents an aggregate demand equation hypothesized to depend on interest rates, exchange rates and demand shocks (ε_t^d). Equation (5) shows the equilibrium of the international capital market. Equation (6) describes the behaviour of the real interest rate which is the difference between expected interest rates and inflation rates. Equation (7) shows the movement of the price index by averages between domestic and international prices. For a detailed discussion of the model used in this paper, see Duguay (1994), Nadal-De Simone et al. (1996), among others.

4. EMPIRICAL RESULTS

4.1. Estimating the Weight of Interest Rate and the Exchange Rates

The monthly and seasonally unadjusted data employed in this paper cover the period from January 1990 to July 1998. They were obtained from the Bank of Thailand Monthly Statistical Bulletin or the Bank's in-house data base.

The econometric results as seen in Fig. 5 suggest that Model A and Model B can explain the movements of the inflation rates fairly well. The estimated coefficients obtained from the two models, especially those associated with the interest rate and inflation variables, have the right signs and are consistent with

	Model A	Model B	Model C
	π_t	π_t	π_t
Constant	0.360121 (8.7)**	0.3616 (8.9)**	0.3867 (9.0)**
$\sum_{i=0}^{4} \Delta r_{t-i}$	− 0.272 (− 1.9)*	− 0.289 (− 2.0)*	− 0.197 (− 1.2)
$\sum_{i=0}^{4} \Delta \ln e_{t-i}$	—	—	− 0.0216 (0.9)
$\sum_{i=0}^{7} \Delta \ln e_{t-i}$	− 0.0818 (− 3.0)*	− 0.0828 (− 3.1)*	—
$\Delta \ln pag_{t-6}$	0.034 (2.19)*	0.0334 (2.1)*	0.0353 (2.1)*
$\Delta \ln pm_{t-8}$	0.0305 (2.0)*	0.030025 (2.0)*	0.0394 (2.6)*
$\sum_{i=5}^{9} \Delta g_{t-i}$	0.0071 (1.2)	—	0.0073 (1.2)
R^2	0.17	0.17	0.09
D.W.	2.0	2.1	2.0
F-statistics	4.5**	5.6**	2.7**
MCI ratio (r:e)	3.3:1	3.5:1	9:1
W_r	0.77	0.78	0.90
W_e	0.23	0.22	0.10

Notes : 1) *Figures in parentheses are t-statistics.*
2) ** and ** represent significance level of 0.05 and 0.01, respectively.*

Fig. 5. The Relative Importance between Interest Rate and the Exchange Rate: The Estimation Results.

economic theories. For instance, the estimated coefficients of the interest rate and the inflation rate calculated from Model A take the values of –0.272 and 0.0818, respectively.

In addition, Model A and B seem to have a similar lag structure. More importantly, the changes in interest rates were found to have significant impacts on the inflation rate after the lag of 4 periods compared to 7 periods in the case of changes in the exchange rates. However, it is important to note that the exchange rate was found to have no significant influence on the movements of the inflation rates when the number of lag variables was adjusted downward from 7 to 4 lags as in the case of Model C.

The finding that interest rate effects the inflation after 4 months and the exchange rate after 7 months is of particular importance. The empirical findings of this kind also lent support to the monetary economic theory which states that changes in the monetary policy actions appear to have a long and variable impact on the ultimate targets of monetary policy.

Comparing the relative importance between interest rate and the exchange rate in influencing the inflation, in each model the coefficients on the interest rates have larger values than those of the exchange rate. Empirical evidence of this kind is considered to be essential information for the construction of the MCI index which is basically derived from the weighted average of the interest rate and the exchange rate as expressed in equation (1).

Note that the R^2 figures calculated from Model A, B, and C have relatively low values. Nonetheless, the low values of the reported R^2 are regarded as the natural result in an econometric analysis in which variables in 'first difference' are used in an empirical analysis. Indeed, the calculated R^2 may have a higher value if the variables used are in 'level' form instead of 'first difference' form. Nonetheless, the use of variables in 'level form may result in a spurious relationship problem, especially in the case where the variables used are associated with time trends. A more detailed explanation on this issue, see Hataiseree (1995c).

4.2. The Construction of the MCI

Judging from the estimated coefficients on interest rate and the exchange rate variables, it is likely that the ratio of the MCI index in the case of Model A, B and C, are in the values of 3.3:1, 3.5:1 and 9:1, respectively. However, the subsequent analysis is based on Model A, as it appears that the estimated coefficients from Model A tended to be relatively statistically more credible when compared to those derived from Model B and C.

On the whole, the empirical findings seem to suggest Thailand's MCI ratio to be roughly 3.3:1. More importantly, the interest rate channel into demand pressure, and thus inflation, is found to be not only more powerful than the exchange rate channel, but it has also a faster impact.

It would be worth noting, however, that the Thai MCI ratio of 3.3:1 seems to be slightly larger than those of other small open economics such as New

Zealand, Sweden, and Canada. The estimated values of the MCI ratio are reported to be an approximate value of 2:1 for New Zealand (Nadal-De Simone et al., 1996); 2:1 for Sweden (Hansson & Lindberg, 1994); 3:1 for Canada (Freedman, 1995). One possible factor attributed to a relatively larger weight on interest rate may be associated with the fact that Thailand has pursued monetary policy using the flexible exchange rate within a much shorter period compared to those countries afore-mentioned. As such, the exchange rate channel on demand pressure, and thus inflation, may have not yet produced a full impact compared with the interest rate channel.

One important implication of the MCI index calculated from Model A is that in the case where the baht is expected to weaken at the average of 3.3% in a period of time, the monetary authorities may have to influence the way of forecasting by pushing up the interest rate averaging 1%, provided that other variables in the model remain unchanged. This sort of operations is needed if the officials concerned do not want the weakening of the baht to affect the forecasting of future inflation.

Nonetheless, considerable care needs to be taken in interpreting the estimated MCI ratio. For example, the authorities concerned are not necessary to automatically push up the short term interest rate to a ratio of 1:3:3 (or by 12%) in the case where the baht has weaken by more than 40%. This seemed to be the case during the first six months after the flotation of the Baht since July 2, 1997. In other words, the authorities concerned need to evaluate and take into account the consideration of the movements of some other factors (e.g. credit expansion or the changes in money supply) that might effect the forecasting of the future inflation rates.

Although the weakening of the baht may have an adverse impact on the forecasted inflation, after an in-depth evaluation of the recent economic and financial developments, there seemed to be insufficient evidence supporting an upward adjustment in the interest rates. For one thing, credit extensions in the private sector tended to decline dramatically following unfavourable economic conditions which do not support commercial banks to extend credit. In effect, it is not always necessary for the monetary authorities to adjust the market interest rates in line with those as implied from the calculated MCI ratio, especially in the case where there are good reasons to believe that some other factors might have exerted the opposite influences on the forecasted inflation.

It is important to note that the calculated MCI ratio may be particularly sensitive to changes in the structure of the economy. As one can see, the MCI ratio calculated in this study is based on data ranging from January 1990 to July 1998, a period in which there were several shifts in the economic system, especially those related to the change in the exchange rate régime, from the

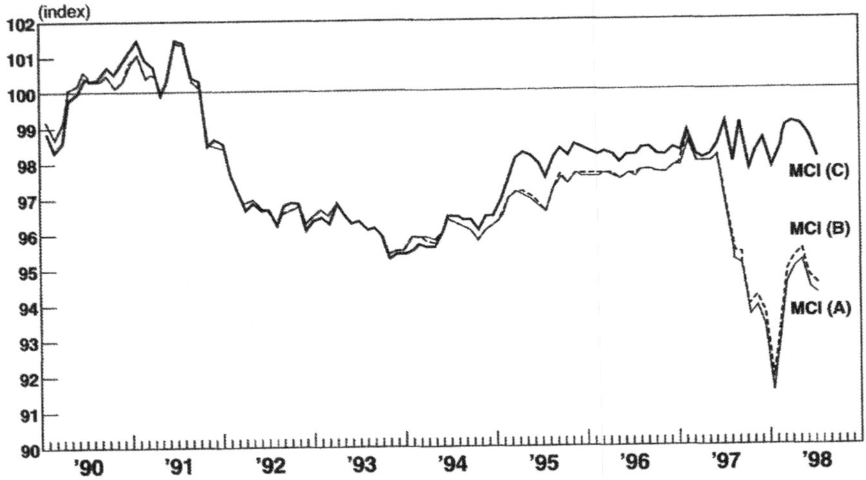

Fig. 6. Different Paths of Monetary Conditions Index for Thailand. (January 1990–July 1998)

basket régime to the managed float régime. For this reason, it is necessary to be cautious in inferring the value of the coefficients derived from the model in this study. In particular, one has to be extremely cautious when opting for using this sort of information in the empirical analysis.

Figure 6 portrays the movements of the estimated MCI using the monthly data from January 1990 to July 1998. As one can see, the movements of the MCI index appeared to be able to reasonably reflect the tightening or relaxing of the stance of monetary policy. More detailed exposition of this is discussed in the following section.

4.3. The Relationship between the MCI and the Inflation Rate

Figure 7 shows the comparison of movements of the interest rate, inflation rate, and the rate of change of the MCI index. Figure 8 demonstrates the movement of the exchange rates, interest rates and the MCI index from January 1990 to July 1998.

From Fig. 8, the monetary policy became less tight on average 3% during the year 1992–1994 compared to the base year 1990. As one can see, the MCI index tended to decline continuously during the mentioned period. However, the monetary policy became more disciplined during the year 1995 until the period prior the flotation of the baht. Thereafter, the overall tightness of

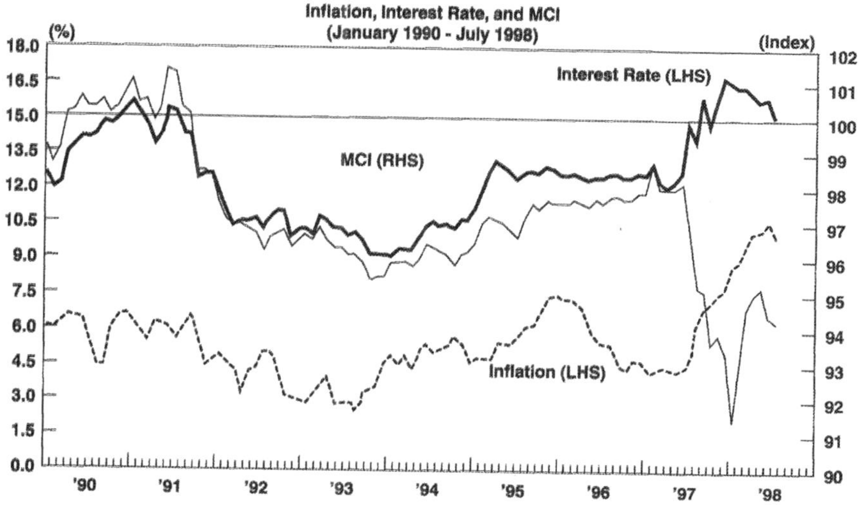

Fig. 7. Inflation, Interest Rate, and MCI.
(January 1990–July 1998)

monetary policy lessened slightly as the rate of increase in the interest rate was less than the rate of depreciation of the baht. Nonetheless, when the baht hit the lowest value in January 1998, the monetary policy became tighter again. This could be noticed from the increase in the rate of interest in the repurchase market particularly in January 1998 when the repurchase rate reached the recorded peak of 24%. However, the monetary policy became more loosened in the second quarter of the year 1998 when the baht continued to show signs of stabilization.

It is important to note, however, that the degree of tightness of the monetary policy showed a significant transformation from mid-1997 to January 1998. The MCI index declined continuously during the mentioned period due to the rate of depreciation of the baht being more than offset by the increase in the interest rate.

The fact that the monetary policy was less tight than it should have been otherwise since the second half of the year 1997 could have been one possible reason for the relatively high inflation rate in that period. This is particularly the case in the environment where the market expected the baht to have been pressured into further depreciation, while the authorities concerned were obliged to alleviate the liquidity problem in the money market.

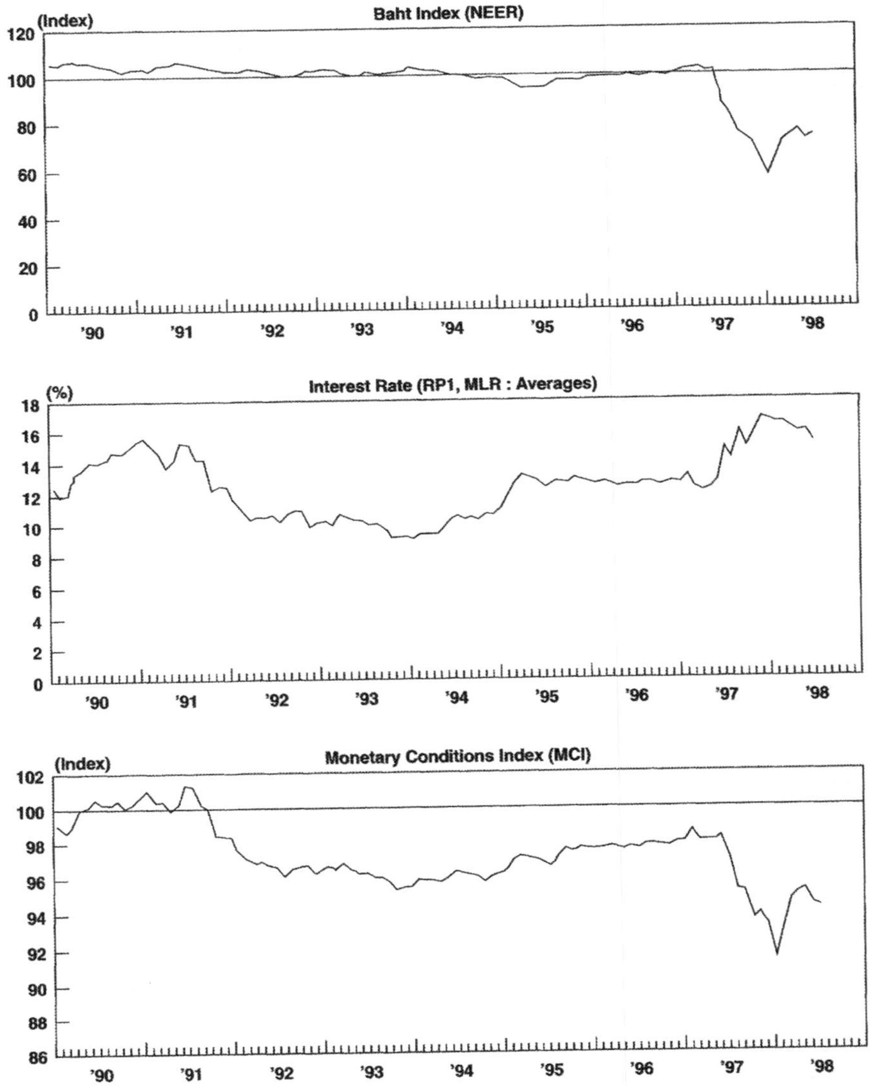

Fig. 8. Nominal Effective Exchange Rate (NEER), Interest Rate, and MCI.

Another noticeable point is the direction of the MCI index (which manifests the degree of tightness of the monetary policy) and the rate of inflation appeared to be in line with the recent movements of the flexible exchange rates. However, one needs to be cautious in assessing the further direction of the baht, as the bath may be adversely affected by an increasing uncertainty of the international economy in the recent periods. Such an unfavourable international environment may cause an depreciation of the Thai baht and, thus, may result in a higher rate of inflation as implied from the relationship mentioned earlier. This is particularly the case where a reduction in the interest rate is chosen with a view to encouraging the resumption of the economy in the real sector.

5. CONCLUSIONS AND IMPLICATIONS FOR THE CONDUCT OF MONETARY POLICY

This paper has established that in small open economies like Thailand monetary policy can be transmitted through both interest rate and the exchange rates.

As the MCI index clearly manifests the comparison of the importance of the interest and the exchange in effecting the aggregate demand and the price stability, which is one of the most important objectives for many central banks, the MCI index can thus be served as an useful indicator for the monetary authorities when making policy decisions. Using an MCI index which characterizes the relative importance of interest rate and the exchange rates on inflationary pressure provides a better indication of the policy stance compared to either interest rate and the exchange rates alone.

The empirical analysis in this study suggests that the relative importance of the interest rate and the exchange rate (MCI ratio) to be 3:3:1. The finding of this kind is essential in improving an assessment of the policy stance of the central bank. In particular, the assessment of economic and monetary conditions using the MCI index was found to be more effective than using either interest rate or the exchange rate alone as an independent variable.

The study also provides an empirical analysis of the relationship between the MCI index and the inflation rate for the Thai economy. The empirical results seem to suggest that the overall movement of the MCI index has some correlation with the movement of the inflation rate.

Another important implication of this finding for the conduct of monetary policy is that the MCI index appears to be a useful indicator for the short term outlook of monetary conditions which can be used to influence the future movements of the inflation rate to be in line with the target. In addition, the finding reflects that the exchange rate plays an increasing role in transmitting the effect of monetary policy to the real economy and, thus, the inflation rate. The development of this kind is deemed to be consistent with the recent economic conditions and developments in Thailand.

Although the above findings seem to provide many positive aspects in using the MCI index as one of the indicative instruments in comparing the impact of interest rate and the exchange rate on the rate of inflation, the actual application of this index in the conduct of monetary policy in Thailand calls for a further study in many important areas. Chief among these are:

Firstly, one can debate whether the Bank's MCI is correctly calibrated, whether a 3.3% fall in the exchange rate is equivalent to a 1% movement in interest rate. Additionally, one can debate whether the Bank's Trade Weighted Index, represently by NEER, correctly measures the movements of the Thai baht.

Secondly, the Thai authorities need to decide between using the MCI index as an operational target (as in the case for New Zealand and Canada) or as a useful indicator for the future assessment of the policy stance (as being the case for those operating in many Scandinavian countries).

Thirdly, the degree of tolerance for the movements of the MCI index needs to be reasonably established. The important message is that in order for the MCI to be effective, the authorities need to be flexible in implementing the policy instruments, thus preventing the use of the MCI index as a fixed formula.

Fourthly, an empirical evaluation of the monetary conditions and its likely outlook, using the MCI index in real or in nominal term as presented in this paper needs to be compared to ascertain whether or not there is any statistically significantly different.

Lastly, price stability has long seen regarded as one of the main objectives for the conduct of monetary policy. However, careful consideration needs to be projected at the choice between using the inflation targeting or monetary targeting régime[3] in order to be most applicable and consistent with the Thai economy. As seen from the empirical analysis, together with the changing of the exchange rate régime from fixed exchange rate régime to the managed-floating régime, the reliance on the fixed exchange rate as a nominal anchor to control the rate of inflation has now been removed.

NOTES

1. For more detail on the implementation of monetary targeting as a monetary policy strategy in major industrialised countries, see Argy et al. (1990) and Neumann (1996). Further information in the case of Thailand may be found in Hataiseree (1993, 1998a, 1998b), and Hataiseree and Phipps (1996b).

2. Empirical evidences obtained thus far have shown that the Thai financial system has become increasingly open. The degree of financial integration has increased from the level of 0.85 during the period of 1980–1992 to 0.89 during the period 1980–1994. Further details on this matter, see Hataiseree (1995a, 1995b), Hataiseree and Phipps (1996a).

3. The empirical findings reported in Hataiseree (1993, 1998a, 1998b) seem to lend supports for the use of the monetary targeting as a potential framework for the conduct of monetary policy in Thailand. In particular, the demand for money M1 was found to be stable and, thus, can be used as a guidepost to control the inflation rate. In addition, M1 was found to have both short term and long-term relationships with the monetary base which is considered the main monetary instrument of the authorities. As such, M1 seems to be a leading indicatior for the future movements and likely outlook of the inflation rate.

ACKNOWLEDGMENTS

We have benefited from comments by Atchana Waikuamdee, Suchada Kirakul, Plearnpit Satsanguan, Sothitorn Mallikgamas, Chayodom Sabhasri, Vimut Vanitchareantham. We are also grateful to Teerapol Rattanalungkarn, Susit Tanarat and Wilatluk Sinswat for invaluable research assistance. The views are those of the authors and are not necessarily those of the Bank of Thailand. The usual disclaimer applies.

REFERENCES

Argy, V., Brennan, A., & Stevens, G. (1990). Monetary Targeting: The International Experience. *Economic Record*, *66*, 37–62.

Battellino, R., Broadbend., J., & Lowe, P. (1997). The Implementation of Monetary Policy in Australia. *Research Bank of Australia*, Discussion Paper 9703.

Brash, D. (1997). Exchange Rates, Export Prices, and the Reserve Bank's Monetary Conditions Index. *RBNZ speech*, *22* (August).

Brash, D. (1998). Monetary Policy in New Zealand: Business as Usual. *Reserve Bank of New Zealand Bulletin*, *61*(2), 154–159.

Debelle, G. (1997). Inflation Targeting in Practice. *IMF Working Papers*, *35* (March).

Duguay, P. (1994). Empirical Evidence on the Strength of the Monetary Transmission Mechanism in Canada. *Journal of Monetary Economics*, *33*, 39–61.

Freedman, C. (1995). The Role of Monetary Conditions and the Monetary Conditions Index in the Conduct of Policy. *Bank of Canada Review*, Autumn, 53–59.

Green, J. H. (1996). Inflation Targeting: Theory and Policy Implications. *IMF Staff Papers*, *43*(4), 779–795.

Hansson, B., & Lindberg, H. (1994). Monetary Conditions Index: A Monetary Indicator. *Quarterly Review* (Sveriges Risksbank), *3*, 12–17.

Hataiseree, R. (1993). The Demand for Money in Thailand: A Cointegration and Error-Correction Approach. *Singapore Economic Review*, *38*(2), 195–230.

Hataiseree, R. (1995a). Degree of Financial Integration: Thailand's Recent Evidence. *Bank of Thailand Monthly Bulletin*, *35*(8), 11–25 (in Thai).

Hataiseree, R. (1995b). Capital Mobility, Sterilization and Monetary Policy: Some Evidence from Thailand since 1980. *Bank of Thailand Papers on Policy Analysis and Assessment*, *3*, 27–51.

Hataiseree, R. (1995c). Cointegration and Error-Correction Approach: An Alternative Application to Thailand's Macroeconomic Modelling. *Thammasat Economic Journal*, *13*(3), 20–55 (in Thai).

Hataiseree, R. (1995d). Monetary Policy in Thailand: Past, Present and Future. *Chulalongkorn Review*, *8*(29), 92–114.

Hataiseree, R. (1995e). Cointegration Tests of Purchasing Power Parity: The Case of Thai Baht. *Asian Economic Journal*, *6*(1), 67–75.

Hataiseree, R. (1995f). Financial Liberalization, Financial Innovation and Its Impacts on Monetary Policy: Thailand's Recent Evidence. *Chulalongkorn Review*, *7*(27), 89–102.

Hataiseree, R. (1996a). Macroeconomic Performance, Monetary and Exchange Rate Policies and Some Salient Features of the Thai Economy Since 1980. *Journal of Business Administration*, *19*(70), 20–41.

Hataiseree, R. (1996c). An Empirical Evaluation of Monetary Control in Thailand: The Post-1980 Experience. *Asian Economies*, *24*(4), 67–113.

Hataiseree, R. (1997a). The Managed Floating Rate Régime: Its Nature and Implications on Thailand's Economy. *Chulalongkorn Review*, *10*(36).

Hataiseree, R. (1997b). Monetary and Fiscal Policy in Thailand: Its Nature and Role as Stabilization Policy. *Journal of Business Administration*, *20*(75), 1–10.

Hataiseree, R. (1997c). Modelling Exchange Rate Policy Reaction Function: Evidence from Thailand Under the Basket Pegging Regime. *Sasin Journal of Management*, *3*(1), 28–40.

Hataiseree, R. (1998a). Strategy of Monetary Targeting: Is It Applicable to Thailand's Case?. *Bank of Thailand Quarterly Bulletin*, *38*(1), 27–56.

Hataiseree, R. (1998b). Monetary Tactics with a Monetary Targeting: A Thai Perspective. *Sasin Journal of Management*, *4*(1), 14–37.

Hataiseree, R., & Phipps, A. (1996a). The Degree of Capital Mobility in Thailand: Some Estimates Using a Cointegration Approach. *Applied Economics Letters*, *3*, 9–12.

Hataiseree, R., & Phipps, A. (1996b). The Relationship between Money and Income in Thailand: Some Evidence for the 1980s Using a Cointegration Approach. *Savings and Development*, *20*, 389–405.

Hataiseree, R. (1999). A Preliminary Analysis of Monetary and Inflation Targeting Frameworks for Thailand. *Bank of Thailand Quarterly Bulletin*, *39*(3), 23–52.

Nadal-De Simone, F., Dennis, R., & Redward, P. (1996). A Monetary Conditions Index for New Zealand. *Reserve Bank of New Zealand*, Dicussion Paper, pp. 96/2

Neumann, M. J. M. (1996). Monetary Targeting in Germany. IMES Discussion Paper 96-E-15, Bank of Japan.

Rattanalungkarn, T. (1995). Core Inflation: Concept and Its Application to Thailand. *Bank of Thailand Paper on Policy Analysis and Assessment*, 97–104 (in Thai).

Svensson, L. (1997). Inflation Forecast Targeting: Implementation and Monitoring Inflation Targets. *European Economic Review*, *41*(6), 1111–1146.

Taylor, J. B. (1996). Policy Rules as a Means to a More Effective Monetary Policy. IMES Discussion Paper 96-E-12, Bank of Japan.

19. THE OPTIMAL-CURRENCY-AREA INDEX: SELECTED ASIAN COUNTRIES

Chayodom Sabhasri and Gene Janevathanavitya

INTRODUCTION

The early literature on optimal currency areas (OCA) are characterized into two approaches: the pre-conditional characteristics of a national economy that desires to participate in a currency union and the costs and benefits to a nation participating in a currency area. The currency area refers to an area in which member countries have their exchange rates irrevocably fixed, or an area in which member countries use a single currency. The monetary cooperation policy, particularly the common monetary authority, is the main ingredient to the success of a currency union. Mundell (1961) discussed that the degree of factor mobility determined the optimal currency area, since the factor mobility led to adjustment of the asymmetric economic shocks across countries. McKinnon (1963) argued that the effectiveness of exchange rates might decline with the openness because changes in exchange rates would destabilize the internal price level. Kenen (1969) suggested that regions with high product diversification would be in a better position to participate in a currency union than those with low diversification, since the former are subject to smaller disturbances. In other words, product diversification insulates an economy from economic shocks. Issues such as the degree of financial market integration, the degree of good market integration, price and wage flexibility, and the degree of fiscal integration have been mentioned as pre-conditional

Restructuring Asian Economics for the New Millennium, Volume 9A, pages 363–373.
2001 by Elsevier Science B.V.
ISBN: 0-7623-0709-9

characteristics for the OCA. The recent work by Tavlas (1993a) raised an issue on the time consistency, which considers the inflation expectations and credible commitment of monetary authorities as a key to success for an OCA.

Several studies, such as Ishiyama (1975), Tavlas (1993b), Canzoneri and Rogers (1990), Frenkel and Rose (1996), and Ricci (1997), discussed issues on the costs and benefits of the OCA. The benefits of the OCA are the elimination in the exchange risk, the elimination of the conversion and transaction costs, the price transparency, the increase in efficiency in both goods and financial markets, the enlargement of foreign exchange market, and the elimination of the speculative financial flows in the area. The costs of participating in a currency union are, for instance, the loss of sovereignty in conducting independent monetary policy and exchange rate policy. Canzoneri and Roger (1990) provided an example on the trade-off between the elimination of the conversion and transaction costs in a single currency area and the independence in managing inflation tax policy.

On the theoretical development of the OCA, Casella and Feinstien (1989) used a general equilibrium approach to compare the cross-country welfare levels between the flexible exchange rate régime and a single currency régime with and without a common central bank. Bayoumi (1994) used a general equilibrium approach with regionally differentiated products to present a model for the OCA. The results show that the choice of a currency union is determined by the size of disturbances, the correlation between disturbances, the transaction costs, the factor mobility, and the interrelationship between demands for different goods. Sabhasri (1994) simulated a general equilibrium with an overlapping generation model to show how differences in the preferential patterns of consumers in private and public goods across countries designate participation in a currency area. Ricci (1997) investigated the circumstances that are beneficial to participate in a currency area: the correlation of the real shocks, the international factor mobility, the fiscal adjustment, the openness, the inflationary biases, and the correlation of monetary shocks. In addition, the European monetary integration has enlightened the advanced research for the OCA. Canzoneri and Roger (1990), Eichengreen (1990), Eichengreen (1997), and Frenkel and Rose (1996 and 1997) are examples of the analysis of the OCA associated with the development of the single currency among the EU members.

In addition to the study of an optimal currency area, the related issues such as economic integration and monetary integration have become increasingly important. Several pre-conditions for an optimal currency area hinge on the degree of integration, for instance, goods and financial market integration, and fiscal integration. The final stage of the Maastricht treaty also deals with the convergence criteria which take into account the convergence in inflation rate, public debt, budget deficit, long-term interest rate, and the stability of the exchange rates among potential members of a single currency area.[1] Bayoumi and Eichengreen (1997) constructed an optimal-currency-area index (OCA index) that showed the relationship between the pre-conditions of countries which the OCA theory states. They also observed the behavior of exchange rates by using a cross-country data of 15 EU members to estimate the OCA index over a period of time and to investigate the relationship between economic integration and the monetary integration.

The objective of this paper is to construct an OCA index for selected Asian countries by using the same estimation and forecasting techniques as Bayoumi and Eichengreen (1997). These selected countries are Indonesia, Japan, Malaysia, Philippines, South Korea, Thailand, and Singapore. The core implications of theory of the OCA are developed in order to find the relationship between the OCA characteristics and the observed behavior of changes in exchange rates. The OCA index for these selected Asian countries will be based on the *Japanese Yen* as a leading currency for the economic and monetary block. The estimated exchange rate equation is also used to predict which countries will be best to support stability of exchange rates and the economic and monetary block.

ANALYSIS OF THE THEORY OF OPTIMAL CURRENCY AREA

The OCA theory focuses on the stability of the exchange rates and monetary unification across the partner countries. According to Bayoumi and Eichengreen (1997), the important factors that influence the stability of the exchange rates are asymmetric shocks to output, trade linkages, usefulness of money for transaction, labor mobility, and automatic stabilizers. Labor mobility and automatic stabilizers among the selected Asian countries do not currently exist or strongly occur even though these two factors have great influences in determining the OCA. Therefore, these two factors will not be included in the construction of the OCA index. There will only be three factors that will be emphasized in the OCA index.

The effects of the asymmetric shocks to output, trade linkages, and the usefulness of money on the variability of exchange rates are hypothesized. The output-related variables that influence the changes in exchange rates can be considered as follows. The standard deviation of the logarithm of relative output in two partner countries is used to measure the asymmetric shocks to output. When the standard deviation is high, the two countries are likely to have different pattern of business cycle. As a result, it is likely to cause fluctuations in their exchange rates. Another proxy that is used to identify the asymmetric shocks, particularly the industry-specific shock, is the dissimilarity of the economic structure. Unlike Bayoumi and Eichengreen (1997), who employed the commodity composition of the exports of the two countries, our study uses the differences in the share of agricultural, industrial, and service sectors as percentage of GDP in two related countries to represent the economic dissimilarity. If a country exports its products in line with the comparative advantage, the definition of our dissimilarity and Bayoumi's and Eichengreen's (1997) dissimilarity will be in the same direction. Since the dissimilarity proxies the asymmetric shocks, the positive relationship with the exchange rate variability is expected. The strong trade linkage between two countries is likely to reduce the instability of the exchange rates for the related countries. The trade linkage is measured by the average value of exports to the partner country as percentage of its GDP. Last but not least, the economic size of the two related countries is determined by their average of the logarithm of real GDPs. Casella and Feinstein (1989) argued that a small country received net benefits from a currency union. Besides, Bayoumi and Eichengreen (1997) also discussed that the benefits of a small country were derived from the role of money as a unit of account, means of payment, and store of value services. Hence, the relationship between the exchange rate stability and the economic size should be negative.

CROSS-COUNTRY DATA, THE ESTIMATING EQUATION AND THE ESTIMATED RESULTS

Cross-country datum for Indonesia, Japan, Korea, Malaysia, Philippines, Thailand, and Singapore are used in our analysis. Figure 1 shows the economic growth of these selected countries. The patterns of the business cycles show the close correlation among these Asian countries from 1985 to 1996. However, due to the economic crisis starting from the middle of 1997, the economic growth rates of these countries have tended to move in less coordinated direction. The economic growth rates then depend on the impacts of the crisis on each economy. Our analysis of the theory of OCA will only use the data from 1990 to 1996 in order to cover the period that the Japanese economic growth has slowed down. The reason is that we assume the Japanese Yen will be the leading currency of the currency block.

Figure 2 exhibits the nominal exchange rates of six candidate countries, i.e. Indonesia, Korea, Malaysia, Philippines, Thailand, and Singapore, vis-à-vis

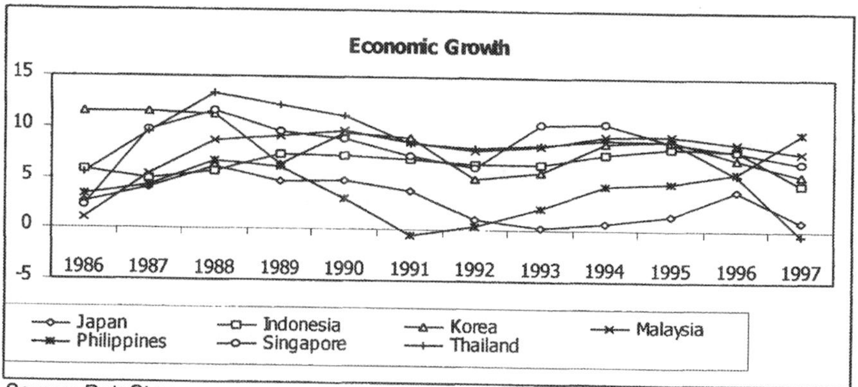

Source: DataStream

Fig. 1. Economic Growth of Selected Asian Countries.

JPY by normalizing the data in 1990Q1 to unity. It can be observed that every currency has depreciated against the JPY after the Asian economic crisis.

Our estimating equation is

$$SD(e_{ij}) = \alpha + \beta_1 SD(\Delta y_i - \Delta y_j) + \beta_2 DISSIM_{ij} + \beta_3 TRADE_{ij} + \beta_4 SIZE_{ij},$$

where

$SD(e_{ij})$ is the standard deviation of the logarithm of the annual average bilateral nominal exchange rate between countries i and j normalized by the mean values,
$SD(\Delta y_i - \Delta y_j)$ is the standard deviation of the differences in the logarithm of real output between countries i and j,
$DISSIM_{ij}$ is the sum of absolute differences in shares of agricultural, industrial and service sectors in relative to domestic GDP for countries i and j,
$TRADE_{ij}$ is the average ratio of bilateral exports to domestic GDP for the two countries i and j, and
$SIZE_{ij}$ is the average of logarithm of the two countries' GDPs.

Indonesia, Japan, Malaysia, Philippines, Korea, Thailand, and Singapore are selected countries; therefore, there are overall 42 observations from the combination of the cross-country data. Observations for each variable are constructed as follows. $SD(e_{ij})$ is calculated by using the bilateral exchange rates for countries i and j from 1990Q1 to 1996Q4. $SD(\Delta y_i - \Delta y_j)$ and $SIZE_{ij}$ are constructed over the same period as $SD(e_{ij})$. $DISSIM_{ij}$ is calculated from the cross-country data of 1996 for Indonesia, Philippines, Korea, Thailand and Singapore, and that of 1995 for Japan and Malaysia due to the availability of

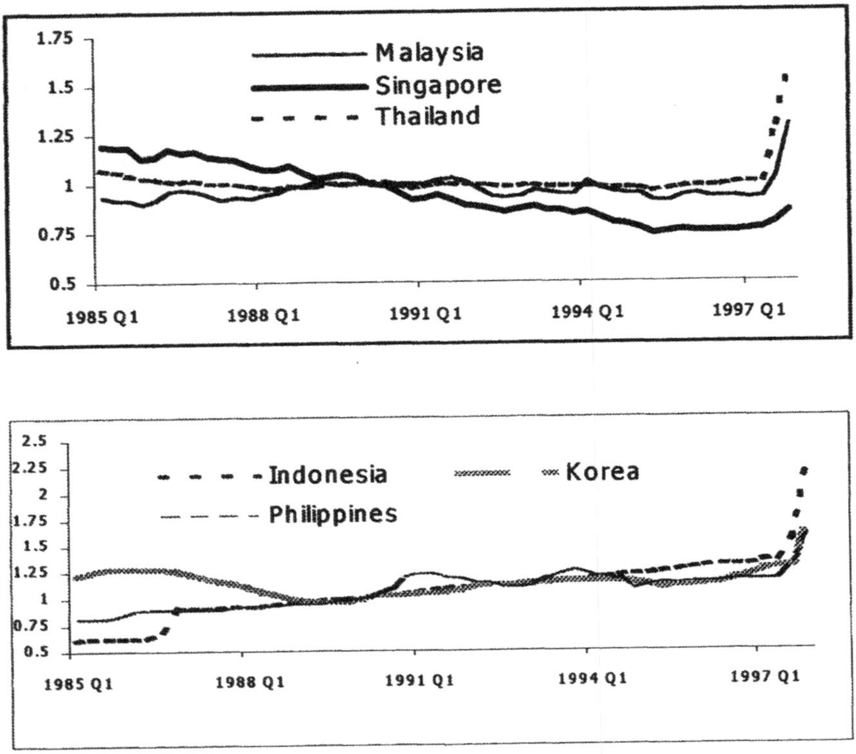

Source: DataStream

Fig. 2. Nominal Exchange Rates vis-à-vis JPY, Normalizing 1990Q1 to Unity. (1990Q1 to 1997Q4).

the data. The results of the estimation should only be minimally affected since the economic structure of a country is likely to remain unaltered during 1990 to 1996. $TRADE_{ij}$is constructed by using the latest data of 1997. Similar to $DISSIM_{ij}$, the trade linkages do not change much during the same period from 1990 to 1996.

The estimation yields the following result with the standard errors in parentheses;

$$\underset{}{SD(e_{ij})} = \underset{(0.12)}{-0.86} \underset{(0.007)}{-0.017} SD(\Delta y_i - \Delta y_j) + \underset{(0.0003)}{0.003} DISSIM_{ij}$$

$$\underset{(0.001)}{-0.004} TRADE_{ij} + \underset{(0.005)}{0.036} SIZE_{ij},$$

where n = 42, adjusted $R^2 = 0.72$, and S. E. = 0.027.

The dissimilarity, trade linkages and economic size have anticipated signs and the estimated coefficients differ from zero at the 1% confidence level. However, the asymmetric shocks to output show unanticipated sign and it is significant at 3% confidence level.

The estimation yields a puzzle for the estimated coefficient of the proxy variable for the asymmetric shocks to output. From the theory of OCA, it is expected that the similar pattern of the business cycles of two related countries lead to exchange rate stability. Bayoumi and Eichengreen (1997) used the data of the 15 European countries, members of the European Monetary System, to estimate the similar equation as above. They found the positive relationship between the standard deviation of nominal exchange rates, $SD(e_{ij})$, and the asymmetric shocks to output, $SD(\Delta y_i - \Delta y\Delta_j)$. One possible explanation of the differences in the results between the relationship of those 15 EU countries and these seven selected Asian countries is that the size of the Japanese economy highly dominates the rest of the Asian countries in the currency block, whereas the German economy does not have an excessive power over EU in economic size. Since the patterns of economic growth of Asian countries are based on high differences in the size of economies, the growth-rate differential may not well represent the asymmetric shocks to output. In our case, it is possible that the government intervention in the foreign exchange market is rather forceful in controlling the fluctuation of nominal exchange rates. Additionally, most of the selected countries run an implicit fixed exchange rate régime with the U.S. dollar or peg its currency with a basket system where the U.S. dollar shares the largest portion. Our ongoing research is to test the similar theoretical concept with the U.S. as a leading country in the currency block.

THE CONSTRUCTION OF THE OCA INDEX

We construct the OCA index for 1994 and 1997 by using Japan as the leader of the economic and monetary block. For 1994 and 1997, the estimated values of $SD(\Delta y_i - \Delta y\Delta_j)$ are calculated and projected over the sample period by centering on the interested years: 1994 and 1997. The values for $SIZE_{ij}$ come from the actual data of 1994 and 1997. The values of $TRADE_{ij}$ and $DISSIM_{ij}$ are kept to remain the same as those values used in the estimation under the assumption that there are no change in the structure of the trade linkages and production, respectively. For a simple interpretation, the estimated dependent variable is transformed to the variance that is reported as the OCA index.

Table 1. OCA Indexes vs. Japan, 1994 and 1997.

	1994	1997
Indonesia	0.007	0.035
Korea	0.053	0.027
Malaysia	0.022	0.027
Philippines	0.411	0.004
Singapore	0.704	0.004
Thailand	0.065	0.002

The OCA index of 1994 is analyzed since this is the year that the selected countries were approximately at the peak of their business cycles, whereas the OCA index of 1997 is used to represent the starting era of economic and financial crisis. Table 1 illustrates the OCA index vis-à-vis Japan for 1994 and 1997. The OCA indexes for Korea, Malaysia, Philippines, Singapore and Thailand for these two periods of time show a convergence towards the exchange rate stability within the Japanese Yen block. The OCA indexes for Korea, Philippines, Thailand and Singapore have shown a dramatic decline in the indexes over the period. Even though Malaysia's index does not significantly change, the indexes are rather small and can be included in the convergence case. Unlike those five countries, the OCA index for Indonesia increased by 5 times in 1997 in comparison to 1994. This result suggests that Korea, Malaysia, Philippines, Singapore and Thailand can be candidates for participating in the currency block, vis-à-vis Japanese Yen, whereas Indonesia has been hit hard with the economic and financial crisis, and its indexes show no sign of convergence towards the currency blocks.

The same approach in constructing the OCA index vis-à-vis Japan is applied in order to observe the bilateral relationships of the countries that will be best

Table 2. OCA Indexes for Specific Relationship, 1994 and 1997.

	1994	1997
Singapore–Malaysia	0.006	0.008
Singapore–Thailand	0.019	0.000
Thailand–Malaysia	0.000	0.001

supporting members of the currency block. In Table 2, the OCA indexes for other bilateral exchange rates are exhibited. These suggest that the likelihood of Malaysia and Thailand taking part in the currency block will depend on Singapore's participation. In our study, the differences in the OCA indexes are mainly driven by the pattern of the asymmetric shocks to output or the business cycles and the size of the economy. The findings from the estimation suggest that the bilateral trade linkages certainly strengthen the exchange rate stability. This result then supports the promotion of free trade policy among the selected countries in order to gain the benefits of the currency area. The relative sizes of the participants and the differences in economic structure are also crucial in determining the currency area. The larger the difference in size and the smaller the difference in economic structure are, the higher the exchange rate stability will be.

CONCLUSION

The objective of this paper is to employ the theory of optimum currency areas by constructing an OCA index using empirical specification. This study is based on Bayoumi and Eichengreen (1997) that operationalize the theory of optimal currency areas by observing the OCA index of the 15 European countries that are preparing for a single-currency area in 1999. Our estimated regression is formed according to the characteristics of pre-conditions for participating in a currency area. The economic disturbances to output, dissimilarity of economic structure, the relative sizes of the candidate countries and the trade linkages are taken into account in determining exchange rate stability. We observe the selected Asian countries: Indonesia, Korea, Japan, Philippines, Malaysia, Thailand, and Singapore. Japanese Yen is chosen to be the leading currency of the currency block. The cross-country data between 1990 and 1996 is used to construct the observations and to estimate the equation determining the OCA index of each country vis-à-vis Japan. By observing the OCA index in 1994 and 1997, the finding suggests that Indonesia is the only country that is not ready for participating in the currency block. Moreover, Thailand's and Malaysia's participation will hinge on Singapore's. The promotion of the free trade area will also foster the currency block.

This study also aims to lighten up the research on the theory of optimal currency areas for Asian countries in particular. There is room for improvement in this field of research. The related issues such as Williamson's target zone and monetary institutional reforms are not yet raised in this study. In case of the

Euro, the development of a single currency in Europe has a long history and has gone through several economic and financial crises before coming to a conclusion in 1999. European Institutional reforms are one of the keys to success for the European monetary integration. The reforms have been on the gradual adjustment path over the past 30 years. Therefore, the formation of a currency block or union in Asia cannot be easily imported from European lessons. Additionally, the Japanese economy is also in a deep crisis. It will be difficult for JPY to be a leading currency of the block for the moment. If the currency block is put into practice in Asia, the promotion of the free- trade area will definitely be a primary stage of monetary integration.

NOTE

1. Empirically, Karras (1997) and Persson (1997) provided simple techniques to perform convergence tests.

REFERENCES

Bayoumi, T. (1994). A formal model of optimum currency areas. *IMF Staff Papers*, *41*, 537–558.

Bayoumi, T., & Eichengreen, B. (1997). Ever closer to heaven? An optimum-currency-area index for European countries. *European Economic Review*, *41*, 761–770.

Canzoneri, M. B., & Roger, C. A. (1990). Is the European Community an optimal currency area? Optimal taxation vesus the costs of multiple currencies. *American Economic Review*, *80* (June).

Casella A., & Feinstein (1989). Participating in a Currency Union, NBER Working paper.

Eichengreen, B. (1997). One money for Europe? Lessons form the U.S. currency union. In: B. Eichengreen (Ed.), *European Monetary Unification: Theory, Practice, and Analysis* (pp. 13–50). Cambridge, MA: The MIT Press.

Eichengreen, B. (1991). Is Europe an optimum currency area? NBER Working papers, No. 3579.

Frenkel, J. A., & Rose, A. K. (1996). Economic structure and the decision to adopt a common currency, Stockholm University, Working Papers.

Frenkel, J. A., & Rose, A. K. (1997). Is EMU more justifiable ex post or ex ante. *European Economic Review*, *41*, 753–760.

Ishiyama, Y. (1975). The theory of optimum currency areas; a survey. *IMF Staff Papers*, *22*.

Karras, G. (1997). Economic integration and convergence: Lessons from Asia, Europe and Latin America. *Journal of Economic Integration*, *12*(4), 419–432.

Kenen, P. (1969). The theory of optimum currency areas: An eclectic view. In: R. A. Mundell & A. K. Swoboda (Eds), *Monetary Problems of the International Economy* (pp. 41–60). Chicago, IL: University of Chicago Press.

McKinnon, R. (1963). Optimum currency areas. *American Economic Review*, *53*, 717–725.

Mundell, R. A. (1961). A theory of optimum currency areas. *American Economic Review, 51*, 657–665.

Persson, J. (1997). Convergence across the Swedish counties, 1911–1993. *European Economic Review, 41*, 1835–1852.

Ricci, L. A. (1997). A model of an optimum currency area, IMF Working Paper.

Sabhasri, C. (1994). Government Management and Monetary Policy Cooperation, Ph.D.thesis, University of Wisconsin-Madison.

Tavlas, G. S. (1993a). The "New" theory of optimum currency areas. *The World Economy, 16*(6).

Tavlas, G. S. (1993b). The theory of optimum currency areas revisited. *Finance and Development*, June.

20. THE ASIAN FINANCIAL CRISIS AND THE BALANCE SHEETS OF BANKING AND CORPORATE SECTORS: AN EMPIRICAL EVALUATION

Sung Yeung Kwack[1]

1. INTRODUCTION

Until the financial crisis in Asia occurred in 1997, economic performance, high saving and investment rates in Asia were envied. Many economic professionals at the time wondered how they achieved such phenomenal growth. This inquiry was dramatically illustrated in the book titled *Asian Miracle*, published by the World Bank in 1993.

The Thailand baht was devalued on July 2, 1997, followed by the Philippine peso, the Malaysian ringgit, and the Indonesian ruphia. The Korean won was floated on December 16, 1997. The value of the currencies of Indonesia, Korea, Malaysia, Philippines, and Thailand fell 52–101% against the U.S. dollar between the end of 1996 and the end of 1997. The Singapore dollar and the Taiwan NT$ depreciated by 20 and 18%, respectively. Many economic analysts continue to question the reasons why the series of crises in Asia have happened and how it can be fixed.

Many studies in recent years have described the factors causing a financial crisis in Asia.[2] They indicated the main causes of a crisis are appreciation in

Restructuring Asian Economics for the New Millennium, Volume 9A, pages 375–393.

ISBN: 0-7623-0709-9

real exchange rates, continuing large current account deficits, high short-term external debts, excessive loans, high leverages of corporations and weaknesses in the balance sheet of financial institutions.

So far, no systematic empirical testing has been performed to identify the causes of the Asian crisis.[3] The purpose of this paper is to set a hypothesis of what causes the Asian crisis and test it. The hypothesis emphasizes the role of non-performing bank loans, which is an indicator for the weakness in the balance sheets of banks.[4] In addition, this paper introduces a hypothesis regarding the determination of non-performing bank loans and tests the hypothesis. Empirical tests are carried out with data on the seven countries in Asia: Indonesia, Korea, Malaysia, Philippines, Singapore, Taiwan, and Thailand.

The rest of the paper is organized as follows. Section 2 sketches financial market conditions in the seven countries of interest. Section 3 specifies the hypotheses to be tested. Section 4 describes the data set and discusses the results of regressions. Section 5 draws the policy implications of empirical results and addresses restructuring of banks and corporations.

2. FINANCIAL MARKET CONDITIONS

Table 1 presents the major indicators for the external and monetary sectors of each economy. Since the figures for 1997 embodied the effects of the financial crisis, our discussions largely focus on the period of three years before the occurrence of the crisis, from 1994 to 1996. During the three years, the average size of the current account deficits of Indonesia, Korea, Malaysia, Philippines and Thailand were in the range from 1% of GDP to 8%. The deficit sizes rose to 3% of GDP or higher in 1996. Hence, the current account deficits of the five countries seem excessive. Taiwan maintained its current account surplus of about 2 to 3% of GDP. The current account surplus of Singapore was about 17% of GDP.

In the 1990s, most of the Asian countries have actively launched financial liberalization, including capital market opening. Recently, there were substantial amounts of capital inflow. The financial account balance per year over the period of three years amounted to \$8 billion, \$17 billion, \$6 billion, \$7 billion, and \$18 billion for Indonesia, Korea, Malaysia, Philippines, and Thailand, respectively (Singapore and Taiwan experienced net capital outflows amounting to \$5 billion and \$6 billion, respectively). These large capital inflows raised their external debt positions.

The external debts of Indonesia, Philippines, and Thailand, reported in *World Financial Markets* by Morgan Guaranty Trust Company, were above 50% of

Table 1. Major Indicator of the External and Monetary Sectors.

	Indonesia	Korea	Malaysia	Phillipines	Singapore	Taiwan	Thailand
Current Account (% of GDP)							
1994	-1.58	-1.01	-6.23	-4.60	17.26	2.70	-5.65
1995	-3.18	-1.81	-8.43	-2.67	16.86	2.10	-8.09
1996	-3.37	-4.75	-5.24	-4.77	15.87	4.05	-8.10
1997	-2.24	-1.85	-6.51	-5.23	15.37	2.74	-1.89
Financial Account of the Balance of Payments($million)							
1994	3839	10610	1288	5120	-11429	-1397	12167
1995	10259	17221	7422	5309	-654	-8190	21909
1996	10847	23924	9200	11277	-2438	-8802	19486
1997	1422	-9195	6800	6396	-4536	-8154	-15441
Total External Debt as % of GDP(J.P. Morgan data)							
1994	58.68	18.25	47.16	66.32	55.61	13.48	48.97
1995	50.26	18.91	42.03	62.32	48.50	15.26	52.24
1996	53.38	29.32	38.92	68.30	45.50	15.72	54.23
1997	61.22	35.09	39.76	71.01	43.81	16.30	66.53
Short-term External Debt as % of Total Debt (Morgan)							
1994	20.62	40.43	34.21	22.82	87.31	77.54	42.80
1995	24.51	45.65	34.60	25.32	88.14	70.60	47.31
1996	23.56	53.20	39.38	25.44	89.81	67.06	38.52
1997	20.24	38.70	36.32	25.51	92.18	63.85	31.15
Total Claims of BIS Reporting Banks as % of GDP							
1994	23.53	16.01	19.97	10.21		9.16	38.03
1995	24.21	18.24	21.49	10.89		8.49	55.04
1996	25.44	22.53	26.13	16.30		8.37	54.71
1997	29.25	23.48	29.57	20.20		7.91	51.76
Short-term Claims of BIS Banks as % of Total Claims							
1994	60.90	70.90	48.80	45.88		90.40	70.60
1995	61.90	70.00	47.20	49.40		87.20	69.40
1996	61.70	67.50	50.30	57.89		86.40	65.20
1997	60.48	62.74	52.75	60.41		81.47	65.86
Claims of BIS Banks(a multiple of international reserves)							
1994	3.15	2.37	0.57	0.92		0.22	1.80
1995	3.31	2.55	0.78	1.04		0.23	2.50
1996	3.00	3.20	0.96	1.15		0.24	2.57
1997	3.61	5.09	1.39	1.90		0.90	14.48
Depreciation of Nominal Exchange Rate(per U.S. $,average)							
1994	3.53	0.10	1.95	-2.59	-5.47	0.23	-0.67
1995	4.07	-4.00	-4.57	-2.66	-7.21	0.15	-0.93
1996	4.15	4.30	0.46	1.93	-0.52	3.66	1.71
1997	24.21	18.25	11.29	12.44	5.30	4.53	23.77
Depreciation of Real Effective Exchange Rate (average)							
1994	2.25	-0.80	4.33	-6.10	-5.22	1.04	0.83
1995	2.24	-2.53	-2.35	-4.44	-1.61	2.68	0.12
1996	-9.85	-4.45	-5.49	-8.70	-3.22	-2.13	-8.54
1997	9.50	9.19	3.38	3.01	-2.54	-0.04	11.52
Banks Claims on the Private Sector(a multiple of GDP)							
1994	0.52	0.57	0.75	0.29	0.84	1.47	0.92
1995	0.53	0.57	0.85	0.38	0.91	1.49	0.98
1996	0.55	0.62	0.93	0.49	0.97	1.44	1.02
1997	0.61	0.70	1.08	0.57	1.00	1.45	1.16

GDP. The external debt of Korea and Malaysia were 22% and 42%, respectively. Taiwan's debt was lower than 15% of GDP. Hence, Indonesia, Korea, and ASEAN countries seem to have large external debt. The shares of short-term external debts to total debt by Korea, Malaysia, Taiwan and Thailand were above 40% in 1996. The shares of short-term claims to total claims by the BIS reporting banks were about 50% or higher. These figures indicate that the short-term debts were excessive for Korea, Malaysia, and Thailand. The claims of the BIS reporting banks as multiples of international reserves were above two for Indonesia, Korea, and Thailand and were about one for Malaysia and Philippines. The short-term obligations of these countries in foreign currencies seem to exceed the amount of foreign currency they can dispose. A high degree of international illiquidity may render a self-fulfilling speculative run possible, as discussed in detail by Chang and Velasco (1998).

During the period from 1994 to 1996, the cumulative depreciation of the nominal exchange rates were 14, 0.4, and 4% for Indonesia, Korea, and Taiwan, respectively. The cumulative appreciations were 2.1, 3.3, and 13.2 for Malaysia, Philippines, and Singapore, respectively. Except for Indonesia and Singapore, nominal exchange rate changes were moderate. The real effective exchange rate indices used here are nominal effective exchange rate indices adjusted for relative consumer price variations.[5] For the three years, the real effective exchange rates appreciated. For 1996, the rates of appreciation were substantial: from 2% for Taiwan to 10% for Indonesia.

The claims of deposit money banks on the private sector as multiples of GDP ranged from 0.5 to 1.5. These figures are comparable to the average figures for Japan, (1.17) and for the United States, (0.64). In recent years, deposit money bank's lending to the private sector has shown a rising trend.

3. SPECIFICATION OF HYPOTHESIS

There are several hypotheses explaining the causes of a crisis. (See Krugman, 1979; Flood & Garber, 1984; Obstfeld, 1986; Krugman, 1998.)[6] We regard a financial crisis as a monetary problem. The hypothesis we introduce below reflects that the crisis in Asia is related, in part, to some structural weaknesses in the financial sector.

3.1. A Hypothesis on Financial Crisis

A financial crisis exerts extreme pressure in the financial markets, including the foreign exchange market. The crisis is signaled by either a sharp exchange rate depreciation and/or a sharp fall in international reserves.[7] The financial crisis variable constructed in this paper is a weighted average of the rate of increase in the holding of international reserves and the rate of appreciation in the nominal exchange rate. It is denoted by CRS. By construction, a higher value means a lesser degree of the financial crisis. When the financial crisis is very severe, the CRS variable has a negative value and is large in the absolute value.

We propose a hypothesis of what factors cause a financial crisis in emerging Asian countries. The first factor is the three-month London Interbank Offer rate on U.S. dollar deposits, denoted by LIBOR. This interest rate represents the cost of borrowing and captures the opportunity cost of lending to Asian countries. The Japanese interest rate was considered as an opportunity cost variable and was not introduced here explicitly because of a high elasticity of substitution between dollar- and yen-denominated assets. Further, LIBOR rate represents output conditions in industrial economies abroad: a rise in the LIBOR rate signals a slowdown in output growth.

The second factor is non-performing bank loans as a percentage of total loans of commercial banks, denoted by NPL, representing the weakness in the balance sheet conditions, poor quality of loan assets and fragility of the financial institutions. The NPL variable is more representative than loan quality. A large non-performing loan ratio indicates high costs in banks in several ways.[8] A high NPL results from poor management in monitoring both the cost conditions of banks and the business conditions of loan customers. It is also caused by events not controllable by the banks such as connected lending and politically directed lending in "crony capitalism". Hence, the NPL variable can be regarded as representing unfavorable micro-economic and socio-political environments under which the financial institutions operate.

In addition to the LIBOR and NPL variables, we introduce a set of variables that have been found important in recent empirical studies on what factors cause a financial crisis, using panel data covering developing countries or both developed and developing countries.[9] The variables include: current account balance as fraction of GDP, CA; the share of short-term external debts to total debts, DTS; and the claims of deposit money banks on the private sector as a multiple of GDP, CL. The hypothesis that deals with the causes of a financial crisis in this paper is summarized by:

$$\text{CRS} = f(\text{NPL, LIBOR, DTS, CL, CA}),\ f_1 < 0,\ f_2 < 0,\ f_3 < 0,\ f_4 < 0,\ f_5 > 0. \quad (1)$$

3.2. A Hypothesis on Non-Performing Bank Loans

Non-performing bank loans represent the balance sheet conditions of commercial banks. Basically, the non-performing bank loan ratios are affected by banks who determine loan supply and firms who determine loan demand. Hence, the willingness and ability of borrowers to meet their payment obligations would have an effect on non-performing bank loans.

A high leverage (debt-to-equity ratio) of firms, denoted by LEV, causes non-performing bank loans to rise, because a high leverage is accompanied by large interest payments. As noted earlier, the CL variable increases non-performing loans. A rise in the profit margin per unit of sales, and thus a rise in corporate cash flows, make payments of debt obligations easier. Hence, the operational margin is defined as the revenue net of cost per unit of sales, OM, is expected to negatively affect non-performing loans. A rise in the foreign interest rate is likely to raise non-performing loans, whereas a large current account surplus is likely to decrease them. On the other hand, banks as lenders extend riskier loans when the risk is borne by a non-bank party, including the government. From this moral hazard view, we expect that low equity-to-assets ratios provide banks an incentive to supply risky loans. The equity capital-to-total assets ratio for banks, denoted by CAP, is expected to have a negative effect on the non-performing bank loans. The hypothesis is specified as follows:

$$\text{NPL} = \text{h}(\text{LEV, LIBOR, DTS, CL, CA, OM, CAP}),\ h_1 > 0,\ h_2 > 0,\ h_3 > 0,\ h_4 > 0,\ h_5 < 0,\ h_6 < 0,\ h_7 < 0. \qquad (2)$$

4. EMPIRICAL TESTS

The Asian crisis started in 1997. Before then, there was no crisis of such a magnitude currently experienced in Asia. To capture more accurately the characteristics of the crisis, we treated the cross country data for 1996 and 1997 as observations for the dependent variable. As a result, the panel data observations are 14. The small panel sample size limited the number of independent variables to be introduced in regression equations, because a large number of independent variables would lower the degrees of freedom. As a consequence, we did not include other variables that would cause the Asian crisis, such as money supply relative to international reserves. Equations (1) and (2) were estimated using the ordinary least squares method on annual observations of the seven Asian countries from 1995 to 1997. We describe the data sources and the results of equations estimates with cross-country time series observations.

4.1. Data Set

Except for the shares of short-term external debts, operational profit margin, corporate leverage ratios, non-performing bank loan ratio, and capital-to-total assets ratio, all data are computed from IMF's *International Financial Statistics*. The financial crisis variable used in the empirical analysis is a weighted average of the rate of nominal exchange rate appreciation against the

U.S. dollar at the end of year and the rate of change in total international reserves, including gold, at the end of the year. The weight assigned to exchange rate changes in the crisis variable is 0.75.[10]

Data on the shares of short-term debt to total debt are computed by making use of data reported in Morgan Guaranty Trust Company's *World Financial Markets*.[11] Corporate operational profit margin and leverages (total debt over equity) ratios of firms are taken from Tables 3 and 6 in "East Asian Corporate: Growth, Financing and Risks over the Last Decade" by Claessen Stijn, Simeon Djankov and Larry Lang (1998).

Data on equity capital-to-total assets ratios for commercial banks are constructed on the basis of information available from the following sources: Bank Supervisory Office (1998), *Financial Statistics Monthly* of Central Bank of China (Taiwan), Corsetti, Giancarlo, P. Pesenti, and Nouriel Roubini (1998a), Statistical Appendix of IMF's Country Report, Goldman Sachs Asia Economic Research Group (1998a), Long-Term Credit Bank of Japan (1998), Monetary Authority of Singapore's *Monthly Statistical Bulletin*, Morgan Guaranty Trust Company (1998) and World Bank (1998).[12] Appendix I provides the figures from the sources used to construct the data on equity-to-total assets ratios for commercial banks and the figures selected to use in regressions.

Data on non-performing bank loans were gathered from BIS (1996, 1997), Bank Supervisory Office (1998), Bank Negara Malaysia (1997), Corsetti et al. (1998a), Goldman Sachs Asia Economic Research Group (1998a, 1998b), and World Bank (1998, Figure 3.3).

Figures on Taiwan's non-performing bank loans from the Central Bank were given to me by Tzong-Shian Yu, and the source of the figures is the Central Bank of China (Taiwan). Appendix II provides the figures from various sources on non-performing bank loans and the figures used in the empirical analysis.

4.2. *Regression Equations*

Equations (1) and (2) were specified in a linear form. We estimated equations using the variables with different time lags, using annual and cross-country data for the seven countries. To capture the effects of a time lag, we used average values over a period of time. The real effective exchange rate variable was tried without success. The data used in the regressions are given in Appendix III. Below are the acceptable specifications for equations (1) and (2), based on earlier experiments:

$$\begin{aligned} CRS = {} & \alpha_1 + \alpha 2_1 NPL + \alpha_3 LIBOR + \alpha_4 0.5*(DTS + DTS_{-1}) \\ & + \alpha_5^* 0.5*(\Delta CL/CL_{-1} + \Delta CL_{-1}/CL_{-2}) + \alpha_6 0.5*(CA + CA_{-1}) \end{aligned} \tag{3}$$

Appendix I: Data on Equity-to-Total Assets for Commercial Banks in Percent

The data on equity-to-total assets for banks were constructed from the following sources: Bank Supervisory Office (1998), Financial Statistics Monthly of Central Bank of China (Taiwan), Statistical Appendix of IMF Country Report, Goldman Sachs Asia Economic Research Group (1998a), Long-Term Credit Bank (1998), Monetary Authority of Singapore's Monthly Statistical Bulletin, Morgan Guaranty Trust Comapany (1998) and World Bank (1998). The figures for 1997 were taken primarily from Goldman Sachs Asian Economic Research Group (1998a) to keep the consistency with the figures on non-performing bank loans for 1997. The figures for other years selected were adjusted to be comparable in their movements.

	Indonesia	Korea	Malaysia	Philippines	Singapore	Taiwan	Thailand
LTC							
1994.00	8.90		6.70	12.50			7.50
1995.00	9.80		6.60	13.50			7.80
1996.00	9.60		6.40				9.00
1997.00	9.40						
BSO							
1994.00		10.62					
1995.00		9.33					
1996.00		9.14					
1997.00		7.04					
WB							
1994	10		6	14			10
1995	10		7	14			10
1996	9		8	13			10
1997							
BS							
1994	21	5	10		8	5	8
1995.00	7.00	4.49	10.24		8.42	6.08	7.83
1996.00	7.45	4.26	10.44		8.30	6.46	8.97
1997.00	6.40	3.20	10.79		8.09	7.01	7.33
GS							
1997.00	8.70	3.40	7.70	13.70	10.90		6.20
JP							
1997.00	8~10	6~10	8~14	15~18	18~22		
SELECTED							
	Indonesia	Korea	Malaysia	Philippines	Singapore	Taiwan	Thailand
1994.00	8.90	5.33	6.70	12.50	7.77	5.30	7.50
1995.00	9.80	4.49	6.60	13.50	8.42	6.08	7.80
1996.00	9.60	4.26	6.40	13.10	8.30	6.46	9.00
1997.00	8.70	3.20	7.70	13.70	10.90	7.01	6.20

LTC refers to Long-Term Credit Bank (1998). BSO refers to Bank Supervisory Office (1998).
BSO refers to Bank Supervisory Office (1998).
WB referes to World bank (1998, Figures 3 and 4). The return-on-assets / the return-on-equity.
BS indicates that the figures are computed by dividing net worth by total assets of commercial banks (deposit money banks in case of Thailand). The sources are Statistical Appendix of IMF's Country Report and Monetary Authority of Singapole's Monthly Statistical Bulletin.
GS refers to Goldman Sachs Asia Economic Research Group (1998a).

Appendix II: Data on Non-Performing Bank Loans as a Percent of Total Loans

We gathered the data on non-performing bank loans from BIS (1996, 1997), Bank Supervisory Office (1998), Bank Negara Malaysia (1997), Corsetti et al (1998a), Goldman Sachs Asia Economic Research Group (1998a, 1998b), and World Bank (1998, Figure 3.3). The figures on Taiwan's non-performing bank loans were given to me by Tzong-Shian Yu, and the source is the Central Bnak of China(Taiwan). I extend my thanks to him. After carefully evaluating the available data for 1997, we selected Goldman Sachs Asia Economic Research Group (1998) as the primary source for the figures of year 1997. The main reason is that the Group provided both non-performing bank loan data and equity-to-toal assets for banks. The figures for other years were selected to make comparable in the movement to the available data.

	Indonesia	Korea	Malaysia	Philippines	Singapore	Taiwan	Thailand
BSO							
1994		5.8					
1995		5.2					
1996		4.1					
1997		6					
BNM							
1994			9.8				
1995			4.9				
1996			3.8				
1997							
BIS							
1994	12	1	8.1			2	7.5
1995	10.4	0.9	5.5			3.1	7.7
1996	8.8	0.8	3.9			3.8	
1997							
CB							
1994						1.85	
1995						2.92	
1996						3.78	
1997						3.82	
WB							
1994		3.2		1.2			1.9
1995		3.5		1			2
1996		4.4		1.9			3.2
CPR							
1996	13	8	10	14	4	4	4
1997	11	16	7.5	5.5	2		15
GS1							
1997	9.2	17	9.1	5.3	3.8		18
GS2							
1997		14.7	12.6	9.8	0.9		44.3
1998		28.6	16.6	28.5	7.7		43.9
SELECTED							
	Indonesia	Korea	Malaysia	Philippines	Singapore	Taiwan	Thailand
1994	12	7.8	11	7.2	3	1.85	7.5
1995	10.4	7.2	8	7	3	2.92	7.7
1996	10.1	6.1	7	8	4	3.78	8.9
1997	9.2	17	9.1	5.3	3.8	3.82	18

BSO refers to Bank Supervisory Office (1998). BNM refers to Bank Negara Malaysia (1997).
BIS refers to BIS (1996, 1997), Annual Report
CB refers to the Central Bank of China (Taiwan). Tzong -Shian Yu provided the data.
WB refers to World Bank (1998, Figure 3.3). The author assigned numbers, based on Figure 3.3.
CPR refers to Corsetti, Giancarlo, P. Pesenti, and N. Roubini (1998a), Tables 21 and 22.
GS1 refers to Goldman Sachs Asia Economic Research Group (1998a).
GS2 refers to Goldman Sachs Asia Economic Research Group (1998b).
SELECTED refers to the figures used in regressions.
CB refers to the Central Bank of China (Taiwan). Tzong -Shian Yu provided the data.
WB refers to World Bank (1998, Figure 3.3). The author assigned numbers, based on Figure 3.3.
CPR refers to Corsetti, Giancarlo, P. Pesenti, and N. Roubini (1998a), Tables 21 and 22.
GS1 refers to Goldman Sachs Asia Economic Research Group (1998a).
GS2 refers to Goldman Sachs Asia Economic Research Group (1998b).

Appendix III: Data Set							
	Indonesia	Korea	Malaysia	Phillipines	Singapore	Taiwan	Thailand
CA, Current Account as a fraction of GDP							
1997	-0.0224	-0.0185	-0.0651	-0.0523	0.1537	0.0274	-0.0169
1996	-0.0337	-0.0475	-0.0524	-0.0477	0.1587	0.0405	-0.0810
1995	-0.0318	-0.0181	-0.0843	-0.0267	0.1686	0.0210	-0.0809
1994	-0.0158	-0.0101	-0.0623	-0.0460	0.1726	0.0270	-0.0565
CAP, Equity-to-Total Asset Ratio for Banks (%)							
1997	8.70	3.20	7.70	13.70	10.90	7.01	6.20
1996	9.60	4.26	6.40	13.10	8.30	6.46	9.00
1995	9.80	4.49	6.60	13.50	8.42	6.08	7.80
1994	8.90	5.33	6.70	12.50	7.77	5.30	7.50
CL, Banks Claims on Private Sector (a multiple of GDP)							
1997	0.61	0.70	1.08	0.57	1.00	1.45	0.92
1996	0.55	0.62	0.93	0.49	0.97	1.44	0.98
1995	0.53	0.57	0.85	0.38	0.91	1.49	1.02
1994	0.52	0.57	0.75	0.29	0.84	1.47	0.92
CRS, Financial Crisis							
1997	-0.39	-0.48	-0.32	-0.32	-0.14	-0.13	-0.42
1996	0.05	-0.05	0.04	0.13	0.04	-0.01	0.00
1995	-0.01	0.08	-0.01	-0.03	0.07	-0.03	0.05
1994	-0.01	0.09	0.02	0.15	0.13	0.04	0.06
DTS, Short-term External Debt as % of Total Debt (Morgan)							
1997	20.24	38.70	36.32	25.51	92.18	63.85	42.80
1996	23.56	53.20	39.38	25.44	89.81	67.06	47.31
1995	24.51	45.65	34.60	25.32	88.14	70.60	47.31
1994	20.62	40.43	34.21	22.82	87.31	77.54	42.80
LEV, Total Debt-to-Equity Ratio for Firms (%)							
1996	187.80	354.50	117.60	128.50	104.90	80.20	236.10
1995	211.50	377.60	110.30	115.00	103.70	79.60	222.40
1994	166.10	353.00	99.10	114.80	86.20	89.40	212.60
LIBOR, Three Month London Interbank Offer Rate ($, %)							
1997	5.76	5.76	5.76	5.76	5.76	5.76	5.76
1996	5.51	5.51	5.51	5.51	5.51	5.51	5.51
1995	6.04	6.04	6.04	6.04	6.04	6.04	6.04
NPL, Non-Performing Bank Loans as percent of Total Loans							
1997	9.20	17.00	9.10	5.30	3.80	3.82	18.00
1996	10.10	6.10	7.00	8.00	4.00	3.78	8.90
1995	10.40	7.20	8.00	7.00	3.00	2.92	7.70
1994	12.00	7.80	11.00	7.20	3.00	1.85	7.50
OM, Operational Margins of Firms (percent of Sales)							
1996	30.60	22.10	25.50	33.30	13.10	21.90	22.70
1995	31.20	21.40	19.50	30.80	13.60	22.30	24.70
1994	32.80	19.60	18.40	27.50	14.10	22.60	24.90
RER, Real Effective Exchange Rate (In Home Currency)							
1997	1.14	1.15	1.02	0.81	0.87	1.07	1.11
1996	1.04	1.05	0.98	0.79	0.89	1.07	1.00
1995	1.16	1.10	1.04	0.86	0.92	1.10	1.09
1994	1.13	1.13	1.07	0.90	0.93	1.07	1.09

$$
\begin{aligned}
NPL = {} & \beta_1 + \beta_2 LEV_{-1} + \beta_3 LIBOR + \beta_4 0.5*(DTS + DTS_{-1}) \\
& + \beta_5^* 0.5*(\Delta CL/CL_{-1} + \Delta CL_{-1}/CL_{-2}) + \beta_6 0.5*(CA + CA_{-1}) \\
& + \beta_7 OM_{-1} + \beta_8 CAP_{-1}.
\end{aligned} \tag{4}
$$

Table 2 contains the estimated coefficients from equation (3). All of the major results are robust, using either average figures during a period of time or figures

Table 2. Estimated Equations.

$$CRS = \alpha_1 + \alpha_2 NPL + \alpha_3 LIBOR + \alpha_4 0.5*(DTS + DTS_{-1}) + \alpha_5^* 0.5*(\Delta CL/CL_{-1} + \Delta CL_{-1}/CL_{-2}) + \alpha_6 0.5*(CA + CA_{-1})$$

Eq.	(5)	(6)	(7)	(8)
α_1	6.961	6.782	6.994	6.771
	(6.06)	(6.43)	(6.41)	(6.96)
α_2	–0.014	–0.014	–0.014	–0.015
	(2.02)	(2.27)	(2.17)	(3.10)
α_3	–1.238	–1.216	–1.237	–1.204
	(6.09)	(6.37)	(6.39)	(6.90)
α_4	–0.001	0.001	–0.001	
	(0.22)	(0.43)	(0.49)	
α_5	0.142	0.189		
	(0.41)	(0.59)		
α_6	0.384		0.459	
	(0.54)		(0.89)	
R^2	0.81	0.83	0.83	0.85
SEE	0.08	0.08	0.08	0.08
DW	1.78	1.84	1.76	1.79

R^2 is the coefficient of determination adjusted for the degree of freedom.
SEE and DW are standard errors of estimates and Durbin Watson statistics.
Figures in parentheses are t-statistic.

in the current period. The coefficient estimates of the LIBOR interest rate and non-performing loan rates (NPL) are significant. Other parameter estimates are not significant. Domestic credit growth relative to output growth is found to have no significant impact on the Asian crisis. The rate of change in the CL variable is viewed by Sachs, Tornell and Velasco (1996) to be an indicator of financial fragility. When bank's lending grows at a rapid pace, a large fraction is likely to become non-performing. In this interpretation, the rate of change in the CL variable serves as a proxy for the NPL variable. Hence, it is expected that the NPL variable is a significant variable, whereas the CL variable is an insignificant variable.

Dropping insignificant variables, we obtain equation (8). In equation (8), the coefficient of the LIBOR interest rate is –1.2 and highly significant at the 1% level. This result is consistent with the findings of Frankel and Rose (1996) and Eichengreen and Ross (1998) that interest rates variables, like U.S. interest rates, are statistically important variables. The impact of a change in the

LIBOR interest rate seems to be very substantial. A one percentage point rise in the LIBOR rate produces a cumulative reduction of 1.2 in the crisis variables. The reduction is about 8 times the mean value. A profound influence of a change in the LIBOR rate in causing the crisis in Asia is not surprising on the following three grounds. First, Calvo et al. (1993) shows that international interest rates determine capital inflows in Latin American countries, followed by the Latin American crisis. Secondly, a small increase in international interest rates would generate larger swings in capital flows and lead to an exchange rate crisis, when thick market externalities exist in the process of intermediation, as analytically discussed in details by Goldfajn and Valdes (1997). Thirdly, the Asian crisis was deepening during the summer of 1998. But, it started to be weaker from late November 1998. During this short period, a significant event in the global financial market was the reduction in the discount rate from 5% to 4.75% on October 17, 1998 and the reduction from 4.75 to 4.5% on November 17, 1998 by the Federal Reserve Board in the United States. This is indirect evidence of the powerful effect of interest rates in international financial centers on the Asian crisis. The reduction in international interest rates reduces the cost of servicing external debts and leads to capital inflows into emerging markets.

The coefficient of the non-performing bank loan ratio, NPL, is –0.015 and is significant at the 1% level. According to the coefficient estimate, without considering the standard errors of estimate, a one percentage point rise in the NPL variable predicts a cumulative reduction in the crisis variable by 0.015, which is about 9% of the mean value during 1996 and 1997. This result shows that the balance sheet conditions of banks are important factors for determining the Asian crisis. This result is consistent with the findings of the study by Corsetti, Pesenti, and Roubini (1998), showing the importance of non-performing bank loans. In contrast to our use, however, the study uses in the regression equations the non-performing bank loan variable modified as follows: the value of non-performing loans as a share of total assets in 1996 is assigned to the original value if the sign of the lending boom in the 1990s is positive, and in the other situation, it is assigned to be zero.

Table 3 contains the coefficient estimates for equation (4) which explains the behavior of non-performing bank loan ratios. As expected, the results were highly robust. Nearly all of the variables appear to be insignificant. In equation (12), the corporate leverage ratio with a lag of one year turns out to be a significant variable at the 5% level. According to the estimate, a 10 percentage point reduction in the leverage ratio decreases the non-performing loan ratio by 0.28, about 3.5% of the mean value during 1996 and 1997.

Table 3. Estimated Equations.

$$NPL = \beta_1 + \beta_2 LEV_{-1} + \beta_3 LIBOR + \beta_4 0.5*(DTS + DTS_{-1}) + \beta_5^* 0.5*(\Delta CL/CL_{-1} + \Delta CL_{-1}/CL_{-2}) + \beta_6 0.5*(CA + CA_{-1}) + \beta_7 OM_{-1} + \beta_8 CAP_{-1}$$

Eq.	(9)	(10)	(11)	(12)
β_1	–39.54	–57.26	–56.84	3.194
	(0.73)	(1.31)	(1.29)	(1.50)
β_2	0.021	0.021	0.028	0.028
	(1.48)	(1.82)	(2.75)	(2.85)
β_3	11.21	11.36	10.55	
	(1.30)	(1.46)	(1.37)	
β_4	–0.174			
	(0.92)			
β_5	–4.562			
	(0.22)			
β_6	–6.820	–25.01		
	(0.15)	(1.28)		
β_7	–0.568	–0.102		
	(1.28)	(0.43)		
β_8	0.36			
	(0.38)			
R^2	0.27	0.37	0.36	0.32
SEE	3.87	3.60	3.64	3.76
DW	1.77	2.06	2.12	2.00

R^2 is the coefficient of determination adjusted for the degree of freedom.
SEE and DW are standard errors of estimates and Durbin Watson statistics.
Figures in parentheses are t-statistic.

Equations (8) and (12) imply that the financial crisis can be explained by the corporate leverage ratio and LIBOR interest rate. We obtained the following reduced form equation:

$$CRS = 7.70 - 0.0007LEV_{-1} - 1.370LIBOR \quad R^2 = 0.85,\ SEE = 0.08,\ DW = 1.76$$
$$(8.25)\ (3.14) \qquad (8.28) \qquad (13)$$

The coefficients of both the corporate leverage ratio and the LIBOR interest rate are consistent with their implied reduced-form coefficients derived from equations (8) and (12). The result of this reduced form regression equation (13) confirms again that both the corporate high leverage ratio and LIBOR interest rates are significant determinants of the Asian crisis and greatly affect the course of the Asian crisis.

5. POLICY IMPLICATIONS: BANK AND CORPORATE RESTRUCTURING

The LIBOR interest rate is, more or less, exogenous to emerging individual Asian countries. However, it is affected by money market conditions in industrial countries. This is why easier monetary policies in the United States and EU countries are expected to be very effective in overcoming the present Asian crisis.

Non-performing bank loans and debt-to-equity ratios are subject to policies of individual Asian countries. Non-performing loans of financial institutions and debt-to-equity ratios of corporations are high. To overcome the present crisis and to reduce the disturbances generated by capital flow variability, it is necessary to take measures that would reduce non-performing assets and debt-to-equity ratios.

A large number of corporations and financial institutions need to deal with non-income-producing loans and high debt-to-equity ratios. Individual cases vary greatly, to which different regulations and laws apply. Conceptually, non-performing loans can be reduced by three means. One way is through bank restructuring. The restructuring of banks refers to the write-off of bad debts, the shutdown of non-viable banks, and the enhancement in banks's efficiency. Banks can reduce their non-performing loans by exchanging non-performing loans for corporate equity. The second way is through corporate restructuring with the intention of lowering their debt-to-equity ratios. This can be done by selling some of their assets including real estates and by raising more equity. The third way is by having governments take a fraction of non-performing loans in exchange for government securities.

The takeover of non-performing bank loans commits the government to making substantial debt-servicing and interest payments. The fiscal burdens are expected to be high. According to IMF (1998a, p. 152), estimated interest costs of bank restructuring are between 0.25 and 3.5% of GDP for Indonesia, Korea, Thailand and Malaysia. Stock of debt are estimated to be between 4 and 32% of GDP.[13] This brings us to the issue of who bears the cost. In bankruptcy cases, shareholders of the banks bear the loss. In restructuring of banks, however, it seems appropriate for the taxpayer to bear the cost on the ground that it is desirable to maintain the safety of deposits against financial panic.

In recent year, most of the Asian countries have actively opened their financial markets and implemented the policies of financial liberalization. The opening of domestic financial markets and a high degree of capital mobility make the existing regulatory financial system less effective. They also make financial institutions and corporations vulnerable to speculative attacks. The broad policy implication emerging from our results is that improving transparency and supervisory oversight and an improvement in governance are essential to avoid speculative attacks on currency and bank run.[14]

NOTES

1. Professor of Economics, Howard University, Washington, DC 20059. skwack@fac.howard.edu.

2. Among many articles and books covering the Asian crisis, see Chang and Velasco (1998), Corsetti et al. (1998a, 1998b), International Monetary Fund (1998), Goldstein (1998), Krugman (1998), Kwack (1998b), Letiche (1998), Moreno et al. (1998), Radelet and Sachs (1998), and World Bank (1998).

3. Judging from the fact that there were no excessive credit growth and government budget deficits, the sources of Asian crisis may be distinguished from the sources of crisis that occurred in other regions, such as the crisis in Latin America. Chang and Velasco (1998) and Dooley (1998) argue that there is no fundamental difference in the causes among the crises occurring in different emerging markets.

4. This point is made by Corsetti et al. (1998a), Moreno et al. (1998) and Sachs et al. (1998). Mishkin (1992, 1996) emphasizes the financial side of view.

5. We constructed nominal and real effective exchange rate indices for each of the seven Asian countries. The nominal effective exchange rate of a country is measured as the exchange rates against the currencies of the other six Asian countries, Japan, and the United States. The weights are the shares of individual trading countries in the export and imports for 1990 and the shares are computed from the figures from IMF's *Direction of Trade*. Real effective exchange rate is nominal effective exchange rate adjusted for relative consumer prices.

6. The first hypothesis on a balance of payments crisis is that of Krugman (1979). If a country with a pegged exchange rate expands domestic credit to finance a budget deficit and investors prefer to hold a foreign currency, international reserves fall. When speculators assume that the country will no longer be able to defend the exchange rate, a speculative attack on the currency will occur. This hypothesis emphasizes macroeconomic policies inconsistent in the long run with the fixed exchange rate and is called the first-generation models, which are surveyed by Agenor, Bhandari and Flood (1992). The second hypothesis is given by Flood and Garber (1984) and Obstfeld (1986). This hypothesis argued that a crisis may develop without a significant change in economic fundamentals. There are two possible equilibria, which depend upon the expectations of market participants about monetary authority's actions in the event of a speculative attack. The state of the economy and the stability of the banking system are the factors that would provide conditions for a self-fulfilling speculative attack. In "What Happened to Asia?," Krugman (1998) noted that Asian banks are managed with implicit government guarantees. The government guarantees can lead to an over-borrowing and an excessively high price of assets including land. As the degree of government guarantees decreases, the price of assets falls, which in turn leads to loan

defaults. This generates self-fulfilling pessimism and hence a financial crisis. For literature review, see Agenor, Bhandari, and Flood (1992), and Flood and Marion (1998).

7. See Kaminsky et al. (1998, p.7) which describes how a "crisis" is defined in the empirical literature. Domestic interest rates are not included in the crisis definition, primarily due to the difficulty in assigning appropriate weights and their sticky responses to market changes, Eichengreen, Ross, and Wyplosz (1995) include domestic interest rates.

8. See Berger and DeYoung (1997) on problem loans and cost efficiency in commercial banks.

9. See Berg and Pattillo (1998), Corsetti et al. (1998a), Domirguc-Kunt et al. (1998), Eichengreen and Rose (1998), Esquivel and Felipe (1998), Frankel and Rose (1996), Kaminsky and Reinhart (1996), and Kaminsky et al. (1998).

10. An alternative crisis variables with the different weight 0.5 was constructed. It did not differ very much from the crisis variable constructed with the 0.75 weight. It did not yield substantially different empirical results. Corsetti et al (1998a) constructed their crisis index, as defined as a weighted average of the percentage rate of exchange rate depreciation relative to the U.S. dollar and the percentage rate of change in foreign reserves between the end of December 1996 and the end of December 1997 and assigned 0.75 to the weight to exchange rate changes.

11. Data on total external debts and short-term debts are also available in *World Debt Tables* by the World Bank and in *The Maturity, Sectoral and Nationality Distribution of International Bank Lending* by Bank for International Settlements. Short-term debt figures by the World Bank are more problematic. See Corsetti et al. (1998a, pp. 30–34) for the nature of the data. Also, Goldman Sachs Asia Economic Research Group (1998a) compares the external debt data by the BIS with the debt data by the World Bank in "Details of Asia's Debt" in *Asia Economics Analyst*, No. 98/17, June 9, 1998. Goldstein (1997) discussed the nature of debt data and capital adequacy of banks.

12. Data on the return-on-assets and the return-on-equity that were used in drawing Figs 3 and 4 of the World Bank (1998) were kindly provided to me by Giovanni Ferri. I wish to express my gratitude. Data on the non-performing bank loans listed under the World Bank (1998, Figure 3.3) are estimated from the Fig. 3.3.

13. For a discussion on the recapitalization, see Daniel (1997). For Korea's on-going restructuring of banks and corporations, see Kang (1998).

14. See Sundararajan (1996) for the discussions on prudential supervision and financial restructuring and Goldstein (1997) for international standard.

ACKNOWLEDGMENTS

The author is grateful to Michael Dooley, John Bennett, Marva Corley, and Jeong-Ho Suh for helpful comments. This paper is a substantially revised version of the paper presented at the conference on "A Macroeconomic Core of Open Economy for Progressive Industrialization and Development in Asian Economies in the New Millennium" sponsored by American Committee on Asian Economic Studies(ACAES) and Chulalongkorn University in Bangkok, Thailand, December 16–18, 1998.

REFERENCES

Agenor, P.-R.Bhandari, J., & Flood, R. (1992). Speculative Attacks and Models of Balance of Payments Crises. *IMF Staff Papers, 39*(2), 357–394.

Bank for International Settlements. *International Banking and Financial Market Development.*

Bank for International Settlements. *The Maturity, Sectoral and nationality distribution of International Bank Lending.*

Bank for International Settlements (1996). *66th Annual Report*, June.

Bank for International Settlements (1997). *67th Annual Report*, June.

Bank Negara Malaysia (1997). *Annual Report 1996.*

Bank Supervisory Office (1998). *Statistics of Bank Management* (in Korean).

Berg, A., & Pattillo, C. (1998). Are Currency Crises Predictable? A Test. *IMF Working Paper, WP/98/154*, November.

Berger, A. N., & DeYoung, R. (1997). Problem Loans and Cost Efficiency in Commercial Banks. *Journal of Banking & Finance, 21*(6), 849–870.

Calvo, G., Leiderman, L., & Reinhart, C. (1993). Capital Inflows and Real Exchange Rate Appreciation in Latin America. *IMF Staff Papers, 40*(1), 108–151.

Central Bank of China (Taiwan). *Financial Statistics Monthly.*

Chang, R., & Velasco, A. (1998). The Asian Liquidity Crisis. *Federal Reserve Bank of Atlanta Working Paper, 98*(11).

Claessens, S., Djankov, S., & King, L. (1998). East Asian Corporate: growth, Financing and Risks over the Last decade. *Draft*, October.

Corsetti, G., Pesenti, P., & Roubini, N. (1998a). What Caused the Asian Currency and Financial Crisis? Part I: A Macroeconomic Overview. *Draft*, September.

Corsetti, G., Pesenti, P., & Roubini, N. (1998b). Paper Tigers? A Model of the Asian Crisis. *Draft*, September.

Daniel, J. A. (1997). Fiscal Aspects of Bank Restructuring. *IMF Working Paper, WP/97/52*, April.

Demirguc-Kunt, A., & Detragiache, E. (1997). The Determinants of Banking Crises in Developing and Developed Countries. *IMF Staff Papers, 40*(1), 81–109.

Dooley, M. P. (1998). A model of Crises in Emerging Markets. *Draft*, September.

Eichengreen, B., & Rose, A. (1998). Staying Afloat When the Wind Shifts: External Factors and Emerging-Market Banking Crises. *NBER Working Paper, 6370* (January).

Eichengreen, B., Rose, A. K., & Wyplosz, C. (1995). Exchange Market Mayhem: the Antecedents and Aftermath of Speculaive Attacks. *Economic Policy, 21* (October), 249–312.

Esquivel, G., & Felipe, L. B. (1998). Explaining Currency Crises. *Draft*, June.

Flood, R., & Garber, R. (1984). Collapsing Exchange Rate Regimes: Some Linear Examples. *Journal of International Economics, 17*(1/2), 1–13.

Flood, R., & Marion, N. (1996). Speculative Attacks:Fundamentals and Self-Fulfilling Prophecies. *NBER Working Paper, 5789* (October).

Flood, R., & Marion, N. (1998). Perspectives on the Recent Currency Crisis Literature. *IMF Working Paper*, WP/98/130, September.

Frankel, J. A., & Rose, A. K. (1996). Currency Crashes in Emerging Markets: An Empirical Treatment. *Journal of International Economics, 41*(3/4), 351–366.

Goldfajan, I., & Valdes, R. O. (1997). Capital Flows and the Twin Crises: The Role of Liquidity. *IMF Working Paper*, WP/97/87, July.

Goldman Sachs Asia Economic Research Group (1998a). Details of Asia's Debt. *Asia Economics Analyst, 98*(17), 1998.

Goldman Sachs Asia Economic Research Group (1998b). *Asian Bank's NPL*, September 16.

Goldstein, M. (1997). *The Case for An International Banking Standard*. Washington, D.C.: Institute for International Economics.

Goldstein, M. (1998). *The Asian Financial Crisis: Causes, Cures, and Systemic Implications*. Washington, D.C.: Institute for International Economics.

International Monetary Fund (1998a). *World Economic Outlook and International Capital Markets-Interim Assessment*, December.

International Monetary Fund (1998b). The Asian Crisis and the Regions's Long-Term Growth Performance. *World Economic Outlook*, October, 82–106.

International Monetary Fund. *International Financial Statistics*.

International Monetary Fund. *Direction of Trade*.

International Monetary Fund. *Statistical Appendix of IMF Staff Country Report*.

Kaminsky, G., & Reinhart, C. M. (1996). The Twin crises: The Causes of Banking and Balance of Payments Problems. *Board of Governors of the Federal Reserve System International Finance Discussion Papers, 544* (March).

Kaminsky, G., Lizondo, S., & Reinhart, C. M. (1998). Leading Indicators of Currency Crises. *IMF Staff Papers, 45*(1), 1–48.

Kang, B. H. (1998). Financial and Corporate Restructuring in Korea. *Journal of Asian Economies, 9*(4).

Krugman, P. (1979). A Model of Balance of Payments Crisis. *Journal of Money, Credit, and Banking, 11*(3), 311–326.

Krugman, P. (1998). What happened to Asia? *Draft*, January.

Kwack, S. Y. (1998a). *The Financial Crisis in Korea: Causes and Cures*. Presented at the Seminar of Research Department at the International Monetary Fund, June 18.

Kwack, S. Y. (1998b). Factors Contributing to the Financial Crisis in Korea. *Journal of Asian Economies, 9*(4), 611–625.

Letiche, J. M. (1998). Causes of the Financial and Economic Crisis in Southeast Asia and the Need for National, Regional, and IMF Structural Reform. *Journal of Asian Economics, 9*(2), 181–191.

Long-Term Credit Bank of Japan (1998). Financial Crisis in ASEAN – Instability in Financial Globalization and System (in Japanese). *LTCBR 89*, June.

Mckinnon, R. I., & Phill, H. (1996). Credible Liberalizations and International Capital Flows: The "Overborrowing Syndrome". In: Ito, Takatoshi & A. O. Krueger (Eds), *Financial Deregulation and Integration in East Asia* (pp. 1–42). Chicago: The University of Chicago Press.

Mishkin, F. S. (1996). Understanding Financial Crises: A Developing Country Perspective. *NBER Working Paper, 5600* (May).

Mishkin, F. S. (1992). Anatomy of A Financial Crisis. *Journal of Evolutionary Economics, 2*(2), 115–130 and *NBER Working Paper, 3934* (December), Monetary Authority of Singapore, *Monthly Statistical Bulletin*.

Moreno, R., Pasadilla, G., & Remolona, E. (1998). Asia's Financial Crisis: Lessons and Policy Responses. *Pacific Basin Working Paper, PB98-02*. Federal Reserve Bank of San Francisco, July.

Morgan Guaranty Trust Company. *World Financial Markets* Morgan Guaranty Trust Company (1998). Financial Restructuring is Key to Asia's Future. *Asian Financial Markets*, Second Quarter 1998, 5–12.

Obstfeld, M. (1986). Rational and Self-Fulfilling Balance of Payments Crises. *American Economic Review, 76*(1), 72–81.

Radelet, S., & Sachs, J. (1998). The Onset of the East Asian Financial Crisis. *NBER Working Paper, 6680* (August).

Sachs, J. D., Tornell, A., & Velasco, A. (1996). Financial Crises in Emerging Markets: The Lessons from 1995. *Brookings Papers on Economic Activity, 1*, 147–215.

Sundararajan, V. (1996). The Role of Prudential Supervision and Financial Restructuring of Banks During Transition to Indirect Instruments of Monetary Control. *IMF Working Paper, WP/96/128* (November).

Weymark, D. N. (1998). A General Approach to Measuring Exchange Market Pressure. *Oxford Economic Papers, 50*(1), 106–121.

World Bank (1993). *East Asian Miracle: Economic Growth and Public Policy.* Washington, D.C.: The World Bank.

World Bank (1998). *East Asia: The Road to Recovery.* Washington, D.C.: The World Bank.

World Bank. *World Debt Tables.*

21. THE THAI FINANCIAL CRISIS AND THE ROLE OF PUBLIC FINANCIAL INSTITUTIONS

Masumi Kishi

1. INTRODUCTION

On August 14, 1998, the Thai government unveiled a comprehensive financial package for dealing with the current financial crisis. This package,[1] which involves merging four banks, including the Bangkok Bank of Thailand, and five finance companies, including Dhana Siam Co., along with investing 300 billion baht of public money to boost other financial institutions' capital, is designed mainly to mobilize savings and stimulate supply of credit to businesses.

Needless to say, this set of measures is intended to remedy the financial crisis that has been ongoing since the collapse of Thailand's bubble economy in 1996. The bubble economy is thought to have its origins in the 1993 launch of Bangkok's offshore market, the Bangkok International Banking Facilities (BIBF). But then inflation broke out, and the tightening of monetary policy that was implemented in response in 1995 triggered a major drop in stock and real estate prices (real estate registration fees). As a result, many financial institutions found themselves in dire straits, and in December 1997, 56 finance companies were shut down, while in February 1998, four commercial banks were placed under government supervision. Meanwhile, Thailand's financial instability manifested itself beyond the country's borders with the introduction of a floating exchange rate for the baht on July 2, 1997.

This predicament was the consequence of large inflows of short-term capital in spite of a decline in Thailand's current account and other real economic

Restructuring Asian Economics for the New Millennium, Volume 9A, pages 395–410.
2001 by Elsevier Science B.V.
ISBN: 0-7623-0709-9

indicators. The launch of the BIBF in particular gave impetus to foreign capital inflows and thus fueled the bubble economy. The current account deficit ballooned during the 1990s, reaching $14.7 billion by 1996.[2]

This growing current account deficit was not, however, the immediate cause of the economic crisis. The current account deficit itself has been chronic since the mid 1970s, when the Thai economy assumed a pattern of export-driven growth. The post-1990 expansion in the deficit is attributable to the fact that deteriorating terms of trade for agricultural and industrial commodities linked to the appreciation of the dollar have placed a damper on growth in exports, while imports have climbed rapidly due to rising incomes and similar factors. Problems in the real economy can presumably be ironed out through the combination of a floating exchange rate and stricter economic policy, but the distortions in funding channels that enabled large inflows of foreign cash to spark off a bubble economy are a more serious conundrum.

The purpose of this paper is to examine the role of public financial institutions in establishing normal funding channels between savings and investment. One can hardly deny that in general short-term inflows of large quantities of private capital exert a distorting effect on funding channels. Here, however, we shall concentrate rather on the Thai financial system itself and its failure to effectively direct foreign capital into productive activities.

Section 2 considers how flow of funds between households and businesses has changed over the course of the 1990s, and analyzes short-term private capital as an external factor in distorting funding channels. Similarly, Section 3 ponders two domestic factors, decline in growth of household savings, and high-risk investment by companies and financial institutions. Then, in Section 4, we discuss policies for strengthening the central bank's monetary controls and more effectively mobilizing domestic savings, and consider the role public financial institutions have to play. Finally, we state our conclusions.

2. IMPACT OF SHORT-TERM PRIVATE CAPITAL INFLOWS

The fundamental cause of the bubble economy was that financial deepening and the inflow of large amounts of short-term capital combined to limit the reach of monetary policy, but financial and capital markets were not yet mature enough to pick up the slack. However, in order to develop markets and enhance operation of the financial system, market mechanisms need to be perfect. The problem is, the distortions in the financial system and funding routes manifested themselves in the context of financial liberalization, increasing internationalization and a greater emphasis on policies for improving living

Table 1. Investment-Savings Gap to GDP (%).

	1989	1990	1991	1992	1993	1994	1995	1996
Household	10.2	5.8	6.0	5.2	5.6	3.1	3.1	2.0
Business sector	–17.8	–21.2	–19.4	–14.1	–13.6	–13.4	–16.3	–12.4
Government sector	3.5	5.2	5.6	3.3	3.3	3.8	4.0	3.0
State enterprise sector	–0.2	–0.5	–1.5	–2.0	–2.4	–2.4	–2.3	–2.6
Financial sector	1.0	2.1	1.5	2.1	2.1	3.4	3.5	1.9
All sectors	–3.5	–8.5	–7.7	–5.5	–5.0	–5.6	–8.0	–8.1

Source: Office of the National Economic and Social Development Board (1998).

standards – all conditions theoretically favorable to enhancing market mechanisms.

First, let us examine how inflows of short-term private capital affected corporate financial liabilities and household financial assets. As indicated in Table 1, which shows the gap between savings and investment as a percentage of GDP, the overall amount of savings in the Thai economy has since 1989 consistently been less than investment. In terms of individual sectors, however, households, the government sector and the financial sector show a surplus, while the business sector and state enterprise sector have a deficit. The monetary surplus in the government sector is a peculiarity of Thailand. That the economy as a whole suffered from a shortage of funds nonetheless was due to two factors: (1) the business sector had a massive deficit; and (2) the household sector's surplus had shrunk. These trends became especially salient after 1993.

In what way has flow of funds changed with regard to how businesses raise money and households invest? First, foreign capital accounts for the largest share of corporate funding needs. As Table 2 illustrates, financial liabilities as of 1996 consisted of 64.9% foreign investment, 13.7% long-term loans, 9.4% commercial bills, and 7.8% stock issues. However, the proportion of commercial bills was down compared to 1992, while by contrast long-term loans had increased and stocks remained at approximately the same level. These figures indicate that payment by commercial bills had decreased, while foreign capital inflows and long-term loans served as the mainstay of both capital investment and financial speculation on the part of Thai firms.

However, usage of foreign funds differs depending on whether they are short- or long-term. Let us examine the breakdown of foreign capital inflows into Thailand as shown in Table 3. The private sector's external debt peaked in the second quarter of 1997 at $75.6 billion. While virtually all the money

Table 2. Financial Liabilities of Corporated Business.
(Millions of Baht)

	1992	1993	1994	1995	1996
Short-term loans	6,620	21,777	60,872	171,229	61,789
	(0.6)	(1.4)	(3.4)	(8.1)	(2.8)
Long-term loans	113,790	265,293	373,806	417,839	299,038
	(9.8)	(17.2)	(20.9)	(20)	(13.7)
Commercial bills	165,881	151,548	143,078	113,818	204,948
	(14.3)	(9.8)	(8)	(5.4)	(9.4)
Share capital	91,647	78,628	161,271	172,234	170,708
	(7.9)	(5.1)	(9)	(8.2)	(7.8)
Debentures	1,890	18,519	44,954	44,168	10,957
	(0.2)	(1.2)	(2.5)	(2.1)	(0.5)
Mortgages	31,921	8,992	27,193	42,736	43,305
	(2.7)	(0.6)	(1.5)	(2)	(2)
Creditors	–1,373	2,488	12,746	12,492	4,394
	(0.1)	(0.2)	(0.7)	(6)	(0.2)
Hire purchase debts	–12,952	–4,577	6,290	773	–1,768
	(1.1)	(0.3)	(0.4)	(0)	(0.1)
Foreign debts and claims	779,677	1,034,785	983,400	1,155,998	1,418,361
	(67.1)	(67)	(54.9)	(55)	(64.9)
Others	–15,679	–32,119	–22,918	–28,578	–27,931
	(1.4)	(2.2)	(1.3)	(7.1)	(1.3)

Note: Figures in parentheses represent percent of total.
Source: Same as Table 1.

absorbed by the public sector was long-term in nature, in the private sector long- and short-term funding was almost evenly split. Long-term funds mainly accompanied direct investment, with the largest portion entered the country via the non-banking sector. Short-term funds, by contrast, may be regarded chiefly as short-term debt invested for reasons of interest arbitrage, and the largest portion came in via the BIBF and commercial banks. These funds, it may be conjectured, were channeled into in stocks and real estate.[3]

Meanwhile, the household sector's surplus dwindled from 10.2% in 1989 to 5.6% in 1993 as a proportion of GDP, then dropped further to 2.0% in 1996. Similarly, the total value of household savings declined slightly from 267.2 billion baht in 1992 to 254.8 billion baht in 1996.[4]

Let us determine why this happened by referring to Table 4, which shows fluctuation in household financial assets. As the table indicates, the two main household assets are deposits and stocks. As of 1996 deposits accounted for the

Table 3. Total Debt Stocks (End of Period).
(Million of Baht)

		1995	1996	1997Q2	1997Q4	1998Q1
Public sector		16,402	16,805	17,354	17,166	17,695
	Long-term	16,317	16,751	17,334	17,146	17,675
	Short-term	85	54	20	20	20
Private sector		66,166	74,731	75,602	67,323	63,168
	Long-term	25,155	36,172	39,066	37,410	37,017
	Short-term	41,011	37,559	36,536	29,913	26,151
Commercial banks		14,436	10,682	11,762	8,164	8,923
	Long-term	4,443	2,314	3,276	2,545	2,835
	Short-term	9,993	8,368	8,486	5,619	6,088
BIBF		27,503	31,187	32,568	30,079	27,819
	Long-term	3,799	10,697	11,399	10,317	9,820
	Short-term	23,704	20,490	21,169	19,762	17,999
Non-bank		24,913	31,862	31,272	29,080	26,426
	Long-term	16,913	23,161	24,391	24,548	24,362
	Short-term	7,314	8,701	6,881	4,532	2,064

Source: BOT (1998c).

largest share of assets at approximately 50%, a level that had remained constant since 1992. Next come stocks, which increased their share from 17.1% in 1992 to 24.4% in 1996. During the same period, foreign debts and claims, and life assurance and pension funds likewise increased their share dramatically. Currency, commercial bills and other types of assets, on the other hand, lost ground.

These developments are a reliable barometer of the progress of financial deepening and diversification: while deposits continued to be the first choice for investment of household assets, ownership of assurance and pensions was also on the rise. Similarly, the rapid increase in ownership of foreign bonds is a gauge of the pace of internationalization, although one factor behind this trend was the desire to hold foreign currency as a hedge against devaluation of the baht. That the savings rate dropped anyway as a percentage of GDP can be attributed to the fact that consumer confidence rose in direct proportion to the decline in amount of cash and bills, which in turn tended to dampen growth in household savings. Another consideration that may be cited is the wealth effect, whereby the accumulation of financial assets drives up stock and real estate prices and thus results in an increase in private wealth, but also leads to a decline in household savings.[5]

Table 4. Main Financial Assets of Household (Flow of Funds).
(Million of Baht)

	1992	1993	1994	1995	1996
Currency	27,296	25,567	30,087	37,485	17,900
	(5.6)	(4.6)	(5.7)	(4.7)	(2.6)
Deposits	244,073	272,849	253,570	440,800	362,469
	(50.1)	(49.3)	(48.3)	(55.3)	(52.1)
Commercial bills	85,737	69,649	83,679	88,747	48,845
	(17.6)	(12.6)	(15.9)	(11.1)	(7)
Share stocks	83,270	69,649	83,679	88,747	48,845
	(17.1)	(13.7)	(28.9)	(19.2)	(24.4)
Life assurance and pension funds	15,580	22,905	23,720	37,489	45,088
	(2.2)	(4.1)	(4.5)	(4.7)	(6.5)
Foreign debts and claims	3,555	5,975	–2,129	27,950	66,763
	(0.7)	(1.1)	(–0.4)	(3.5)	(9.6)
Others	18,886	80,636	–15,429	11,487	–14,692
	(3.9)	(14.6)	(–2.9)	(1.4)	(–2.1)

Note: Figures in parentheses represent percentage of total.
Source: Same as Table 2.

Alongside financial deepening, a second factor can be identified behind the decline in household savings as a percentage of GDP: the impact of foreign capital inflows. Financial liberalization attracted foreign investment, which in turn reduced the need to mobilize household savings. In other words, once cheap money became available from overseas, incentives to mobilize household savings were weakened. In the case of Thailand, therefore, one may conclude that the funding channels linking household savings to corporate investment became distorted under the weight of a number of factors: financial deepening and the wealth effect, which first came into the limelight in the late 1980s; financial liberalization starting circa 1990; and the rapid expansion of foreign capital inflows from 1993.

3. DOMESTIC FACTORS IN THE FORMATION OF THE BUBBLE ECONOMY

There were several domestic factors behind the formation of Thailand's bubble economy. First, commercial banks and finance companies, buoyed up by infusions of foreign cash, overlent to the housing and real estate industries. Let

Table 5. Housing Credits Extended by Financial Institutions.

	1989	1990	1991	1992
Commercial banks	100,624.90	177,699.00	207,140.30	251,124.90
	(85)	(76.6)	(16.6)	(21.2)
Government Housing Bank	3,221.00	2,640.70	2,938.00	3,075.50
	(3.2)	(−18.0)	(11.3)	(−4.7)
Finance and securities companies	42,073.40	72,247.30	99,453.80	127,926.00
	(84.5)	(71.7)	(37.7)	(28.6)
Credit foncier companies	—	—	—	1,095.20
Government Savings Bank	—	—	—	—
Total	145,919.30	252,587.00	309,532.10	383,221.60
	(81.7)	(73.1)	(22.5)	(23.8)

(Million of Baht)

	1993	1994	1995	1996
Commercial banks	303,801.40	364,208.00	400,184.30	426,101.10
	(21)	(19.9)	(9.9)	(6.5)
Government Housing Bank	1,857.10	1,837.60	2,156.20	2,142.10
	(−39.6)	(1.1)	(17.3)	(−0.7)
Finance and securities companies	163,757.10	239,925.80	325,923.30	362,842.10
	(28)	(46.5)	(35.8)	(11.3)
Credit foncier companies	1,385.00	1,974.70	2,289.30	2,372.60
	(26.5)	(42.6)	(15.9)	(3.7)
Government Savings Bank	—	—	—	—
Total	470,800.70	607,946.10	730,553.10	793,456.90
	(22.9)	(29.1)	(20.2)	(8.6)

Note: Figures in parentheses are percentage changes from previous year.
Source: Bangkok Bank (1997).

us examine Thailand's economic cycle in terms of fluctuations in stock and real estate prices.[6] The SET index, which went through one boom in 1977–1979 and a second in 1986–1989, climbed from 1984.1 points in 1975 to 1,682.9 points at the peak of the third boom in 1993. But then it reversed into steady decline, plummeting to 372.7 points at the end of 1997. The housing and real estate industries have experienced a similar cycle to that of the stock market. There was a boom in 1989–1990, followed by a slowdown in 1991–1992, the bubble of 1993–1995, and then collapse in 1996. Inextricably bound up in all this were loans by commercial banks and finance companies. As Table 5 illustrates, housing loans by commercial banks shot up 4.2 times between 1989 and 1996. But finance companies lent out even more feverishly, chalking up an increase of 8.6 times over the same period. That growth is all the more impressive considering the Government Housing Bank and credit foncier companies were also expanding their financing, while the loans involved were relatively small. Moreover, in the boom years of 1989, 1990, 1994 and 1995, the rate of increase in housing loans was dramatically higher for finance companies than for other types of financial institutions. During the bubble years of 1994 and 1995 in particular, even commercial banks increased their loans at a modest annual rate of 19.9% and 9.9% respectively. Loans by finance companies, by contrast, skyrocketed by 46.5% and 35.8%, eloquent testimony to their deep involvement in generating the bubble economy.

The second factor was the way this increase in lending limited the effectiveness of monetary policy. As financial deepening progresses and household and corporate ownership of securities rises, that alone is enough to make it more difficult for the central bank to control the monetary supply. In addition, as private-sector financial institutions lend out more money, the upshot is the proliferation of another financial product in the form of credit, which competes with the money supply. Even then, so long as complete information is available and the market functions perfectly, a reduction in money supply is still effective in triggering a rise in short- and long-term interest rates, driving up the cost of money and cooling down overall demand. Thus the central bank can easily retain control by adjusting money supply and manipulating interest rates. But where asymmetry of information exists, it becomes difficult for the central bank to steer policy once the lending behavior of private-sector financial institutions begins to exert considerable impact.[7]

Figure 1 shows the correlation between investment and various monetary policy instruments. With the exception of 1997, when the economy was in the adjustment phase, rate of investment growth has since around 1992 been more sensitive to growth in credit than growth in the money supply. Similarly, while there appears to have been almost no relationship between investment and interbank interest rates during the years 1987–1992, in 1993–1996 interest rate fluctuations begin to make themselves felt. On this basis three observations can be made: (1) The influence of the money supply has weakened. (2) Financial liberalization has started to bolster the role of interest rates. (3) Lending by private-sector financial institutions is beginning to exert a powerful impact.

The third factor that may be cited is the management strategy of commercial banks and finance companies. Private firms got into the habit of using short-term funds to make long-term investments in housing and real estate. As explained by the principal-agent theory,[8] such high-risk investment occurs

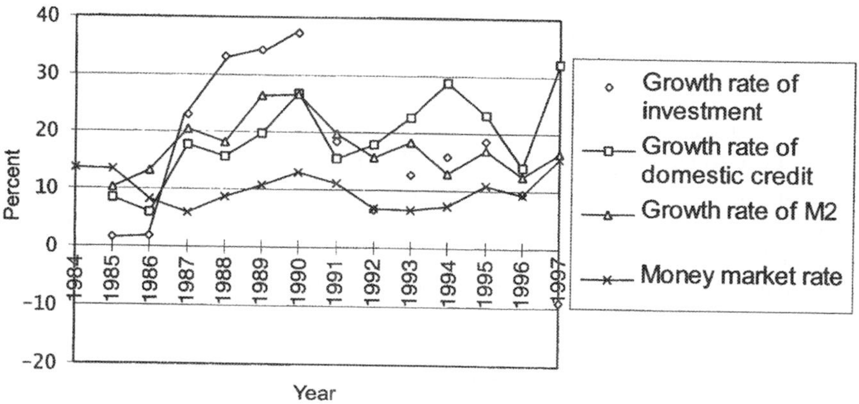

Fig. 1. Domestic Credit, M2, Money Market Rate and Investment.

because the contract with the bank works in the borrower's favor. Say for example a firm takes only 1% of the funds it needs for an investment out of its own assets, and raises the remaining 99% by issuing stocks, offering high dividends to boot. Since virtually all the corporate profit will end up in the pockets of shareholders anyway, company managers will feel little need to boost profits unless shareholders keep a close eye on their behavior. If, on the other hand, the company raises money by borrowing from a bank, it gets to keep the lion's share of any profit that is generated; hence a predilection for bank loans. The catch is that managers will tend to channel these funds that they can control themselves into high-risk investments.

In a country like Thailand where commercial banks and finance companies enjoy an overwhelming share of savings, loans and overall assets, there is a real danger of such high-risk investment taking place. For example, since 1980 the commercial banks' share of financial assets held by all types of financial institution has remained a consistent 70%. Finance companies for their part have generally maintained a 20% share, although there have been slight fluctuations, with the figure declining in the late 1980s and picking up again from around 1992. Government financial institutions, on the other hand, account for a paltry 10% share over the same period.[9]

In such a climate, unless lending by private-sector financial institutions is properly screened, there is a heightened risk of moral hazard. Normally, when a bank lends a firm money, it obtains appropriate information on the probability of success or riskiness of the investment project in question; if the risk is high, it will raise interest on the loan proportionately. However, where asymmetry of information exists – in other words, where a firm possesses information that the bank does not – the greater the potential return on a project, even though the risk may be considerable, the more tempted the firm will be to borrow money, even at high rates of interest. But if commercial banks and finance companies still choose to lend money to high-risk, high-return projects in spite of the fact they have accurate information on them by virtue of the considerable influence they exercise over corporate management, then the result will be the emergence of a bubble economy. Such behavior constitutes moral hazard, and is ultimately detrimental to the interests of depositors. By acting in this fashion, financial institutions abdicate what is supposed to be their proper function: generating information on behalf of savers who do not possess adequate information and lack the technical expertise to monitor borrowers. This is where the central bank and government financial institutions have a role to play, the former as watchdog over the actions of commercial banks and finance companies, the latter as a counterbalance to them.

4. PUBLIC FINANCIAL INSTITUTIONS AS A COUNTERBALANCE TO PRIVATE-SECTOR FINANCE

Financial liberalization and internationalization were both supposed to improve the operation of market mechanisms and lead to greater affluence. Instead, they have left behind a mountain of bad loans in the aftermath of the bubble economy's collapse, and given rise to factors like moral hazard and rent seeking that rather disrupt market mechanisms. As we have already seen, this was the consequence of a weakening of the reach of monetary policy in combination with risky behavior on the part of financial institutions and companies. Reinforcing market mechanisms and disposing of bad debt will therefore require an overhaul of the financial system. The government's role must be threefold: (1) providing public goods; (2) setting national minimum standards; and (3) developing financial infrastructure. The central bank and government financial institutions have a similar responsibility to develop the financial system and establish financial infrastructure. Two functions will be particularly crucial: setting up a money market capable of accommodating inflows of overseas capital and fluctuations in exchange rate, and establishing funding channels whereby household savings can be mobilized.

First, expanding the money market is essential for both sterilization and implementation of monetary policy, because all other tools of monetary policy with the exception of open market operations via the money market largely lack effectiveness. For example, adjusting the bank rate accomplishes little, since private financial institutions' dependence on the Bank of Thailand is minimal.

Thus it becomes all the more important to ensure monetary policy for regulating effective demand can percolate through the system. That means adjusting supply and demand for call money and BOT bonds etc. such as to affect money market interest rates and by extension influence lending by financial institutions.

However, the Thai money market is exclusively an interbank market, consisting of a call money market and a repurchase market. Let us therefore first examine the feasibility of policy transmission via the call money market. For instance, if the Bank of Thailand decides to implement a fastening of economic policy, it has to lower call rates. That means it will need to induce private financial institutions to raise their reserves so as to lesson demand for funds on the call money market. This increment in reserves consists of the following elements:

Increment in reserves = inflow of Bank of Thailand bank notes
+ government's payments
+ supply of Bank of Thailand credits.

The first item in the above formula refers to circulation of funds between households and businesses on the one hand and financial institutions on the other. For example, if a company repays its loan to a commercial bank, the result is a extension in the bank's reserves. However, this behavior may not be suitable at this liquidity shortage situation Government payment and/or provision of Bank of Thailand credits are required in order to increase bank's reserves without inducing the contraction of bank lending.

Then the second item refers to circulation of funds between households and businesses and the government. When the government adopts a tax reduction policy, household's and/or businesses' deposits at banks extend, and commercial banks reserves are increased. Similarly, if the Bank of Thailand intervenes on the foreign exchange market by buying dollars and selling baht , baht flow from the foreign exchange account into the private sector, and commercial bank reserves are again increased, although this policy will not be taken at the present situation.

The third item refers to transfer of funds between private financial institutions and the Bank of Thailand by means of the latter's lending activities and market operations. When the Bank of Thailand powers funds in the call money market, a increment in commercial banks' reserves takes place.

According to the above formula, consolidation of the call money market will be indispensable for business adjustment and foreign exchange market intervention. However, this assumes that money can circulate unimpeded between households, businesses and the Bank of Thailand. In order to expand the market, therefore, measures will be needed to enhance the effectiveness of policy guidance by the Bank of Thailand, e.g. beefing up requirements for private banks to accurately report fluctuations in their reserves.

Like the call money market, the repurchase market is growing in importance. The financial products offered on the repurchase market comprise government bonds, state enterprise bonds and BOT bonds. However, while outstanding issues of private securities totaled 1342.6 billion baht as of the end of 1997 (at market price), the corresponding figure for public bonds was a mere 336.9 billion baht, partially because there have been no government bond issues.[10]

However, as indicated in Fig. 2, while there was considerable discrepancy between the bank rate, interbank lending rate and repurchase rate around 1989 and 1993, the variation between the three rates narrowed in the years 1994–1996. Interest rates are exerting an increasing impact on investment, as we have already seen from Fig 1. Leaving aside the period of economic confusion in 1997, it is fair to conclude that the practicality of directly controlling corporate investment and household spending on housing by means of a policy based primarily on repurchase market interest rates has increased. Moreover, the importance of open market operations involving the issue of government and BOT bonds is growing, since call money market rates are susceptible to the influence of lending by commercial banks and finance companies.

Second, the role of such organizations as the Government Housing Bank and Government Savings Bank needs to be beefed up as a way to counteract the

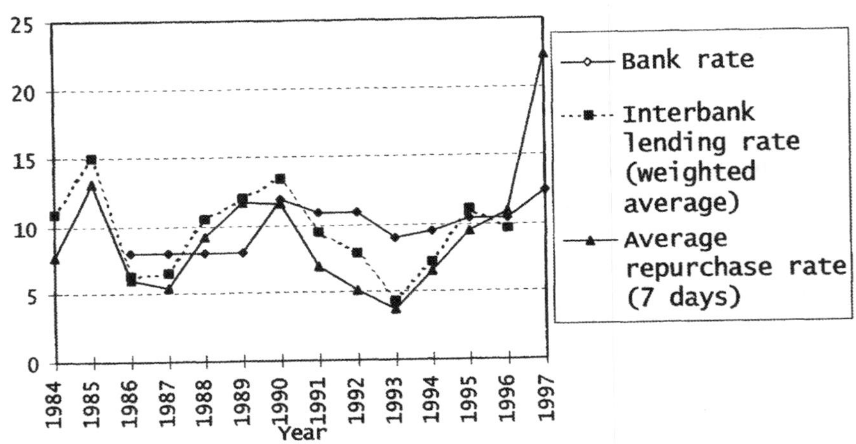

Fig. 2. Interest Rate Policy (1984–1997).

influence of private financial institutions and enhance channels for mobilizing savings. The establishment of efficient routes of funding linking savings and investment assumes access to complete information. However, in the case of Thailand, the information available to economic agents that utilize foreign capital, like branches of overseas banks and local financial institutions or the large corporations with which they deal, does not necessarily coincide with that shared by provincial households and small businesses. One way to alleviate asymmetry of information, surely, is to establish a direct link between investment objectives and savings objectives. With this goal in mind, the government in 1987 imposed a rule requiring commercial banks to lend no less than 20% of the total value of their deposits to farmers and small businesses.[11] However, ignoring as it does the problem of cost, this type of control can hardly be described as efficient.

Another option is to ensure rural districts and small businesses greater access to credit by expanding the branch network of government financial institutions like the Government Savings Bank and the Government Housing Bank. The Government Savings Bank[12] provides financing for social infrastructure improvement projects in such areas as transportation, communications, public works and education, as well as for promoting welfare and building housing. It also supplies loans under the Credit Program for Rural Development, which is designed to foster self-sustained regional development through local networks. The Bank is funded from small deposits consisting specifically of savings deposits, premium savings certificates and life insurance deposits.

Similarly, the Government Housing Bank[13] provides financing for construction and renovation of homes and acquisition of land. It also offers refinancing loans. Its chief sources of funding are public deposits and promissory notes, which at the end of 1997 accounted for 60.7% of all the moneys it raised. The rest of its funding comes from financing from the Government Savings Bank and Bank of Thailand, loans from the offshore market, and bond issues. The Bank's lending activities are characterized by a preponderance of loans to middle income borrowers with a monthly income of 20,000–50,000 baht (who accounted for 41.1% of loans to all income brackets as of the end of 1997).

However, while in 1985 the Government Savings Bank accounted for 9.2% of household savings mobilized,[14] by 1996 its share had dropped to 5.2%. In the same period, the equivalent figure for the Government Housing Bank edged up slightly from 1.2% to 1.5%. The contrast is striking. One reason that can be cited for it is the difference in ratio of loans to deposits between the two institutions. The Government Housing Bank had a ratio of loans to deposits of 160.0% in 1985 and 195.0% in 1996. The ratio for the Government Savings Bank, by contrast, was a mere 21.7% and 26.9% respectively for those two years. Since 1987 a policy to increase the ratio of loans to deposits has been undertaken in an effort to get around this situation. Nonetheless, the Government Savings Bank continues to make little headway in more effectively mobilizing household savings, not least of all because it has a relatively small branch network that is growing only very slowly (between 1985 and 1996 the number of branches increased just slightly from 415 to 548). The Government Housing Bank has given priority to outlying areas of the country in developing its branch network, and it is important that the Government Savings Bank likewise expand more aggressively into outlying regions. In addition, the two institutions need to increase market-based lending by for instance pooling information, since both provide housing loans.

5. CONCLUSION

The massive inflows of short-term private capital sparked by the launch of the BIBF fueled the bubble economy and precipitated a financial crisis. By the second quarter of 1997, Thai companies were burdened with external debt totaling \$75.6 billion, approximately half of which consisted of short-term capital loans. But another side of the story as far as flow of funds goes is the failure of household savings to grow.

The intent of the present paper has been to shift the spotlight onto the problem of distortions in domestic funding channels. It can be argued that over-lending to the housing and real estate industries, weakening of the central bank's monetary controls, and commercial banks' and finance companies' management strategies all conspired to produce the bubble economy.

In order to deal effectively with these internal and external problems, it is urgent above all that Thailand develop a money market capable of handling foreign capital inflows and exchange rate fluctuations, as well as establish funding channels through which to mobilize household savings. The Bank of Thailand, the Government Savings Bank and the Government Housing Bank all have a key role to fulfill in this regard by countervailing risky behavior on the part of private-sector financial institutions and companies.

However, if the reforms to the financial system that are currently underway are to succeed, and a direct link is to be established between investment and savings, then cooperation with the United States, Japan and other supporters from the sidelines will be essential. One example it has not been possible to discuss here: a policy to adjust exchange rates based on current account trends could be adopted should major fluctuations in exchange rates conflict with Thai monetary policy.

Establishing financial infrastructure is also indispensable. Relaxation of economic regulations has so far been successful in Thailand, but there is still a great deal to be desired in terms of developing financial infrastructure, which entails specifically strengthening prudential regulation and the information regulation upon which it is based. The most serious problem of all is the inadequacy of legislation to protect lenders, as became all too evident with the collapse of the bubble economy, when inaccurate finance reports, low levels of disclosure and faulty settlement procedures all came into the spotlight.

NOTES

1. In addition, the First Bangkok City Bank and the Union Bank was slated to be merged with the state-owned Krung Thai Bank, while the country's smallest bank, Laem Thong, was scheduled to be amalgamated with the Radhanasin Bank. Moreover, the largest of the finance companies, Dhana Siam Co., was earmarked for a merger with Krung Thai Thanakit Co. The Nation, August 15, 1998.
2. IMF (1998).
3. On the shift from direct investment to short-term capital debt and its impact on productive activity, see Jansen (1997), pp. 147–198.
4. Office of The National Economic and Social Development Board (1998).
5. See BOT (1998a), pp. 34–35.
6. Real estate price trends (real estate registration fees) are based on BOT (1997). SET index figures are taken from SET (1996).
7. On funding channels under conditions of market imperfection, see Kishi (1996), pp. 229–255 and (1998), pp. 420–422.
8. See Bernanke (1993), pp. 55–60.
9. This share is the total for the Government Savings Bank, the Bank for Agriculture and Agricultural Cooperatives, Industrial Finance and Finance & Finance Corporation of Thailand, the Government Housing Bank, the Export-Import Bank of Thailand, and Finance. For this figures, see BOT (1998b).
10. BOT (1998c).
11. The Bank for Agriculture and Agricultural Cooperatives (1985), pp. 6–17.
12. The Government Savings Bank (1998).
13. The Government Housing Bank (1997).
14. BOT (1998d).

ACKNOWLEDGMENT

An earlier version of this paper was presented at the Bangkok Conference on "A Macroeconomic Core of an Open Economy for Progressive Industrialization and Development in Asia in the New Millennium", sponsored by the American Committee on Asian Economic Studies, and the Chulalongkorn University, December 16–18,1998. I am indebted to the distinguished conferees, particularly to Prof. M. Jan Dutta and Prof. Kazuo Sato for their precious comments. I am, of course, responsible for all remaining errors.

REFERENCES

BAAC (1985). *BAAC and Agricultural Development in Thailand.*

Bangkok Bank (1997). *Monthly Review, 38*(8).

Bernanke, B. S. (1993). Credit in the Macroeconomy. *FRBNY Quarterly Review,* Spring 1992–1993.

BOT (1997). *Annual Economic Report 1996.*

BOT (1998a). *Annual Economic Report 1997.*

BOT (1998b). *Quarterly Bulletin, 38*(1).

BOT (1998c). *Quarterly Bulletin, 38*(2).

BOT (1998d). Interim Source.

Government Housing Bank (1998). *Annual Report 1997.*

Government Savings Bank (1998). *Annual Report 1997.*

IMF (1998). *International Financial Statistics Yearbook 1997.*

Jansen, K. (1997). *External Finance in Thailand's Development.* Macmillan Press and St.Martin's Press, 147–198.

Kishi, M. (1996). Kinyu Kaikaku to Jutaku (Financial Reform and Housing). In: K. Hayakawa & K. Yokota (Eds), *Gendai Kyoju 4 (Modern Housing 4)* (pp. 229–255). Tokyo University Press.

Kishi, M. (1998). Tai no Kinyu-Shihon Shijyo no Hatten Katei (The Development Process of the Thai Financial and Capital Markets). Edited by Institute of Monetary and Fiscal Policies, Ministry of Finance, 407–426.

Office of The National Economic and Social Development Board (1998). *Flow-of-Funds Accounts of Thailand.*

SET (1996). Securities Market in Thailand 1997.